Praise for *Consciousness and the Brain*

"Brilliant . . . Dehaene's special contribution is his global-workspace theory, the first step in a complete account of why some neural processes lead to conscious experience. . . . Dehaene's account is the most sophisticated story about the neural basis of consciousness so far." —Chris Frith, *Nature*

"In *Consciousness and the Brain*, [Dehaene] summaries the fruits of two decades of vigorous experimentation and modeling. . . . The book introduces the methods that acted as midwife at the birth of a science of consciousness . . . postulating that global availability of information is what we subjectively experience as a conscious state begets the question of why . . . answering such questions requires an information-theoretical account of what type of data, communicated within what system, gives rise to conscious experience in biological or artificial organisms. Dehaene's well-written and well-sourced book avoids this, as he opts to restrict it to behavioral and neuronal observables." —Christof Koch, *Science*

"Consciousness tomes have become a dime a dozen over the past decade or so, with every last researcher feeling the need to join the fray. But Stanislas Dehaene is one of the few at the top of the disciplines involved—philosophy, history, cognitive psychology, brain imaging, computer modeling—to add something new." —*New Scientist*

"An excellent teacher with a gift for vivid analogies, Dehaene writes that 'consciousness is like the spokesperson in a large institution . . . with a staff of a hundred billion neurons' issuing briefs that tell us what we need to know moment by moment. He then explains his and his colleagues' groundbreaking theory about the 'global neuronal workspace,' where information is made 'available to the rest of the brain,' wowing us with descriptions of our pyramidal neurons and their spiny dendrites and the discovery that each neuron 'cares' about such specific stimuli as 'faces, hands, objects.' A stunning delineation of the intricate biological machinery' that has made us an animal unlike any —*Booklist* (starred review)

"Dehaene's knack for explaining complex terms in interesting, understandable phrases is bolstered by accompanying images that enhance the basic comprehension of the material. . . . In all respects, this book will bring the brain's marvelous mechanisms into clearer focus."   —*Publishers Weekly*

"Stanislas Dehaene's remarkable book is the best modern treatment of consciousness I have read to date. Dehaene, a world-class scientist, has pioneered the development of a set of experiments for studying consciousness that have revolutionized the field and given us the first direct approach to its biology. Simply stated this book is a tour de force. It opens up a whole new world of intellectual exploration for the general reader."
      —Eric Kandel, author of *In Search of Memory* and *The Age of Insight*,
      and winner of the Nobel Prize in Physiology or Medicine

PENGUIN BOOKS

CONSCIOUSNESS AND THE BRAIN

Stanislas Dehaene trained as a mathematician and psychologist before becoming one of the world's most active researchers on the cognitive neuroscience of language and number processing in the human brain. He is the director of the Cognitive Neuroimaging Unit in Saclay, France, a professor of experimental cognitive psychology at the College de France, and a member of the French Academy of Sciences and the Pontifical Academy of Sciences. He has published extensively in peer-reviewed scientific journals and is the author of *The Number Sense* and *Reading in the Brain*. He lives in France.

CONSCIOUSNESS
AND THE **BRAIN**

Deciphering How the Brain

Codes Our Thoughts

STANISLAS DEHAENE

PENGUIN BOOKS

PENGUIN BOOKS

Published by the Penguin Group
Penguin Group (USA) LLC
375 Hudson Street
New York, New York 10014

USA | Canada | UK | Ireland | Australia | New Zealand | India | South Africa | China
penguin.com
A Penguin Random House Company

First published in the United States of America by Viking Penguin, a member of Penguin Group (USA) LLC, 2014
Published in Penguin Books 2014

THE LIBRARY OF CONGRESS HAS CATALOGED THE HARDCOVER EDITION AS FOLLOWS:
Dehaene, Stanislas.
Consciousness and the brain : deciphering how the brain codes our thoughts / Stanislas Dehaene.
pages   cm
Includes bibliographical references and index.
ISBN 978-0-670-02543-5 (hc.)
ISBN 978-0-14-312626-3 (pbk.)
1. Consciousness.   2. Brain.   3. Cognitive neuroscience.   I. Title.
QP411.D44 2014
612.8'2—dc23
2013036814

Printed in the United States of America
ScoutAutomatedPrintCode

Set in Warnock with Gotham Display
Designed by Carla Bolte

To my parents,

and to Ann and Dan, my American parents

Consciousness is the only real thing in the world
and the greatest mystery of all.

—Vladimir Nabokov, *Bend Sinister* (1947)

The brain is wider than the sky,
For, put them side by side,
The one the other will include
With ease, and you beside.

—Emily Dickinson (ca. 1862)

CONTENTS

CONSCIOUSNESS
AND THE **BRAIN**

INTRODUCTION: THE STUFF OF THOUGHT

Deep inside the Lascaux cave, past the world-renowned Great Hall of the Bulls, where Paleolithic artists painted a colorful menagerie of horses, deer, and bulls, starts a lesser-known corridor known as the Apse. There, at the bottom of a sixteen-foot pit, next to fine drawings of a wounded bison and a rhinoceros, lies one of the rare depictions of a human being in prehistoric art (figure 1). The man is lying flat on his back, palms up and arms extended. Next to him stands a bird perched on a stick. Nearby lies a broken spear that was probably used to disembowel the bison, whose intestines are hanging out.

The person is clearly a man, for his penis is fully erect. And this, according to the sleep researcher Michel Jouvet, illuminates the drawing's meaning: it depicts a dreamer and his dream.[1] As Jouvet and his team discovered, dreaming occurs primarily during a specific phase of sleep, which they dubbed "paradoxical" because it does not look like sleep; during this period, the brain is almost as active as it is in wakefulness, and the eyes ceaselessly move around. In males, this phase is invariably accompanied by a strong erection (even when the dream is devoid of sexual content). Although this weird physiological fact became known to science only in the twentieth century, Jouvet wittily remarks that our ancestors would easily have noticed it. And the bird seems the most natural metaphor for the dreamer's soul: during dreams, the mind flies to distant places and ancient times, free as a sparrow.

This idea might seem fanciful were it not for the remarkable recurrence of imagery of sleep, birds, souls, and erections in the art and symbolism of all sorts of cultures. In ancient Egypt, a human-headed bird, often depicted with an erect phallus, symbolized the Ba, the immaterial soul. Within every human being, it was said, dwelled an immortal Ba that upon death took flight to seek the afterworld. A conventional depiction of the

1

FIGURE 1. The mind may fly while the body is inert. In this prehistoric drawing, dated approximately 18,000 years ago, a man lies supine. He is probably asleep and dreaming, as hinted by his strong erection, characteristic of the phase of rapid-eye-movement sleep, during which dreams are most vivid. Next to him, the artist painted a disemboweled bison and a bird. According to the sleep researcher Michel Jouvet, this may be one of the first depictions of a dreamer and his dream. In many cultures, the bird symbolizes the mind's ability to fly away during dreams—a premonition of dualism, the misguided intuition that thoughts belong to a different realm from the body.

great god Osiris, eerily similar to Lascaux's Apse painting, shows him lying on his back, penis erect, while Isis the owl hovers over his body, taking his sperm to engender Horus. In the Upanishads, the Hindu sacred texts, the soul is similarly depicted as a dove that flies away at death and may come back as a spirit. Centuries later doves and other white-winged birds came to symbolize the Christian soul, the Holy Spirit, and the visiting angels. From the Egyptian phoenix, symbol of resurrection, to the Finnish Sielulintu, the soul bird that delivers a psyche to newborn babies and takes it away from the dying, flying spirits appear as a universal metaphor for the autonomous mind.

Behind the bird allegory stands an intuition: the stuff of our thoughts differs radically from the lowly matter that shapes our bodies. During dreams, while the body lies still, thoughts wander into the remote realms of imagination and memory. Could there be a better proof that mental activity cannot be reduced to the material world? That the mind is made of a distinct stuff? How could the free-flying mind ever have arisen from a down-to-earth brain?

Descartes's Challenge

The idea that the mind belongs to a separate realm, distinct from the body, was theorized early on, in major philosophical texts such Plato's *Phaedo* (fourth century BC) and Thomas Aquinas's *Summa theologica* (1265–74), a foundational text for the Christian view of the soul. But it was the French philosopher René Descartes (1596–1650) who explicitly stated what is now known as dualism: the thesis that the conscious mind is made of a nonmaterial substance that eludes the normal laws of physics.

Ridiculing Descartes has become fashionable in neuroscience. Following the publication of Antonio Damasio's best-selling book *Descartes' Error* in 1994,[2] many contemporary textbooks on consciousness have started out by bashing Descartes for allegedly setting neuroscience research years behind. The truth, however, is that Descartes was a pioneering scientist and fundamentally a reductionist whose mechanical analysis of the human mind, well in advance of his time, was the first exercise in synthetic biology and theoretical modeling. Descartes's dualism was no whim of the moment—it was based on a logical argument that asserted the impossibility of a machine ever mimicking the freedom of the conscious mind.

The founding father of modern psychology, William James, acknowledges our debt: "To Descartes belongs the credit of having first been bold enough to conceive of a completely self-sufficing nervous mechanism which should be able to perform complicated and apparently intelligent acts."[3] Indeed, in visionary volumes called *Description of the Human Body, Passions of the Soul,* and *L'homme* (Man), Descartes presented a resolutely mechanical perspective on the inner operation of the body. We are sophisticated automata, wrote this bold philosopher. Our bodies and brains literally act as a collection of "organs": musical instruments comparable to those found in the churches of his time, with massive bellows forcing a special fluid called "animal spirits" into reservoirs, then a broad variety of pipes, whose combinations generate all the rhythms and music of our actions.

> I desire that you consider that all the functions that I have attributed to this machine, such as the digestion of food, the beating of the heart and the arteries, the nourishment and growth of the bodily parts, respiration, waking and sleeping; the reception of light, sounds, odours, smells, heat, and other such qualities by the external sense organs; the impression of the ideas of them in the organ of common sense and the imagination, the retention or imprint of these ideas in the memory; the internal movements of the appetites and the passions; and finally the external movements of all the bodily parts that so aptly follow both the actions of objects presented to the senses. . . . These functions follow in this machine simply from the disposition of the organs as wholly naturally as the movements of a clock or other automaton follow from the disposition of its counterweights and wheels.[4]

Descartes's hydraulic brain had no difficulty moving his hand toward an object. The object's visual features, impinging on the inner surface of the eye, activated a specific set of pipes. An inner decision-making system that was located in the pineal gland then leaned in a certain direction, thus sending the spirits flowing, to cause precisely the appropriate movement of the limbs (figure 2). Memory corresponded to the selective reinforcement of some of these pathways—an insightful anticipation of the contemporary idea that learning relies on changes in the brain's connections ("neurons that fire together wire together"). Descartes even presented an explicit mechanical model of sleep, which he theorized as a reduced pressure of the spirits. When the source of animal spirits was abundant, it circulated through every nerve, and this pressurized machine, ready to respond to any stimulation, provided an accurate model of the wake state. When the pressure weakened, making the lowly spirits capable of moving only a few threads, the person fell asleep.

Descartes concluded with a lyrical appeal to materialism—which was

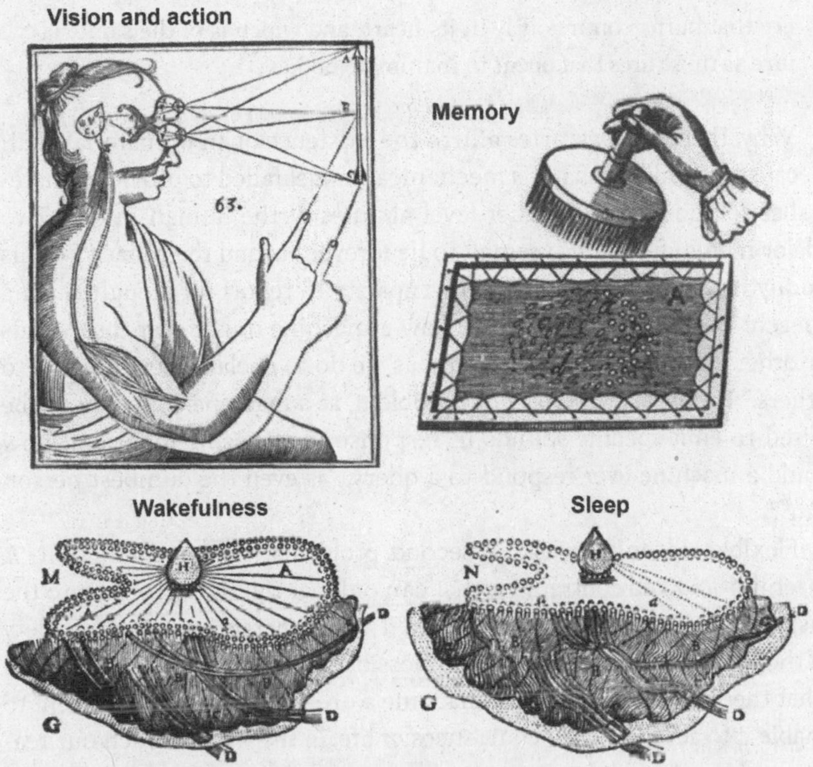

Vision and action

Memory

Wakefulness

Sleep

FIGURE 2. René Descartes's theory of the nervous system stopped short of a fully materialistic conception of thought. In *L'homme*, published posthumously in 1664, Descartes foresaw that vision and action could result from a proper arrangement of the connections between the eye, the pineal gland inside the brain, and the arm muscles. He envisaged memory as the selective reinforcement of these pathways, like the punching of holes in cloth. Even fluctuations in consciousness could be explained by variations in the pressure of the animal spirits that moved the pineal gland: high pressure led to wakefulness, low pressure to sleep. In spite of this mechanistic stance, Descartes believed that the mind and the body were made of different kinds of stuff that interacted through the pineal gland.

quite unexpected, coming from the pen of the founder of substance dualism:

> To explain these functions, then, it is not necessary to conceive of any vegetative or sensitive soul, or any other principle of movement or life, other than its blood and its spirits which are agitated by the heat of the

fire that burns continuously in its heart, and which is of the same nature as those fires that occur in inanimate bodies.

Why, then, did Descartes affirm the existence of an immaterial soul? Because he realized that his mechanical model failed to provide a materialist solution for the higher-level abilities of the human mind.[5] Two major mental functions seemed to lie forever beyond the capacity of his bodily machine. The first was the capacity to report its thoughts using speech. Descartes could not see how a machine might ever "use words or other signs by composing them, as we do to declare our thoughts to others." Reflexive cries posed no problem, as a machine could always be wired to emit specific sounds in response to a specific input; but how could a machine ever respond to a query, "as even the dumbest person can"?

Flexible reasoning was the second problematic mental function. A machine is a fixed contraption that can only act rigidly, "according to the disposition of its organs." How could it ever generate an infinite variety of thoughts? "It must be morally impossible," our philosopher concluded, "that there should exist in any machine a diversity of organs sufficient to enable it to act in all the occurrences of life, in the way in which our reason enables us to act."

Descartes's challenges to materialism stand to this very day. How could a machine like the brain ever express itself verbally, with all the subtleties of human language, and reflect upon its own mental states? And how might it make rational decisions in a flexible manner? Any science of consciousness must address these key issues.

The Last Problem

As humans, we can identify galaxies light years away, study particles smaller than an atom. But we still haven't unlocked the mystery of the three pounds of matter that sits between our ears.

—Barack Obama announcing the BRAIN initiative (April 2, 2013)

Thanks to Euclid, Karl Friedrich Gauss, and Albert Einstein, we possess a reasonable understanding of the mathematical principles that govern the physical world. Standing as we do on the shoulders of such giants as Isaac

Newton and Edwin Hubble, we understand that our earth is just a speck of dust in one of a billion galaxies that originated from a primeval explosion, the big bang. And Charles Darwin, Louis Pasteur, James Watson, and Francis Crick showed us that life is made of billions of evolved chemical reactions—just plain physics.

Only the story of the emergence of consciousness seems to remain in medieval darkness. How do I think? What is this "I" that seems to be doing the thinking? Would I be different if I had been born at a different time, in another place, or in another body? Where do I go when I fall asleep, and dream, and die? Does it all arise from my brain? Or am I in part a spirit, made of distinct stuff of thought?

These vexing questions have perplexed many a bright mind. Writing in 1580, the French humanist Michel de Montaigne, in one of his famous essays, lamented that he could find no coherence in what past thinkers had written about the nature of the soul—they all disagreed, both on its nature and on its seat within the body: "Hippocrates and Hierophilus lodge it in the ventricle of the brain; Democritus and Aristotle, throughout the body, Epicurus in the stomach, the Stoics within and around the heart, Empedocles, in the blood; Galen thought that each part of the body had its own soul; Strato lodged it between the eyebrows."[6]

Throughout the nineteenth and twentieth centuries, the question of consciousness lay outside the boundaries of normal science. It was a fuzzy, ill-defined domain whose subjectivity put it forever beyond the reach of objective experimentation. For many years, no serious researcher would touch the problem: speculating about consciousness was a tolerated hobby for the aging scientist. In his textbook *Psychology, the Science of Mental Life* (1962), George Miller, the founding father of cognitive psychology, proposed an official ban: "Consciousness is a word worn smooth by a million tongues. . . . Maybe we should ban the word for a decade or two until we can develop more precise terms for the several uses which 'consciousness' now obscures."

And banned it was. When I was a student in the late 1980s, I was surprised to discover that, during lab meetings, we were not allowed to use the C-word. We all studied consciousness in one way or another, of course, by asking human subjects to categorize what they had seen or to form mental images in darkness, but the word itself remained taboo: no serious scientific publication used it. Even when experimenters flashed brief pictures at the threshold of participants' conscious perception, they did not care to report whether the participants saw the stimuli or not. With a few major exceptions,[7] the general feeling was that using the term *consciousness* added nothing of value to psychological science. In the emerging positive science of cognition, mental operations were to be solely described in terms of the processing of information and its molecular and neuronal implementation. Consciousness was ill defined, unnecessary, and passé.

And then in the late 1980s everything changed. Today the problem of consciousness is at the forefront of neuroscience research. It is an exciting field with its own scientific societies and journals. And it is beginning to address Descartes's major challenges, including how our brain generates a subjective perspective that we can flexibly use and report to others. This book tells the story of how the tables have turned.

Cracking Consciousness

In the past twenty years, the fields of cognitive science, neurophysiology, and brain imaging have mounted a solid empirical attack on consciousness. As a result, the problem has lost its speculative status and become an issue of experimental ingenuity.

In this book, I will review in great detail the strategy that has turned a philosophical mystery into a laboratory phenomenon. Three fundamental ingredients have made this transformation possible: the articulation of a better definition of consciousness; the discovery that consciousness can be experimentally manipulated; and a new respect for subjective phenomena.

The word *consciousness*, as we use it in everyday speech, is loaded with fuzzy meanings, covering a broad range of complex phenomena. Our first task, then, will be to bring order to this confused state of affairs. We will have to narrow our subject matter to a definite point that can be subjected to precise experiments. As we will see, the contemporary science of consciousness distinguishes a minimum of three concepts: vigilance—the state of wakefulness, which varies when we fall asleep or wake up; attention—the focusing of our mental resources onto a specific piece of information; and conscious access—the fact that some of the attended information eventually enters our awareness and becomes reportable to others.

What counts as genuine consciousness, I will argue, is conscious access—the simple fact that usually, whenever we are awake, whatever we decide to focus on may become conscious. Neither vigilance nor attention alone is sufficient. When we are fully awake and attentive, sometimes we can see an object and describe our perception to others, but sometimes we cannot—perhaps the object was too faint, or it was flashed too briefly to be visible. In the first case, we are said to enjoy *conscious access*, and in the second we are not (and yet as we shall see, our brain may be processing the information unconsciously).

In the new science of consciousness, conscious access is a well-defined phenomenon, distinct from vigilance and attention. Furthermore, it can be easily studied in the laboratory. We now know of dozens of ways in which a stimulus can cross the border between unperceived and perceived, between invisible and visible, allowing us to probe what this crossing changes in the brain.

Conscious access is also the gateway to more complex forms of conscious experience. In everyday language, we often conflate our consciousness with our sense of self—how the brain creates a point of view, an "I" that looks at its surroundings from a specific vantage point. Consciousness can also be recursive: our "I" can look down at itself, comment on its own performance, and even know when it does not know something. The good news is that even these higher-order meanings of consciousness are no longer inaccessible to experimentation. In our laboratories, we have learned to quantify what the "I" feels and reports, both about the external environment and about itself. We can even manipulate the sense of self, so that people may have an out-of-body experience while they lie inside a magnetic resonance imager.

Some philosophers still think that none of the above ideas will suffice to solve the problem. The heart of the problem, they believe, lies in another sense of consciousness, which they call "phenomenal awareness": the intuitive feeling, present in all of us, that our internal experiences possess exclusive qualities, unique qualia such as the exquisite sharpness of tooth pain or the inimitable greenness of a fresh leaf. These inner qualities, they argue, can never be reduced to a scientific neuronal description; by nature, they are personal and subjective, and thus they defy any exhaustive verbal communication to others. But I disagree, and I will argue that the notion of a phenomenal consciousness that is distinct from conscious access is highly misleading and leads down a slippery slope to dualism. We should start simple and first study conscious access. Once we clarify how any piece of sensory information can gain access to our mind and become reportable, then the insurmountable problem of our ineffable experiences will disappear.

To See or Not to See

Conscious access is deceptively trivial: we lay our eyes on an object, and seemingly immediately, we become aware of its shape, color, and identity. Behind our perceptual awareness, however, lies an intricate avalanche of brain activity that involves billions of visual neurons and that may take nearly half a second to complete before consciousness kicks in. How can we analyze this long processing chain? How can we tell which part corresponds to purely unconscious and automatic operations, and which part leads to our conscious sense of seeing?

This is where the second ingredient of the modern science of consciousness kicks in: we now have a strong experimental handle on the mechanisms of conscious perception. In the past twenty years, cognitive scientists have discovered an amazing variety of ways to manipulate consciousness. Even a minuscule change in experimental design can cause us to see or not to see. We can flash a word so briefly that viewers will fail to notice it. We can create a carefully cluttered visual scene, in which one item remains wholly invisible to a participant because the other items always win out in the inner competition for conscious perception. We can also distract your attention: as any magician knows, even an obvious gesture can become utterly invisible if the watcher's mind is drawn to another train of thought. And we can even let your brain do the magic: when two distinct images are presented to your two eyes, the brain will spontaneously oscillate and let you see one picture, then the other, but never both at the same time.

The perceived image, the one that makes it into awareness, and the losing image, which vanishes into unconscious oblivion, may differ minimally on the input side. But within the brain, this difference must be amplified, because ultimately you can speak about one but not about the other. Figuring out exactly where and when this amplification occurs is the object of the new science of consciousness.

The experimental strategy of creating a minimal contrast between conscious and unconscious perception was the key idea that cracked wide open the doors to the supposedly inaccessible sanctuary of consciousness.[8] Over the years, we discovered many well-matched experimental contrasts in which one condition led to conscious perception while the other did not. The daunting problem of consciousness was reduced to the experimental issue of deciphering the brain mechanisms that distinguish two sets of trials—a much more tractable problem.

Turning Subjectivity into a Science

This research strategy was simple enough, yet it relied on a controversial step, one that I personally view as the third key ingredient to the new science of consciousness: taking subjective reports seriously. It was not enough to present people with two types of visual stimuli; as experimenters, we had to carefully record what they thought of them. The participant's introspection was crucial: it defined the very phenomenon that we aimed to study. If the experimenter could see an image but the subject denied seeing it, then it was the latter response that counted—the image had to be scored as invisible. Thus, psychologists were forced to find new ways of monitoring subjective introspection, as accurately as possible.

This emphasis on the subjective has been a revolution for psychology. At the beginning of the twentieth century, behaviorists such as John Broadus Watson (1878–1958) had forcefully ousted introspection from the science of psychology:

> Psychology as the behaviorist views it is a purely objective experimental branch of natural science. Its theoretical goal is the prediction and control of behaviour. Introspection forms no essential part of its methods, nor is the scientific value of its data dependent upon the readiness with which they lend themselves to interpretation in terms of consciousness.[9]

Although behaviorism itself was also eventually rejected, it left a lasting mark: throughout the twentieth century, any recourse to introspection remained highly suspicious in psychology. However, I will argue that this dogmatic position is dead wrong. It conflates two distinct issues: introspection as a research method, and introspection as raw data. As a research method, introspection cannot be trusted.[10] Obviously, we cannot count on naïve human subjects to tell us how their mind works; otherwise our science would be too easy. And we should not take their subjective experiences too literally, as when they claim to have had an out-of-body experience and flown to the ceiling, or to have met their dead grandmother in a dream. But in a sense, even such bizarre introspections must be trusted: unless the subject is lying, they correspond to genuine mental events that beg for an explanation.

The correct perspective is to think of subjective reports as raw data.[11] A person who claims to have had an out-of-body experience genuinely *feels* dragged to the ceiling, and we will have no science of consciousness unless we seriously address why such feelings occur. In fact, the new science of consciousness makes an enormous use of purely subjective phenomena, such as visual illusions, misperceived pictures, psychiatric delusions, and other figments of the imagination. Only these events allow us to distinguish objective physical stimulation from subjective perception, and therefore to search for brain correlates of the latter rather than the former. As consciousness scientists, we are never as pleased as when we discover a new visual display that can be subjectively either seen or missed, or a sound that is sometimes reported as audible and sometimes as inaudible. As long as we carefully record, on every trial, what our participants feel, we are in business, because then we can sort the trials into conscious and unconscious ones and search for brain activity patterns that separate them.

Signatures of Conscious Thoughts

These three ingredients—focusing on conscious access, manipulating conscious perception, and carefully recording introspection—have transformed the study of consciousness into a normal experimental science. We can probe the extent to which a picture that people claim not to have seen is in fact processed by the brain. As we will discover, a stagger-ing amount of unconscious processing occurs beneath the surface of our conscious mind. Research using subliminal images has provided a strong platform to study the brain mechanisms of conscious experience. Mod-ern brain imaging methods have given us a means of investigating how far an unconscious stimulus can travel in the brain, and exactly where it stops, thus defining what patterns of neural activity are exclusively asso-ciated with conscious processing.

For fifteen years now, my research team has been using every tool at its disposal, from functional magnetic resonance imaging (fMRI), to electro- and magnetoencephalography, and even electrodes inserted deep in the human brain, to try to identify the cerebral underpinnings of consciousness. Like many other laboratories throughout the world, ours is engaged in a systematic experimental search for patterns of brain ac-tivity that appear if and only if the scanned person is having a conscious experience—what I call the "signatures of consciousness." And our search has been successful. In one experiment after another, the same signatures show up: several markers of brain activity change massively whenever a person becomes aware of a picture, a word, a digit, or a sound. These signatures are remarkably stable and can be observed in a great variety of visual, auditory, tactile, and cognitive stimulations.

The empirical discovery of reproducible signatures of consciousness, which are present in all conscious humans, is only a first step. We need to work on the theoretical end as well: How do these signatures origi-nate? Why do they index a conscious brain? Why does only a certain type of brain state cause an inner conscious experience? Today no scien-tist can claim to have solved these problems, but we do have some strong and testable hypotheses. My collaborators and I have elaborated a theory that we call the "global neuronal workspace." We propose that conscious-ness is global information broadcasting within the cortex: it arises from a neuronal network whose raison d'être is the massive sharing of pertinent information throughout the brain.

The philosopher Daniel Dennett aptly calls this idea "fame in the brain." Thanks to the global neuronal workspace, we can keep in mind any idea that makes a strong imprint on us, for however long we choose, and make sure that it gets incorporated into our future plans, whatever they might be. Thus consciousness has a precise role to play in the computational economy of the brain—it selects, amplifies, and propagates relevant thoughts.

What circuit is responsible for this broadcasting function of consciousness? We believe that a special set of neurons diffuses conscious messages throughout the brain: giant cells whose long axons crisscross the cortex, interconnecting it into an integrated whole. Computer simulations of this architecture have reproduced our main experimental findings. When enough brain regions agree about the importance of incoming sensory information, they synchronize into a large-scale state of global communication. A broad network ignites into a burst of high-level activation—and the nature of this ignition explains our empirical signatures of consciousness.

Although unconscious processing can be deep, conscious access adds an additional layer of functionality. The broadcasting function of consciousness allows us to perform uniquely powerful operations. The global neuronal workspace opens up an internal space for thought experiments, purely mental operations that can be detached from the external world. Thanks to it, we can keep important data in mind for an arbitrarily long duration. We can pass it on to any other arbitrary mental process, thus granting our brains the kind of flexibility that Descartes was looking for. Once information is conscious, it can enter into a long series of arbitrary operations—it is no longer processed in a reflexive manner but can be pondered and reoriented at will. And thanks to a connection to language areas, we can report it to others.

Equally fundamental to the global neuronal workspace is its autonomy. Recent studies have revealed that the brain is the seat of intense spontaneous activity. It is constantly traversed by global patterns of internal activity that originate not from the external world but from within, from the neurons' peculiar capacity to self-activate in a partly random fashion. As a result, and quite opposite to Descartes's organ metaphor, our global neuronal workspace does not operate in an input-output manner, waiting to be stimulated before producing its outputs. On the contrary, even in full darkness, it ceaselessly broadcasts global patterns of neural activity, causing what William James called the "stream of consciousness"—an uninterrupted flow of loosely connected thoughts, primarily shaped by our current goals and only occasionally seeking information in the senses.

René Descartes could not have imagined a machine of this sort, where intentions, thoughts, and plans continually pop up to shape our behavior. The outcome, I argue, is a "free-willing" machine that resolves Descartes's challenge and begins to look like a good model for consciousness.

The Future of Consciousness

Our understanding of consciousness remains rudimentary. What does the future hold in store? At the end of this book, we will return to the deep philosophical questions, but with better scientific answers. There I will argue that our growing understanding of consciousness will help us not only resolve some of our deepest interrogations about ourselves but also face difficult societal decisions and even develop new technologies that mimic the computational power of the human mind.

To be sure, many details remain to be nailed down, but the science of consciousness is already more than a mere hypothesis. Medical applications now lie within our grasp. In countless hospitals throughout the world, thousands of patients in a coma or a vegetative state lie in terrible isolation, motionless, speechless, their brains destroyed by a stroke, a car accident, or a transient deprivation of oxygen. Will they ever regain consciousness? Might some of them already be conscious but fully "locked in" and unable to let us know? Can we help them by turning our brain-imaging studies into a real-time monitor of conscious experience?

My laboratory is now designing powerful new tests that begin to reliably tell whether a person is or is not conscious. The availability of objective signatures of consciousness is already helping coma clinics worldwide and will soon also inform the related issue of whether and when infants are conscious. Although no science will ever turn an *is* into an *ought*, I am convinced that, once we manage to objectively determine whether subjective feelings are present in patients or in infants, we will make better ethical decisions.

Another fascinating application of the science of consciousness involves computing technologies. Will we ever be able to imitate brain circuits *in silico*? Is our current knowledge sufficient to build a conscious computer? If not, what would it take? As consciousness theory improves, it should become possible to create artificial architectures of electronic chips that mimic the operation of consciousness in real neurons and

circuits. Will the next step be a machine that is aware of its own knowledge? Can we grant it a sense of self and even the experience of free will?

I now invite you to take a journey into the cutting-edge science of consciousness, a quest that will guarantee deeper meaning to the Greek
motto "Know thyself."

1

CONSCIOUSNESS ENTERS THE LAB

How did the study of consciousness become a science? First, we had to focus on the simplest possible definition of the problem. Shelving for later the vexing issues of free will and self-consciousness, we concentrated on the narrower issue of conscious access—why some of our sensations turn into conscious perceptions, while others remain unconscious. Then many simple experiments allowed us to create minimal contrasts between conscious and unconscious perception. Today we can literally render an image visible or invisible at will, under full experimental control. By identifying threshold conditions, in which the very same image is perceived consciously only half the time, we can even keep the stimulus constant and let the brain do the switching. It then becomes crucial to collect the viewer's introspection, because it defines the contents of consciousness. We end up with a simple research program: a search for objective mechanisms of subjective states, systematic "signatures" in brain activity that index the transition from unconsciousness to consciousness.

Take a look at the visual illusion in figure 3. Twelve dots, printed in light gray, surround a black cross. Now stare intently at the central cross. After a few seconds, you should see some of the gray dots fade in and out of existence. For a few seconds, they vanish from your awareness; then they pop back in. Sometimes the entire set goes away, temporarily leaving you with a blank page—only to return a few seconds later with a seemingly darker shade of gray.

An objectively fixed visual display can pop in and out of our subjective awareness, more or less at random. This profound observation forms the basis of the modern science of consciousness. In the 1990s, the late Nobel Prize winner Francis Crick and the neurobiologist Christof Koch

FIGURE 3. A visual illusion called "Troxler fading" illustrates one of the many ways in which the subjective content of consciousness can be manipulated. Stare intently at the central cross. After a few seconds, some of the gray dots should vanish, then return at random moments. The objective stimulus is constant, but its subjective interpretation keeps changing. Something must be changing inside your brain—can we track it?

jointly realized that such visual illusions gave scientists a means to track the fate of conscious versus unconscious stimuli in the brain.[1]

Conceptually at least, this research program poses no major difficulty. During the experiment with the twelve dots, for instance, we can record the discharges of neurons from different places in the brain during moments in which the dots are seen, and compare these recordings with those made during moments in which the dots are not seen. Crick and Koch singled out vision as a domain ripe for such investigations, not only because we are beginning to understand in great detail the neural pathways that carry visual information from the retina to the cortex, but also because there are myriad visual illusions that can be used to contrast visible and invisible stimuli.[2] Do they share anything? Is there a single pattern of brain activity that underlies all conscious states and that provides a unifying "signature" of conscious access in the brain? Finding such a signature pattern would be a major step forward for consciousness research.

In their down-to-earth manner, Crick and Koch had cracked the problem open. Following their lead, dozens of laboratories started studying consciousness through elementary visual illusions such as the one you just experienced. Three features of this research program suddenly put conscious perception within experimental reach. First, the illusions did not require an elaborate notion of consciousness—just the simple act of seeing or not seeing, what I have called conscious access. Second, a great many illusions were available for study—as we shall see, cognitive scientists have invented dozens of techniques to make words, pictures, sounds, and even gorillas disappear at will. And third, such illusions are eminently subjective—only you can tell when and where the dots disappear in your mind. Yet the results are reproducible: anyone who watches the figure reports having the same kind of experience. There is no point in denying it: we all agree that something real, peculiar, and fascinating is going on in our awareness. We have to take it seriously.

I argue that those three crucial ingredients have brought consciousness within the reach of science: focusing on conscious access; using a panoply of tricks to manipulate consciousness at will; and treating subjective reports as genuine scientific data. Let us now consider each of these points in turn.

The Many Facets of Consciousness

Consciousness: the having of perceptions, thoughts and feelings; awareness. The term is impossible to define except in terms that are unintelligible without a grasp of what consciousness means. . . . Nothing worth reading has been written about it.

—Stuart Sutherland, *International Dictionary of Psychology* (1996)

Science often progresses by carving out new distinctions that refine the fuzzy categories of natural language. In the history of science, a classic example is the separation of the concepts of heat and temperature. Everyday intuition treats them as one and the same. After all, adding heat to something will increase its temperature, right? Wrong—a block of ice, when heated, will melt while staying at a fixed temperature of zero degrees Celsius. A material may have a high temperature (e.g., a firework spark, which may reach a few thousand degrees Celsius) but have so little heat that it won't burn the skin (because it has very little mass). In the nineteenth century, distinguishing heat (the amount of energy transferred) from temperature (the average kinetic energy in a body) was key to making progress in thermodynamics.

The word *consciousness*, as we use it in daily conversation, is similar to the layman's *heat*: it conflates multiple meanings that cause considerable confusion. In order to bring order to this field, we first need to sort them out. In this book, I argue that one of them, *conscious access*, denotes a well-defined question, one that is sufficiently focused to be studied with modern experimental tools, and that has a good chance of shedding light on the entire problem.

So what do I mean by conscious access? At any given time, a massive flow of sensory stimulation reaches our senses, but our conscious mind seems to gain access to only a very small amount of it. Every morning as I drive to work, I pass the same houses without ever noticing the color of their roof or the number of their windows. As I sit at my desk and concentrate on writing this book, my retina is bombarded with information about the surrounding objects, photographs, and paintings, their shapes and colors. Simultaneously, my ears are stirred with music, birdsong, noise from the neighbors—and yet all these distracting bits remain in the unconscious background while I focus on writing.

Conscious access is, at once, extraordinarily open and inordinately selective. Its *potential* repertoire is vast. At any given moment, with a switch of my attention, I can become conscious of a color, a scent, a sound, a lost memory, a feeling, a strategy, an error—or even the multiple meanings of the word *consciousness*. If I make a blunder, I may even become *self-conscious*—which means that my emotions, strategies, errors, and regrets will enter my conscious mind. At any moment, however, the *actual* repertoire of consciousness is dramatically limited. We are fundamentally reduced to just about one conscious thought at a time (although a single thought can be a substantial "chunk" with several subcomponents, as when we ponder the meaning of a sentence).

Because of its limited capacity, consciousness must withdraw from one item in order to gain access to another. Stop reading for a second, and notice the position of your legs; perhaps you feel a pressure here or a pain there. This perception is now conscious. But a second earlier it was *preconscious*—accessible but not accessed, it lay dormant amid the vast repository of unconscious states. It did not necessarily remain unprocessed: you constantly adjust your posture unconsciously in response to such bodily signals. However, conscious access made it available to your mind—all at once, it became accessible to your language system and to many other processes of memory, attention, intention, and planning. It is precisely that switch from preconscious to conscious, suddenly letting a piece of information into awareness, that I will discuss in the next chapters. Exactly what happens then is the question that I hope to clarify in this book: the brain mechanisms of conscious access.

To do so, we will need to further distinguish conscious access from mere attention—a delicate but indispensable step. What is attention? In his landmark opus *The Principles of Psychology* (1890), William James proposed a famous definition. Attention, he said, is "the taking possession by the mind, in clear and vivid form, of one out of what seem several simultaneously possible objects or trains of thought." Unfortunately, this definition actually conflates two different notions with distinct brain mechanisms: *selection* and *access*. William James's "taking possession by the mind" is essentially what I have called conscious access. It is the bringing of information to the forefront of our thinking, such that it becomes a conscious mental object that we "keep in mind." That aspect of attention, almost by definition, coincides with consciousness: when an object takes possession of our mind such that we can report it (verbally or by gesturing), then we are conscious of it.

However, James's definition also includes a second concept: the isolation of one out of many possible trains of thought, which we now call "selective attention." At any moment, our sensory environment is buzzing with myriad potential perceptions. Likewise, our memory is teeming with knowledge that could, in the next instant, surface back into our consciousness. In order to avoid information overload, many of our brain systems therefore apply a selective filter. Out of countless potential thoughts, what reaches our conscious mind is *la crème de la crème*, the outcome of the very complex sieve that we call attention. Our brain ruthlessly discards the irrelevant information and ultimately isolates a single conscious object, based on its salience or its relevance to our current goals. This stimulus then becomes amplified and able to orient our behavior.

Clearly, then, most if not all of the selective functions of attention have to operate outside our awareness. How could we ever think, if we first had to consciously sift through all the candidate objects of our thoughts? Attention's sieve operates largely unconsciously—attention is dissociable from conscious access. True enough, in everyday life, our environment is often clogged with stimulating information, and we have to give it enough attention to select which item we are going to access. Thus attention often serves as the gateway for consciousness.[3] However, in the lab, experimenters can create situations so simple that only one piece of information is present—and then selection is barely needed before that information gets into the subject's awareness.[4] Conversely, in many cases attention operates sub rosa, covertly amplifying or squashing incoming information even though the final outcome never makes it into our awareness. In a nutshell, selective attention and conscious access are distinct processes.

There is a third concept that we need to carefully set apart: vigilance, also called "intransitive consciousness." In English, the adjective *conscious* can be transitive: we can be conscious *of* a trend, a touch, a tingle, or a toothache. In this case, the word denotes "conscious access," the fact that an object may or may not enter our awareness. But *conscious* can also be intransitive, as when we say "the wounded soldier remained conscious." Here it refers to a *state* with many gradations. In this sense, consciousness is a general faculty that we lose during sleep, when we faint, or when we undergo general anesthesia.

To avoid confusion, scientists often refer to this sense of consciousness as "wakefulness" or "vigilance." Even these two terms should probably be separated: *wakefulness* refers primarily to the sleep-wake cycle, which arises from subcortical mechanisms, whereas *vigilance* refers to the level of excitement in the cortical and thalamic networks that support conscious states. Both concepts, however, differ sharply from conscious access. Wakefulness, vigilance, and attention are just enabling conditions for conscious access. They are necessary but not always sufficient to make us aware of a specific piece of information. For instance, some patients, following a small stroke in the visual cortex, may become color-blind. These patients are still awake and attentive: their vigilance is intact, and so is their capacity to attend. But the loss of a small circuit specialized in color perception prevents them from gaining access to this aspect of the world. In Chapter 6 we will meet patients in a vegetative state who still awaken in the morning and fall asleep at night—yet do not seem to access any information consciously during their waking time. Their wakefulness is intact, yet their impaired brain no longer seems able to sustain conscious states.

In most of this book, we will be asking the "access" question: What happens during consciousness *of* some thought? In Chapter 6, however, we will return to the "vigilance" meaning of consciousness and consider the applications of the growing science of consciousness to patients in a coma or a vegetative state, or with related disorders.

The word *consciousness* has still other meanings. Many philosophers and scientists believe that consciousness, as a subjective state, is intimately related to the sense of self. The "I" seems an essential piece of the puzzle: How can we ever understand conscious perception without first figuring out who is doing the perceiving? In a standard cliché, the first words that a hero utters upon recovering from a knockout blow are "Where am I?" My colleague the neurologist Antonio Damasio defines consciousness as "the self in the act of knowing"—a definition that implies that we cannot solve the riddle of consciousness until we know what a self is.

The same intuition underlies Gordon Gallup's classic mirror self-recognition test, which probes whether children and animals recognize themselves in a mirror.[5] Self-awareness is attributed to a child who uses the mirror to gain access to hidden parts of his body—for instance, to spot a red sticker surreptitiously placed on his forehead. Children gain the ability to detect the sticker by use of a mirror, typically between eighteen and twenty-four months. Chimpanzees, gorillas, orangutans, and even dolphins, elephants, and magpies have been said to pass this test[6]— leading a group of colleagues to bluntly assert, in the Cambridge Declaration on Consciousness (July 7, 2012), that "the weight of evidence indicates that humans are not unique in possessing the neurological substrates that generate consciousness."

Once again, however, science requires that we refine the concepts. Mirror recognition need not indicate consciousness. It could be accomplished by an utterly unconscious device that merely predicts how the body should look and move and that adjusts its movements based on a comparison of these predictions with the actual visual stimulation—as when I thoughtlessly use a mirror to shave. Pigeons can be conditioned to pass the test—although only after considerable training that essentially turns them into mirror-using automata.[7] The mirror recognition test may just be measuring the extent to which an organism has learned enough about its own body to develop expectations of what it looks like, and enough about mirrors to use them to compare expectation with reality—an interesting competence without doubt, but far from a litmus test for possession of a self-concept.[8]

Most important, the link between conscious perception and self-knowledge is unnecessary. Attending a concert or watching a gorgeous sunset can put me in a heightened state of consciousness without requiring that I constantly remind myself that "*I* am in the act of enjoying *myself*." My body and self remain in the background, like recurrent sounds or backdrop illumination: they are potential topics for my attention, lying outside my awareness, that I can attend to and bring into focus whenever needed. In my view, self-consciousness is much like consciousness of color or sound. Becoming conscious of some aspect of myself could just be another form of conscious access in which the information being accessed is not sensory in nature but concerns one of the various mental representations of "me"—my body, my behavior, my feelings, or my thoughts.

What is special and fascinating about self-consciousness is that it seems to include a strange loop.[9] When I reflect upon myself, the "I" appears twice, both as the perceiver and as the perceived. How is this possible? This recursive sense of consciousness is what cognitive scientists call *metacognition*: the capacity to think about one's own mind. The French positivist philosopher Auguste Comte (1798–1857) considered this a logical impossibility. "The thinking individual," he wrote, "could not divide into two, one reasoning, the other watching the reasoning. The observed organ and the observing organ being identical in this case, how could the observation be made?"[10]

Comte was wrong, however: as John Stuart Mill immediately noted, the paradox dissolves when the observing and the observed are encoded at different times or within different systems. One brain system may notice when another fails. We do it all the time, as when we experience a word on the tip of our tongue (we know we should know), notice a reasoning error (we know we erred), or brood over a failed exam (we know we studied, we thought we knew the answers, and we cannot imagine why we failed). Some areas of the prefrontal cortex monitor our plans, attach confidence to our decisions, and detect our errors. Working as a closed-loop simulator, in tight interaction with our long-term memory and imagination, they support an internal soliloquy that lets us reflect upon ourselves without external help. (The very word *reflection* hints at the mirroring function whereby some brain areas "re-present" and evaluate the operation of others.)

All in all, as scientists, we are better off starting with the simplest notion of consciousness: conscious access, or how we become aware of a specific piece of information. The thornier issues of self and recursive consciousness should best be kept for a later time. Maintaining a focus on conscious access, carefully setting it apart from the related concepts of attention, wakefulness, vigilance, self-consciousness, and metacognition, is the first ingredient in our contemporary science of consciousness.[11]

Minimal Contrasts

The second ingredient that makes the science of consciousness possible is the panoply of experimental manipulations that affects the contents of our consciousness. In the 1990s cognitive psychologists suddenly realized that they could fiddle with consciousness by contrasting conscious and unconscious states. Pictures, words, and even movies could be made invisible. What happened to those images at the brain level? By carefully delimiting the powers and limits of unconscious processing, one could begin to delineate, as in a photographic negative, the contours of consciousness itself. Combined with brain imaging, this simple idea provided a solid experimental platform for studying the cerebral mechanisms of consciousness.

In 1989 the psychologist Bernard Baars, in his important book ambitiously called *A Cognitive Theory of Consciousness*,[12] forcefully argued that there are, in fact, dozens of experiments that provide direct forays into the nature of consciousness. Baars added a crucial observation: many of these experiments provide a "minimal contrast": a pair of experimental situations that are minimally different but only one of which is consciously perceived. Such cases are ideal, because they allow scientists to treat conscious perception as an experimental variable that changes considerably even though the stimulus remains virtually constant. By concentrating on such minimal contrasts, and trying to understand what changes in the brain, researchers could get rid of all the irrelevant brain operations that are common to conscious and unconscious processing and concentrate solely on the brain events that track the switch from the unaware to the aware mode.

Consider, for instance, the acquisition of a motor activity such as typing. When we first learn to type, we are slow, attentive, and painfully self-conscious of every move we make. But after a few weeks of practice, typing becomes so fluent that we can do it automatically, while talking or thinking of something else, and without consciously remembering the locations of the keys. For scientists, studying what happens as behavior automatizes sheds light on the transition from conscious to unconscious. It turns out that this very simple contrast identifies a major cortical network, particularly including regions of the prefrontal lobe that activate whenever conscious access occurs.[13]

Studying the converse transition, from unconscious to conscious, is now equally feasible. Visual perception affords experimenters plenty of opportunities for creating stimuli that come in and out of conscious experience. One example is the illusion with which we opened this chapter (see figure 3). Why do the fixed dots occasionally vanish from sight? We still don't fully understand the mechanism, but the general idea is that our visual system treats a constant image as a nuisance rather than as a genuine input.[14] As we keep our eyes perfectly still, each spot creates a constant, motionless stain of blurry gray on our retina—and at some point, our visual system decides to get rid of this constant blot. Our blindness to such spots may reflect an evolved system that filters out the defects of our eyes. Our retina is full of imperfections, such as blood vessels running in front of the photoreceptors, which we must learn to interpret as coming from inside rather than from outside. (Imagine how horrible it would be to be constantly distracted by wiggly bloody curves barring our gaze.) An object's perfect immobility is a cue that our visual system uses in order to decide to fill in the missing information using the nearby texture. (Such "filling in" explains why we fail to notice the "blind spot" in our retina, at the place occupied by the visual nerve and therefore devoid of light receptors.) When we move our eyes, even by a very small amount, the spots drift slightly on the retina. The visual system therefore realizes that they must come from the external world rather than the eye itself—and it immediately lets them pop back into awareness.

Filling in blind spots is just one of the many visual illusions that let us study the transition from unconscious to conscious. Let us take a quick tour of the many other paradigms available in the cognitive scientist's toolkit.

Rival Images

Historically, one of the first productive contrasts between conscious and unconscious vision came from the study of "binocular rivalry," the curious tug-of-war that occurs, inside our brains, when distinct images are shown to the two eyes.

Our consciousness is entirely oblivious to the fact that we have two eyes that constantly move around. While our brain lets us see a stable three-dimensional world, it hides from our sight the amazingly complex operations that underlie this feat. At any given time, each of our eyes receives a slightly different image of the external world—yet we do not experience double vision. Under natural conditions, we typically fail to notice the two images and simply fuse them together into a single homogeneous visual scene. Our brain even takes advantage of the slight space between our two eyes, which induces a relative shift in the two images. As first observed by the English scientist Charles Wheatstone in 1838, it exploits this disparity to locate objects in depth, thus giving us a vivid sense of the third dimension.

But what would happen, Wheatstone wondered, if the two eyes received completely different images, such as a picture of a face in one eye and of a house in the other? Would the images still be fused? Could we see two unrelated scenes at once?

To find out, Wheatstone built a device that he dubbed the stereoscope. (It quickly started a craze for stereo pictures, from landscapes to pornography, that lasted throughout the Victorian era and beyond.) Two mirrors, placed in front of the left and right eyes, allowed the presentation of distinct pictures to the two eyes (figure 4). To Wheatstone's amazement, when the two pictures were unrelated (such as a face and a house), vision became utterly unstable. Instead of fusing the scene, the viewer's perception ceaselessly alternated between one image and the other, with only brief transitions between them. For a few seconds, the face would appear; then it would break down and vanish to reveal the house; and so on in a alternation created solely by the brain. As Wheatstone noted, "It does not appear to be in the power of the will to determine the appearance" of either image. Rather the brain, when confronted with an utterly implausible stimulus, seems to waver between two interpretations: face or house. The two incompatible images seem to fight for conscious perception— hence the term *binocular rivalry*.

Binocular rivalry is an experimenter's dream because it provides a pure test of subjective perception: although the stimulus is constant, the viewer reports that his vision changes. Furthermore, across time, the very same image changes in status: sometimes it is fully visible, while at other times it vanishes completely from conscious perception. What happens to it then? By recording data from neurons in monkeys' visual cortex, the neurophysiologists David Leopold and Nikos Logothetis were the first to observe the cerebral fate of seen and unseen visual images.[15] They trained the monkeys to report their perception by using a lever, then showed that monkeys experienced semirandom alternations of the two images, just as we do; they finally tracked the responses of single neurons as the monkeys' preferred image faded in and out of conscious experience. The results were clear. At the earliest stage of processing, in the primary visual cortex that acts as the visual gateway into the cortex, many cells reflected the objective stimuli: their firing simply depended on which images were presented to the two eyes, and it did not change when the animal reported that his perception had switched. As visual processing progressed to a more advanced level, within so-called higher visual areas such as area V4 and the inferotemporal cortex, more and more neurons began to agree with the animal's report: they fired strongly when the animal reported seeing its preferred image, and much less or

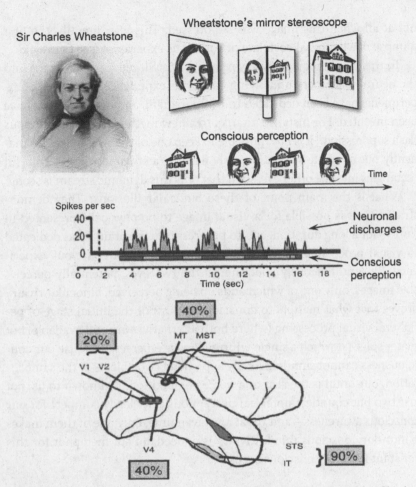

Sir Charles Wheatstone

Wheatstone's mirror stereoscope

Conscious perception

Time

Spikes per sec

Neuronal discharges

Conscious perception

Time (sec)

20%

V1 V2

40%

MT MST

V4

40%

STS

IT

90%

FIGURE 4. Binocular rivalry is a powerful visual illusion discovered by Charles Wheatstone in 1838. A distinct image is presented to each eye, but at any given time we see only one image. Here, a face is presented to the left eye, and a house to the right eye. Rather than seeing two fused images, we see endless alternations of the face, the house, the face again, and so on. Nikos Logothetis and David Leopold trained monkeys to use a joy stick to report what they saw. The researchers showed that monkeys too experience this illusion, and went on to record the activity of neurons in the animals' brains. The illusion was not present at the earliest stages of visual processing, in areas V1 and V2, where most neurons encoded both images equally well. However, at higher levels of the cortical hierarchy, particularly the brain areas IT (inferotemporal cortex) and STS (superior temporal sulcus), most cells correlated with subjective awareness: their discharge rate predicted which image was subjectively seen. Numbers indicate the fraction of such cells in different brain regions. This pioneering research suggests that conscious perception relies predominantly on higher-level association cortex.

not at all when this image was suppressed. This was, literally, the first glimpse of a neuronal correlate of conscious experience (see figure 4).

To this day, binocular rivalry remains a privileged mode of access into the neural machinery underlying conscious experience. Hundreds of experiments have been dedicated to this paradigm, and many variants have been invented. For instance, thanks to a new method called "continuous flash suppression," it is now possible to keep one of the two images permanently out of sight, by continuously flashing a stream of bright colorful rectangles into the other eye, such that only this dynamic stream is seen.[16]

What is the main point of these binocular illusions? They demonstrate that it is possible for a visual image to be physically presented in the eye for a long duration, and to progress into the brain areas dedicated to visual processing, yet be totally suppressed from conscious experience. By simultaneously injecting, in the two eyes, potentially perceivable images, only one of which ends up being perceived, binocular rivalry proves that what matters to consciousness is not the initial stage of peripheral visual processing (where both alternatives are still available) but a later stage (at which a single winning image emerges). Because our consciousness cannot simultaneously apprehend two objects at the same location, our brain is the seat of a fierce competition. Unknown to us, not just two but countless potential perceptions ceaselessly compete for our conscious awareness—and yet at any given time, only one of them makes it into our conscious mind. Rivalry is, indeed, an apt metaphor for this constant fight for conscious access.

When Attention Blinks

Is this rivalry a passive process, or can we consciously decide which image will be the winner of the fight? When we perceive two competing images, our subjective impression is that we are passively submitted to these ceaseless alternations. That impression is false, however: attention does play an important role in the cortical competition process. First of all, if we try hard to attend to one of the two images—for instance, the face rather than the house—its perception lasts a little bit longer.[17] That effect, though, is weak: the fight between the two images starts at stages that are not in our control.

But most important, the very existence of a single winner depends on our giving it our attention; the fighting arena itself, as it were, is made up of the conscious mind.[18] When we remove our attention from the location where the two images are presented, they cease to compete.

The reader may ask, How do we know this? We cannot ask a distracted person what she sees, and whether she still perceives the images as alternating—because in order to respond, she would have to attend to that location. At first sight, the task of determining how much you perceive without attending smacks of circularity, much like trying to monitor how your eyes move in a mirror: no doubt your eyes move constantly, but whenever you watch them in a mirror, that very act forces them to remain still. For a long time, trying to study rivalry without attention seemed a self-defeating strategy, like asking what sound a falling tree makes when no one is around to hear it, or how we feel at the precise moment when we fall asleep.

But science often achieves the impossible. Peng Zhang and his collaborators at the University of Minnesota realized that they did not have to ask the viewer whether the images were still alternating when she was not attending.[19] All they had to do was to find brain markers of rivalry, which would indicate whether the two images still competed with each other. They already knew that, during rivalry, neurons alternatively fire for one or the other image (see figure 4)—so could they still measure such alternations in the absence of attention? Zhang used a technique called "frequency tagging," whereby each image is "tagged" by flickering at its own specific rhythm. The two frequency tags can then be easily picked up on an electroencephalogram, recorded by electrodes placed on the head. Characteristically, during rivalry, the two frequencies exclude each other: if one oscillation is strong, the other is weak, reflecting the fact that we perceive only one image at a time. As soon as we cease to attend, however, these alternations stop, and the two tags co-occur independently of each other: inattention prevents rivalry.

Another experiment confirms this conclusion by pure introspection: when attention is removed from rival images for a fixed duration, the image that is perceived upon return differs from what it should have been, had the images continued to alternate during the inattention period.[20] Thus binocular rivalry depends on attention: in the absence of a consciously attending mind, the two images are jointly processed and no longer compete. Rivalry requires an active, attentive observer.

Attention thus imposes a sharp limit on the number of images that can be simultaneously attended. This limit, in turn, leads to new minimal contrasts for conscious access. One method, aptly called the "attentional blink," consists of creating a brief period of invisibility of an image by temporarily saturating the conscious mind.[21] Figure 5 illustrates the typical conditions under which this blink occurs. A stream of symbols appears at the same location on a computer screen. Most of the symbols are digits, but some are letters, which the participant is told to remember. The first letter is easily remembered. If a second letter occurs half a second or more after the first, it too is accurately committed to memory. If the two letters appear in close succession, however, the second one is often completely missed. The viewer reports seeing only one letter and is quite surprised to learn that there were two of them. The very act of attending to the first letter creates a temporary "blink of the mind" that annihilates the perception of the second.

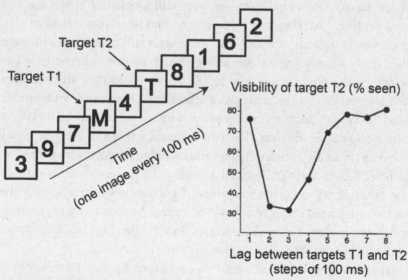

FIGURE 5. The attentional blink illustrates the temporal limitations of conscious perception. When we view a stream of digits interspersed with an occasional letter, we easily identify the first letter (here an M) but not the second (here a T). While we are committing the first letter to memory, our consciousness temporarily "blinks," and we fail to perceive a second stimulus presented within the next instant.

NOTE: ms = milliseconds throughout.

Using brain imaging, we see that all letters, even the unconscious ones, enter the brain. They all reach early visual areas and may even proceed quite deeply into the visual system, to the point of being classified as a target: part of the brain "knows" when a target letter has been presented.[22] But somehow this knowledge never makes it into our conscious awareness. To be consciously perceived, the letter must reach a stage of processing that registers it into our awareness.[23] This registering appears tightly limited: at any given time, only one chunk of information can go through it. Meanwhile everything else in the visual scene remains unperceived.

Binocular rivalry reveals a competition between two simultaneous images. During the attentional blink, a similar competition occurs across time, between two images that are successively presented at the same location. Our consciousness is often too slow to keep up with a fast rate of image presentation on screen. Although we seem to "see" all the digits and letters if we just passively look at them, the act of committing a letter to memory suffices to tie up our conscious resources long enough to create a temporary period of invisibility for the others. The fortress of the conscious mind possesses a small drawbridge that forces mental representations to compete with one another. Conscious access imposes a narrow bottleneck.

The reader may object that we sometimes see two successive letters (about one-third of the time in the data from figure 5). Furthermore, in many other real-life situations, we seem to have no problem perceiving two things that appear almost simultaneously. For instance, we can hear a car's horn while attending to a picture. Psychologists call such situations "dual tasks," because the person is asked to do two things at once. So what happens then? Does dual task performance refute the idea that our conscious awareness is structurally limited to one chunk at a time? No. The evidence shows that, even in such cases, we are still tightly limited. We never really process two unrelated items consciously at exactly the same moment. When we attempt to attend to two things at once, the impression that our consciousness is immediate and "online" with both stimuli is just an illusion. In truth, the subjective mind does not perceive them simultaneously. One of them gets accessed and enters awareness, but the second must wait.

This bottleneck creates an easily measurable processing delay, which is aptly called the "psychological refractory period."[24] While the conscious mind is processing a first item at a conscious level, it appears to be temporarily refractory to further inputs—and therefore to be very late in processing them. While it is processing the first item, the second one lingers in an unconscious buffer. It stays there until the processing of the first item is complete.

We remain oblivious to this unconscious waiting period. But how could it be otherwise? Our consciousness is occupied elsewhere, so we have no means of stepping outside the system and realizing that our conscious perception of the second item is delayed. As a consequence, whenever we are mentally preoccupied, our subjective perception of the timing of events can be systematically wrong.[25] Once we are engaged in a first task, then are asked to estimate *when* a second item appeared, we wrongly postdate it to the moment when it entered our consciousness. Even when two inputs are objectively simultaneous, we fail to perceive their simultaneity and systematically feel that the first one that we attended to appeared earlier than the other. In truth, this subjective delay arises solely from the sluggishness of our consciousness.

The attentional blink and the refractory period are deeply related psychological phenomena. Whenever the conscious mind is occupied, all other candidates for awareness have to wait in an unconscious buffer—and the wait is risky: at any time, due to internal noise, distracting thoughts, or other incoming stimuli, a buffered item may be erased and vanish from awareness (the blink). Experiments indeed confirm that, during a dual task, both refractoriness and blinking occur. Conscious perception of the second item is always delayed, and the probability of a complete blackout increases with the duration of the delay.[26]

During most dual task experiments, the blink lasts only a fraction of a second. Committing a letter to memory, indeed, requires only a brief moment. What happens, however, when we perform a much longer distracting task? The surprising answer is that we can become totally oblivious to the external world. Avid readers, concentrated chess players, and focused mathematicians know all too well that intellectual absorption can create long periods of mental isolation, during which we lose all awareness of our surroundings. The phenomenon, dubbed "inattentional blindness," is easily demonstrated in the lab. In one experiment,[27] participants gaze at the center of a computer screen but are told to attend to the top side. A letter will soon appear there, they are told, and they will have to remember it. They train on this task for two trials. Then on the third, simultaneously with the peripheral letter, an unexpected shape also appears at the center. It may be a large dark spot, a digit, or even a word—and it may last for nearly a second. But amazingly, up to two-thirds of the participants fail to notice it. They report seeing the peripheral letter and nothing else. Only when the trial is rerun do they realize, to their utter surprise, that they missed a major visual event. In brief, inattention breeds invisibility.

For another classic demonstration, consider Dan Simons and Christopher Chabris's extraordinary experiment known as "the invisible gorilla" (figure 6).[28] A film shows two teams—one wearing white T-shirts, one wearing black—practicing basketball. Viewers are asked to count the passes made by the team wearing white. The movie lasts about thirty seconds, and with a little concentration, nearly everyone counts fifteen passes. And then the experimenter asks, "Did you see the gorilla?" Of course not! The tape is rewound, and there it is: in the middle of the film, an actor dressed in a gorilla suit enters the stage, bangs his chest several times in full sight, then leaves. A majority of the viewers fail to detect the gorilla in the first showing: they swear that there never was one. They are so sure of themselves that they accuse the experimenter of showing a different movie the second time! The very act of concentrating on the players wearing white T-shirts makes a black gorilla vanish into oblivion.

In cognitive psychology, the gorilla study is a landmark. At about the same time, researchers discovered dozens of similar situations in which inattention leads to temporary blindness. People turn out to be terrible witnesses. Simple manipulations can make us unconscious of even the most blatant parts of a visual scene. Kevin O'Regan and Ron Rensink discovered "change blindness,"[29] a striking inability to detect which part of a picture has been erased. Two versions of the picture, with or without a deletion, alternate on screen every second or so, with just a short blank between them. Viewers swear that the two pictures are identical—even when the change is huge (a jet loses its engine) or highly relevant (in a driving scene, the central road line changes from broken to continuous).

Dan Simons demonstrated change blindness in a staged experiment using live actors. An actor asks a student for directions on the Harvard campus. The conversation is briefly interrupted by passing workers, and

FIGURE 6. Inattention may cause blindness. Our conscious perception is tightly limited, so the very act of attending to an item can prevent us from perceiving others. In the classic gorilla movie (above), viewers are asked to count how many times the players who are wearing white pass a basketball. As they concentrate on the white-clad team, they fail to notice that an actor, dressed as a gorilla, enters the stage and forcefully bangs his chest before leaving. In another movie (below), no fewer than twenty-one major items in the crime scene change without viewers noticing. How many "gorillas in our midst" do we miss in our everyday lives?

once it resumes, two seconds later, the original actor has been replaced by a second. Even though the two people have different hair and clothing styles, most students fail to notice the swap.

An even more remarkable case is Peter Johansson's study of "choice blindness."[30] In this experiment, a male subject is shown two cards, each with a picture of a female face, and chooses which he prefers. The card bearing the chosen picture is passed to him, but while it is briefly held face down, the experimenter surreptitiously swaps the two cards. The participant ends up holding a picture of the face that he did *not* choose. Half of the participants are oblivious to this manipulation. They happily proceed to comment on the choice they never made and readily invent explanations for why this face is definitely more attractive than the other!

For the most spectacular demonstration of visual unawareness, connect to YouTube and search for *Whodunnit?*, a brief detective movie commissioned by the London transportation department.[31] A distinguished British detective grills three suspects and ends up arresting one of them. Nothing suspicious . . . until the movie rewinds, the camera backs up, and we suddenly realize that we missed massive anomalies. Within one minute, no fewer than twenty-one elements of the visual scene were incoherently changed, right in front of our eyes. Five assistants swapped the furniture, replaced a huge stuffed bear with a medieval suit of armor, and helped the actors change their coats and trade the objects that they held. A naïve spectator misses it all.

The impressive change blindness movie ends with the mayor of London's moralizing words: "It's easy to miss something you're not looking for. On a busy road, this could be fatal—look out for cyclists!" And the mayor is right. Flight-simulation studies have shown that trained pilots, when communicating with traffic control, become so oblivious to other events that they may even crash into a plane that they failed to detect.

The lesson is clear: inattention can make virtually any object vanish from our consciousness. As such, it provides an essential tool for contrasting conscious and unconscious perception.

Masking Conscious Perception

In the laboratory, testing inattentional blindness has a problem: experiments require replication over hundreds of trials, but inattention is a very labile phenomenon. On the first trial, most naïve viewers miss even a massive change, but the slightest hint of the manipulation is enough to make them become watchful. As soon as they are on the alert, the change's invisibility is compromised.

Furthermore, although unattended stimuli can create a powerful subjective feeling of unconsciousness, scientists find it quite hard to prove, beyond a reasonable doubt, that participants are truly unaware of the changes that they claim not to have seen. One may question them after every trial, but this procedure is slow and puts them on the lookout. Another possibility is to postpone the questioning until the end of the whole experiment, but this is equally problematic because forgetting then becomes an issue: after a few minutes, viewers may underestimate what they had been aware of.

Some researchers suggest that, during change blindness experiments, participants are always aware of the whole scene but simply fail to commit most of the details to memory.[32] Thus, change blindness may arise not from a lack of awareness but from an inability to compare the old scene with the new one. Once motion cues are eliminated, even one second of delay may make it difficult for the brain to compare two pictures. By default, the participant would respond that nothing has changed; according to this interpretation, they consciously perceived all the scenes and merely failed to notice that they differ.

I personally doubt that the forgetting explanation accounts for all inattention and change blindness—after all, a gorilla in a basketball game or a stuffed bear in a crime scene should be rather memorable. But a lingering doubt remains. For an unquestionably scientific study, what is needed is a paradigm in which the image is 100 percent invisible—and no matter how informed the participants are, no matter how hard they try to discern it, and no matter how many times they view the film, they still do not see it. Fortunately, such a complete form of invisibility exists. Psychologists call it "masking"; the rest of the world knows it as "subliminal images." A subliminal image is one that is presented below the threshold of consciousness (literally—*limen* means "threshold" in Latin), such that nobody can see it, even with considerable effort.

How does one create such an image? One possibility is to make it very faint. Unfortunately, that solution typically degrades the image so much that it produces very little brain activity. A more interesting method is to flash the image for a brief moment, sandwiched between two other pictures. Figure 7 shows how we can "mask" an image of the word *radio*. First, we flash the word for a short duration of 33 milliseconds, about the length of one movie frame. By itself, this duration does not suffice to

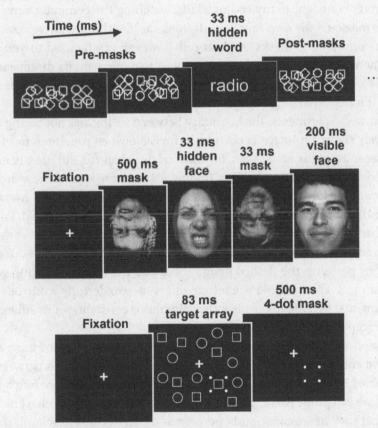

FIGURE 7. Masking can make an image invisible. This technique consists of flashing a picture, surrounded in time by other similar shapes that act as masks and prevent its conscious perception. In the top example, a single word briefly flashed within a series of random geometric shapes remains invisible to the viewer. In the middle, a flashed face, even if it carries a strong emotion, can be made unconscious by surrounding it with random pictures: the viewer sees only the masks and the final face. In the bottom case, a whole array of shapes serves as the target. Paradoxically, the only shape that cannot be perceived is the one that is signaled by four surrounding dots. By extending beyond the duration of the initial array, the four dots act as masks.

induce invisibility—in complete darkness, even a microsecond-long flash of light will illuminate a scene and freeze it. What makes the image of *radio* invisible, however, is a visual illusion called "masking." The word is preceded and followed by geometric shapes that appear at the same location. When the timing is right, the viewer sees only the flickering patterns. Sandwiched between them, the word becomes utterly invisible.

I devise many subliminal masking experiments myself, and although I am pretty confident in my coding skills, watching the computer screen makes me doubt my own eyes. It really looks as if nothing at all is present in between the two masks. A photocell, however, can be used to verify that the word is indeed flashed for an objective moment: its disappearance is a purely subjective phenomenon. The word invariably reappears when it is displayed long enough.

In many experiments, the boundary between seeing and not seeing is relatively sharp: an image is downright invisible when presented for 40 milliseconds, but is easily seen, on most trials, when the duration is increased to 60 milliseconds. This finding justifies the use of the words *subliminal* (below threshold) and *supraliminal* (above threshold). Metaphorically, the gateway to consciousness is a well-defined threshold, and a flashed image is either in or out. The length of the threshold varies across subjects, but it always falls close to 50 milliseconds. At this duration, one perceives the flashed image about half the time. Presenting visual stimuli at threshold therefore offers a wonderfully controlled experimental paradigm: the objective stimulus is constant, yet its subjective perception varies from trial to trial.

Several variants of masking can be used to modulate consciousness at will. An entire picture may vanish from sight when sandwiched between scrambled images. When the picture is a smiling or a fearful face (see figure 7), we can probe participants' subliminal perception of a hidden emotion that they never consciously perceived—at an unconscious level, the emotion shines through. Another version of masking involves flashing an array of shapes and cueing one of them by surrounding it with four long-lasting dots (see figure 7).[33] Surprisingly, only the cued shape vanishes from conscious experience; all the others remain clearly visible. Because they last longer than the array, the four dots and the white space that they enclose appear to replace and wipe out any conscious perception of a shape at that location; hence this method is called "substitution masking."

Masking is a great laboratory tool because it allows us to study the fate of an unconscious visual stimulus with high temporal precision and with complete control over experimental parameters. The best conditions involve flashing a single target stimulus followed by a single mask. At a precise moment, we "inject" into the viewer's brain a well-controlled dose of visual information (say, a word). In principle, this dose should suffice for the viewer to consciously perceive the word, because if we remove the trailing mask, he or she always sees it. But when the mask is present, it somehow overrides the prior image and is the only thing that the viewer perceives. A strange race must be happening in the brain: although the word enters first, the subsequent mask seems to catch up and abolish it from conscious perception. One possibility is that the brain behaves like a statistician weighing the evidence before deciding whether one item or two were present. When the word presentation is short enough, and the mask strong enough, then the viewer's brain receives overwhelming evidence in favor of the conclusion that only the mask was present—and it becomes oblivious to the word.

Primacy of the Subjective

Can we guarantee that masked words and pictures are truly unconscious? In my lab's latest experiments, we simply ask the participants, after each trial, whether they saw a word or not.[34] Several of our colleagues quibble over this procedure, which they judge "too subjective." But such skepticism seems off the mark: by definition, in consciousness research, subjectivity is at the heart of our subject matter.

Fortunately, we also have other means of convincing the skeptics. First, masking is a subjective phenomenon that induces considerable agreement among viewers. Below a duration of about 30 milliseconds, all participants, in every trial, deny seeing a word; only the minimal duration that they require before perceiving something varies somewhat.

Most important, it is easy to verify that during masking, subjective invisibility has objective consequences. In trials where subjects report seeing nothing, they usually cannot name the word. (Only when forced to respond do they perform slightly above chance—a finding that indicates a degree of subliminal perception, and to which we will return in the next chapter.) A few seconds later, they fail to make even the simplest judgments, such as deciding whether a masked digit is larger or smaller than the number 5. In one of my lab's experiments,[35] we repeatedly presented the same list of thirty-seven words up to twenty times, but with masks that made them invisible. At the end of the experiment, we asked viewers to select these old words from among new ones that had not been presented. They were utterly unable to do so, suggesting that the masked words had left no trace in their memory.

All this evidence points to an important conclusion, the third key ingredient in our budding science of consciousness: *subjective reports can and should be trusted*. Although invisibility caused by masking is a subjective phenomenon, it has very real consequences for our capacity to process information. In particular, it drastically reduces our naming and memory abilities. Near the masking threshold, the trials that a viewer labels as "conscious" are accompanied by a massive change in the amount of available information, which is reflected not only in a subjective feeling of being aware but also in a host of other improvements in processing the stimulus.[36] Whatever information we are conscious of, we can name it, rate it, judge it, or memorize it much better than we can when it is subliminal. In other words, human observers are neither random nor whimsical about their subjective reports: when they report an honest-to-god feeling of seeing, such conscious access corresponds to a massive change in information processing, which almost always results in an enhanced performance.

In other words, contrary to a century of behaviorist and cognitive suspicion, introspection is a respectable source of information. Not only does it provide valuable data, which can often be confirmed objectively, by behavioral or brain-imaging measures, it also *defines* the very essence of what a science of consciousness is about. We are looking for an objective explanation of subjective reports: signatures of consciousness, or sets of neuronal events that systematically unfold in the brain of a person whenever she experiences a certain conscious state. By definition, only she can tell us about it.

In a 2001 review that became a manifesto of the field, my colleague Lionel Naccache and I summarized this position as follows: "Subjective reports are the key phenomena that a cognitive neuroscience of consciousness purports to study. As such, they constitute primary data that need to be measured and recorded along with other psychophysiological observations."[37]

This being said, we should not be naïve about introspection: while it certainly provides raw data for the psychologist, it is not a direct window into the operations of the mind. When a neurological or psychiatric patient tells us that he sees faces in the dark, we do not take him literally—but neither should we deny that he has had this experience. We just need to explain *why* he has had it—perhaps because of a spontaneous, possibly epileptic activation of the face circuits in his temporal lobe.[38]

Even in normal people, introspection can be demonstrably wrong.[39] By definition, we have no access to our many unconscious processes—but this does not prevent us from making up stories about them. For instance, many people think that when they read a word, they recognize it instantaneously and "as a whole," based on its overall shape; but actually a sophisticated series of letter-based analyses occurs in their brain, of which they are completely unaware.[40] As a second example, consider what happens when we try to make sense of our past actions. People often invent all sorts of contorted, after-the-fact interpretations for their decisions—oblivious to their true unconscious motivations. In a classic experiment, consumers were shown four pairs of nylon stockings and asked to judge which pair was the best quality. In fact, all the stockings were identical, but people nevertheless showed a strong preference for whichever pair was presented on the right side of the shelf. When asked to explain their choice, none of them ever mentioned the role of shelf location; instead they commented at some length on the quality of the fabric! In this instance, introspection was demonstrably delusional.

In that sense, the behaviorists were right: as a method, introspection provides a shaky ground for a science of psychology, because no amount of introspection will tell us how the mind works. However, as a measure, introspection still constitutes the perfect, indeed the only, platform on which to build a science of consciousness, because it supplies a crucial half of the equation—namely, how subjects feel about some experience (however wrong they are about the ground truth). To attain a scientific understanding of consciousness, we cognitive neuroscientists "just" have to determine the other half of the equation: Which objective neurobiological events systematically underlie a person's subjective experience?

Sometimes, as we just saw for masking, subjective reports can be immediately corroborated by objective evidence: a person says that she saw a masked word, and she immediately proves it by accurately naming it aloud. Consciousness researchers should not be wary, however, of the many other cases in which subjects report on a purely internal state that, superficially at least, seems utterly unverifiable. Even in such cases, there must be objective neural events that explain the person's experience—and since this experience is detached from any physical stimulus, it may actually be easier for researchers to isolate its cerebral source, because they will not confound it with other sensory parameters. Thus contemporary consciousness researchers are constantly on the hunt for "purely subjective" situations, in which sensory stimulation is constant (sometimes even absent), yet subjective perception varies. These ideal cases turn conscious experience into a pure experimental variable.

A case in point is the Swiss neurologist Olaf Blanke's beautiful series of experiments on out-of-body experiences. Surgery patients occasionally report leaving their bodies during anesthesia. They describe an irrepressible feeling of hovering at the ceiling and even looking down at their inert body from up there. Should we take them seriously? Does out-of-body flight "really" happen?

In order to verify the patients' reports, some pseudoscientists hide drawings of objects atop closets, where only a flying patient could see them. This approach is ridiculous, of course. The correct stance is to ask how this subjective experience could arise from a brain dysfunction. What kind of brain representation, Blanke asked, underlies our adoption of a specific point of view on the external world? How does the brain assess the body's location? After investigating many neurological and surgery patients, Blanke discovered that a cortical region in the right temporoparietal junction, when impaired or electrically perturbed, repeatedly caused a sensation of out-of-body transportation.[41] This region is situated in a high-level zone where multiple signals converge: those arising from vision; from the somatosensory and kinesthetic systems (our brain's map of bodily touch, muscular, and action signals); and from the vestibular system (the biological inertial platform, located in our inner ear, which monitors our head movements). By piecing together these various clues, the brain generates an integrated representation of the body's location relative to its environment. However, this process can go awry if the signals disagree or become ambiguous as a result of brain damage. Out-of-body flight "really" happens, then—it is a real physical event, but only in the patient's brain and, as a result, in his subjective experience. The out-of-body state is, by and large, an exacerbated form of the dizziness that we all experience when our vision disagrees with our vestibular system, as on a rocking boat.

Blanke went on to show that *any* human can leave her body: he created just the right amount of stimulation, via synchronized but delocalized visual and touch signals, to elicit an out-of-body experience in the normal brain.[42] Using a clever robot, he even managed to re-create the illusion in a magnetic resonance imager. And while the scanned person experienced the illusion, her brain lit up in the temporoparietal junction—very close to where the patient's lesions were located.

We still do not know exactly how this region works to generate a feeling of self-location. Still, the amazing story of how the out-of-body state moved from parapsychological curiosity to mainstream neuroscience gives a message of hope. Even outlandish subjective phenomena can be traced back to their neural origins. The key is to treat such introspections with just the right amount of seriousness. They do not give direct insights into our brain's inner mechanisms; rather, they constitute the raw material on which a solid science of consciousness can be properly founded.

. . .

At the end of this brief review of the contemporary approach to consciousness, we thus reach an optimistic conclusion. In the past twenty years, many clever experimental tools have emerged, with which researchers may manipulate consciousness at will. Using them, we can make words, pictures, and even entire movies vanish from awareness—and then, with minimal changes or sometimes none at all, make them visible again.

With these tools in hand, we can now ask all the burning questions that René Descartes would have loved to raise. First, what happens to an unseen image? Is it still processed in the brain? For how long? How far does it go into the cortex? Do the answers depend on how the stimulus was made unconscious? [43] And then, second, what changes when a stimulus becomes consciously perceived? Are there unique brain events that appear only when an item makes it into conscious awareness? Can we identify these signatures of consciousness and use them to theorize what consciousness is?

In the next chapter, we begin with the first of these questions: the fascinating issue of whether subliminal images deeply influence our brains, thoughts, and decisions.

2

FATHOMING UNCONSCIOUS DEPTHS

How deep can an invisible image travel into the brain? Can it reach our higher cortical centers and influence the decisions we make? Answering these questions is crucial to delineating the unique contours of conscious thought. Recent experiments in psychology and brain imaging have tracked the fate of unconscious pictures in the brain. We recognize and categorize masked images unconsciously, and we even decipher and interpret unseen words. Subliminal pictures trigger motivations and rewards in us—all without our awareness. Even complex operations linking perception to action can unfold covertly, demonstrating how frequently we rely on an unconscious "automatic pilot." Oblivious to this boiling hodgepodge of unconscious processes, we constantly overestimate the power of our consciousness in making decisions—but in truth, our capacity for conscious control is limited

Time past and time future allow but a little consciousness.

—T. S. Eliot, *Burnt Norton* (1935)

During the 2000 presidential campaign, a nasty commercial concocted by George W. Bush's team featured a caricature of Al Gore's economic plan, accompanied by the word *RATS* in huge capital letters (figure 8). Although not strictly subliminal, the image went largely unnoticed, for it flew by inconspicuously at the end of the word *bureaucrats*. The offending epithet stirred a debate: Did the viewer's brain register the hidden meaning? How far did it travel in the brain? Could it reach the voter's emotional center and influence an electoral decision?

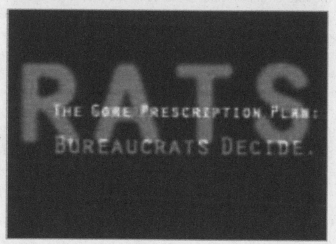

FIGURE 8. Subliminal images are occasionally used in the media. During the 1988 French presidential campaign, the face of president and candidate François Mitterrand was briefly flashed within the logo of the major public TV program. In 2000, in one of George W. Bush's commercials, Al Gore's economic plan was surreptitiously labeled with the word *RATS*. Are such unconscious images processed by the brain, and do they influence our decisions?

The French elections, twelve years earlier, had been the theater of an even more controversial use of subliminal images. The face of presidential candidate François Mitterrand was briefly flashed within the logo of the main state television program (figure 8). This invisible image appeared daily at the opening of the eight p.m. news broadcast, a popular program for French viewers. Did it bias the votes? Even a very small shift, in a nation of fifty-five million, would mean thousands of ballots.

The mother of all subliminal manipulations is the (in)famous insertion of a frame with the words *Drink Coca Cola* into a 1957 movie. Everybody knows the story and its outcome: a massive increase in sales of soft drinks. Yet this foundational myth of subliminal research was a complete fabrication. James Vicary made up the story and later admitted that the experiment was a hoax. Only the myth persists, and so does the scientific question: Can unseen images influence our thoughts? This is not just an important issue for freedom and mass manipulation, but also a key interrogation for our scientific understanding of the brain. Do we need to be conscious of an image in order to process it? Or can we perceive, categorize, and decide without awareness?

The issue has become all the more pressing now that a variety of methods exists for presenting information to the brain in an unconscious manner. Binocular images, inattention, masking, and many other situations render us oblivious to many aspects of our surroundings. Are we just blind to them? Whenever we attend to a given object, do we cease to perceive all the unattended surroundings? Or do we continue to process them, but in a subliminal manner? And if we do, how far can they progress into the brain without receiving the beam of consciousness?

Answering those questions is particularly crucial for our scientific goal of identifying the brain signatures of conscious experience. If subliminal processing is deep, and if we can fathom that depth, then we will understand the nature of consciousness much better. Once we know, for instance, that the early stages of perception can operate without awareness, we will be able to exclude them from our search for consciousness. By extending this process of elimination to higher-level operations, we will learn more and more about the specifics of the conscious mind. Delineating the contours of the unconscious will progressively print a negative photograph of the conscious mind.

Pioneers of the Unconscious

The discovery that a dramatic amount of mental processing occurs outside our awareness is generally credited to Sigmund Freud (1856–1939). However, this is a myth, crafted in large part by Freud himself.[1] As noted by the historian and philosopher Marcel Gauchet, "When Freud declares, in substance, that prior to psychoanalysis the mind was systematically identified with consciousness, we have to declare this statement rigorously false."[2]

In truth, the realization that many of our mental operations occur sub rosa, and that consciousness is only a thin veneer lying atop sundry unconscious processors, predates Freud by decades or even centuries.[3] In Roman antiquity, the physician Galen (ca. 129–200) and the philosopher Plotinus (ca. 204–270) had already noticed that some of the body's operations, such as walking and breathing, occur without attention. Much of their medical knowledge was in fact inherited from Hippocrates (ca. 460–377 BC), a keen observer of diseases whose name remains an emblem of the medical profession. Hippocrates wrote an entire treatise on epilepsy, called *The Sacred Disease*, in which he noted that the body suddenly misbehaves against its owner's will. He concluded that the brain constantly controls us and covertly weaves the fabric of our mental life:

> Men ought to know that from the brain, and from the brain alone, arise our pleasures, joys, laughter and jests, as well as our sorrows, pains, grieves and tears. Through it, in particular, we think, see, hear and distinguish the ugly from the beautiful, the bad from the good, the pleasant from the unpleasant.

During the Dark Ages, which followed the fall of the Roman Empire, Indian and Arab scholars preserved some of antiquity's medical wisdom. In the eleventh century, the Arab scientist known as Alhazen (Ibn al-Haytham, 965–1040) discovered the main principles of visual perception. Centuries before Descartes, he understood that the eye operates as a camera obscura, a receiver rather than an emitter of light, and he foresaw that various illusions could fool our conscious perception.[4] Consciousness was not always in control, Alhazen concluded. He was the first to postulate an automatic process of unconscious inference: unknown to us, the brain jumps to conclusions beyond the available sense data, sometimes causing us to see things that are not there.[5] Eight centuries later the physicist Hermann von Helmholtz, in his 1867 book, *Physiological Optics*, would use the very same term, *unconscious inference*, to describe how our vision automatically computes the best interpretation compatible with incoming sense data.

Beyond the issue of unconscious perception lay the greater issue of the origins of our deepest motives and desires. Centuries before Freud, many philosophers—including Augustine (354–430), Thomas Aquinas (1225–74), Descartes (1596–1650), Spinoza (1632–77), and Leibniz (1646–1716)—noted that the course of human actions is driven by a broad array of mechanisms that are inaccessible to introspection, from sensorimotor reflexes to unaware motives and hidden desires. Spinoza cited a hodgepodge of unconscious drives: a child's desire for milk, an injured person's will for revenge, a drunkard's craving for the bottle, and a chatterbox's uncontrollable speech.

During the eighteenth and nineteenth centuries, the first neurologists discovered proof after proof of the omnipresence of unconscious circuits in the nervous system. Marshall Hall (1790–1857) pioneered the concept of a "reflex arc," linking specific sensory inputs to particular motor outputs, and he emphasized our lack of voluntary control over basic movements that originate in the spinal cord. Following in his footsteps, John Hughlings Jackson (1835–1911) underscored the hierarchical organization of the nervous system, from the brain stem to the cerebral cortex and from automatic operations to increasingly voluntary and conscious ones. In France, the psychologists and sociologists Théodule Ribot (1839–1916), Gabriel Tarde (1843–1904), and Pierre Janet (1859–1947) stressed the broad range of human automatisms, from practical knowledge stored in our action memory (Ribot) to unconscious imitation (Tarde) and even to subconscious goals that date from early childhood and become defining facets of our personality (Janet).

French scientists were so advanced that when the ambitious Freud published his first claims to fame, Janet protested that he owned the paternity of many of Freud's ideas. As early as 1868, the British psychiatrist Henry Maudsley (1835–1918) had written that "the most important part of mental action, the essential process on which thinking depends, is unconscious mental activity."[6] Another contemporary neurologist, Sigmund Exner, who was Freud's colleague in Vienna, had stated in 1899: "We shouldn't say 'I think,' 'I feel,' but rather 'it thinks in me' [*es denkt in mir*], 'it feels in me' [*es fühlt in mir*]"—a full twenty years prior to Freud's reflections in *The Ego and the Id* (*Das Ich und das Es*), published in 1923.

At the turn of the century, the ubiquity of unconscious processes was so well accepted that in his major treatise *The Principles of Psychology* (1890), the great American psychologist and philosopher William James could boldly state: "All these facts, taken together, form unquestionably the beginning of an inquiry which is destined to throw a new light into the very abysses of our nature. . . . They prove one thing conclusively, namely, that we must never take a person's testimony, however sincere, that he has felt nothing, as proof positive that no feeling has been there."[7] Any human subject, he surmised, "will do all sorts of incongruous things of which he remains quite unaware."

Relative to this flurry of neurological and psychological observations, clearly demonstrating that unconscious mechanisms drive much of our lives, Freud's own contribution appears speculative. It would not be a huge exaggeration to say that in his work, the ideas that are solid are not his own, while those that are his own are not solid. In hindsight, it is particularly disappointing that Freud never tried to put his views to an empirical test. The late nineteenth and early twentieth centuries saw the birth of experimental psychology. New empirical methods flourished, including the systematic collection of precise response times and errors. But Freud seemed content with proposing metaphorical models of the mind without seriously testing them. One of my favorite writers, Vladimir Nabokov, had no patience with Freud's method and nastily barked: "Let the credulous and the vulgar continue to believe that all mental woes can be cured by a daily application of old Greek myths to their private parts. I really do not care."[8]

The Seat of Unconscious Operations

In spite of the major medical advances of the nineteenth and twentieth centuries, only twenty years ago, in the 1990s, when my colleagues and I started to apply brain-imaging techniques to subliminal perception, an enormous amount of confusion still surrounded the issue of unseen pictures in the brain. Many conflicting accounts of a division of labor were being proposed. The simplest idea was that the cortex—the folded sheets of neurons that form the surface of our two cerebral hemispheres—was conscious while all the other circuits were not. The cortex, the most evolved part of the brain in mammals, hosts the advanced operations that underlie attending, planning, and speaking. Thus, it was a fairly natural hypothesis to consider that whatever information reached the cortex had to be conscious. Unconscious operations, by contrast, were thought to take place solely within specialized brain nuclei such as the amygdala or the colliculus, which had evolved to perform dedicated functions such as the detection of fearful stimuli or the movement of the eyes. These groups of neurons form "subcortical" circuits, so called because they lie underneath the cortex.

A different but equally naïve proposal introduced a dichotomy between the two hemispheres of the brain. The left hemisphere, which hosts the language circuits, could report on what it was doing. Therefore it would be conscious, while the right wouldn't.

A third hypothesis was that some cortical circuits were conscious, while others were not. Specifically, whatever visual information is carried through the brain by the ventral route, which recognizes the identity of objects and faces, would necessarily be conscious. Meanwhile information carried by the dorsal visual route, which goes through the parietal cortex and uses object shape and location to guide our actions, would forever lie on the unconscious dark side.

None of these simplistic dichotomies held up to scrutiny. Based on what we now know, virtually all the brain's regions can participate in both conscious and unconscious thought. To get to this conclusion, however, clever experiments were needed to progressively expand our understanding of the range of the unconscious.

Initially, simple experiments in patients with brain injuries suggested that unconscious operations brooded in the hidden basement of the brain, beneath the cortex. The amygdala, for instance, an almond-shaped group of neurons located beneath the temporal lobe, flags important, emotionally laden situations of everyday life. It is particularly crucial for coding fear; frightening stimuli, such as the sight of a snake, can activate it on a fast track from the retina, well before we register the emotion at a conscious cortical level.[9] Many experiments have indicated that such emotional appraisals are made extraordinarily quickly and unconsciously, mediated by the fast circuitry of the amygdala. Early in the 1900s, the Swiss neurologist Édouard Claparède demonstrated an unconscious emotional memory: while he was shaking the hand of an amnesic patient, he pricked her with a pin, and the next day, while her amnesia prevented her from recognizing him, she emphatically refused to shake his hand. Such experiments provided a first proof that complex emotional operations could unfold below the level of awareness, and they always seemed to arise from a set of subcortical nuclei specialized for emotional processing.

Another source of data on subliminal processing was "blindsight" patients, those with lesions of the primary visual cortex, the main source of visual inputs into the cortex. The oxymoronic term *blindsight* may seem bizarre, but it accurately describes these individuals' Shakespearean condition: to see, but not to see. A lesion in the primary visual cortex should make a person blind, and it does deprive such patients of their *conscious* vision—they assure you that they cannot see anything in a specific part of the visual field (which corresponds precisely to the destroyed area of cortex), and they behave as if they were blind. Incredibly enough, however, when an experimenter shows them objects or flashes of light, they accurately point to them.[10] In a zombielike manner, they unconsciously guide their hand to locations that they do not see—blindsight indeed.

Which intact anatomical pathways support unconscious vision in blindsight patients? Clearly, in these patients, some visual information still makes it through from the retina to the hand, bypassing the lesion that makes them blind. Because the entry point into the patients' visual cortex had been destroyed, the researchers initially suspected that their unconscious behavior arose entirely from subcortical circuits. A key suspect was the superior colliculus, a nucleus in the midbrain that specializes in the cross-registration of vision, eye movements, and other spatial responses. Indeed, the first functional MRI study of blindsight demonstrated that unseen targets triggered a strong activation in the superior colliculus.[11] But that study also contained evidence that the unseen stimuli evoked activations in the cortex—and sure enough, later research confirmed that invisible stimuli could still activate both the thalamus and higher-level visual areas of the cortex, somehow bypassing the damaged primary visual area.[12] Clearly, the brain circuits that take part in our unconscious inner zombie and that guide our eye and hand movements include much more than just old subcortical routes.

Another patient, studied by the Canadian psychologist Melvyn Goodale, strengthened the case for a cortical contribution to unconscious processing. At the age of thirty-four, D.F. suffered carbon monoxide intoxication.[13] Lack of oxygen caused dramatic and irreversible damage to her left and right lateral visual cortexes. As a result, she lost some of the most basic aspects of conscious vision, developing what neurologists call "visual form agnosia." For purposes of shape recognition, D.F. was essentially blind—she could not tell a square from an elongated rectangle. Her deficit was so severe that she failed to recognize the orientation of a slanted line (vertical, horizontal, or oblique). Yet her gesture system was still remarkably functional: when asked to post a card through a slanted slit, whose orientation she consistently failed to perceive, her hand behaved with perfect accuracy. Her motor system always seemed to unconsciously "see" things better than she could consciously. She also adapted the size of her grasp to the objects that she reached for—yet she was utterly unable to do so voluntarily, using the finger-to-thumb distance as a symbolic gesture for perceived size.

D.F.'s unconscious ability to perform motor actions seemed to vastly exceed her capacity for consciously perceiving the same visual shapes. Goodale and his collaborators argued that her performance could not be explained solely by subcortical motor pathways but had to involve the cortex of the parietal lobes as well. Although D.F. was unaware of it, information about the size and orientation of objects was still proceeding unconsciously down her occipital and parietal lobes. There, intact circuits extracted visual information about size, location, and even shape that she could not consciously see.

Since then, severe blindsight and agnosia have been studied in a host of similar patients. Some of them could navigate a busy hallway without bumping into objects, all the while claiming total blindness. Other patients experienced a form of unconsciousness called "spatial neglect." In this fascinating condition, a lesion to the right hemisphere, typically in the vicinity of the inferior parietal lobe, prevents a patient from attending to the left side of space. As a result, he or she often misses the entire left half of a scene or an object. One patient forcefully complained about not being given enough food: he had eaten all the food on the right side of his plate but failed to notice that the left half was still full.

Spatial neglect patients, while dramatically impaired in their conscious judgments and reports, are not truly blind in the left visual field. Their retinas and early visual cortex are perfectly functional, yet somehow a higher-level lesion prevents them from attending this information and registering it at a conscious level. Is the unattended information totally lost? The answer is no: the cortex still processes the neglected information, but at an unconscious level. John Marshall and Peter Halligan elegantly made this point by showing a spatial neglect patient pictures of two houses, one of which was on fire on the left side (figure 9).[14] The patient forcefully denied seeing any difference between them—he claimed that the houses were identical. But when asked to choose which one he would prefer to live in, he consistently avoided picking the one on fire. Obviously, his brain was still processing visual information deeply enough that it could categorize the fire as a danger to be avoided. A few years later, brain-imaging techniques showed that in spatial neglect patients, an unseen stimulus could still activate the regions of the ventral visual cortex that respond to houses and faces.[15] Even the meaning of neglected words and numbers invisibly made its way into the patients' brain.[16]

The Brain's Dark Side

All this evidence initially arose from patients with severe and often massive brain lesions that had arguably altered the separation between conscious and unconscious operations. Do normal brains, in the absence of a lesion, also process images unconsciously at a deep visual level? Can our cortex operate without our awareness? Might even the sophisticated functions that we acquire at school, such as reading or arithmetic, execute unconsciously? My laboratory was among the first to provide a positive answer to these important questions; we used brain imaging to demonstrate that invisible words and digits reach quite deep in the cortex.

As I explained in Chapter 1, we can flash a picture for several dozen milliseconds, yet keep it unseen. The trick is to mask the critical event that we wish to hide from consciousness with other shapes just before and after it (see figure 7). But how far does such a masked picture travel in the brain? My colleagues and I got an indication by using the clever technique of "subliminal priming." We briefly flashed a subliminal word or picture (dubbed the prime) and immediately followed it with another visible item (the target). On successive trials, the target might be identical to the prime or different from it. For example, we flashed the prime word *house* so briefly that the participants did not see it, and then the target

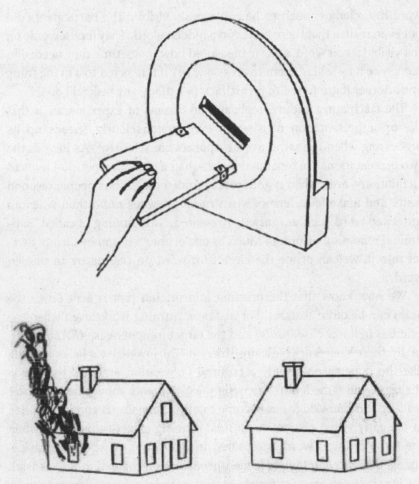

FIGURE 9. Patients with brain lesions provided the first solid evidence that unconscious images are processed in the cortex. Following a brain lesion, Goodale and Milner's (1991) patient D.F. lost all visual recognition ability and became utterly unable to perceive and describe shapes, even one as simple as a slanted slit (above). Nevertheless, she could accurately post a card through it, suggesting that complex hand movements can be guided unconsciously. Marshall and Halligan's (1988) patient P.S., who suffered from massive neglect of the left side of space, failed to consciously perceive any difference between the two houses below. Yet when asked which one he would prefer to live in, he consistently avoided the house on fire, suggesting that he unconsciously understood the meaning of the drawing.

word *radio* long enough to be consciously visible. The participants did not even realize that there had been a hidden word. They focused only on the visible target word, and we measured how much time they needed to recognize it by asking them to press one key if it referred to a living thing and another if it referred to an artifact. (Virtually any task will do.)

The fascinating finding, replicated in dozens of experiments, is that the prior presentation of a word, even unconsciously, speeds up its processing when the same word reappears consciously.[17] As long as the two presentations are separated by less than a second, repetition leads to facilitation—even when it goes totally undetected. Thus people respond faster and make fewer errors when *radio* precedes *radio* than when an unrelated word such as *house* is presented. This finding is called "subliminal repetition priming." Much as one primes a pump by flushing water into it, we can prime the circuit for word processing by an unseen word.

We now know that the priming information that is sent down the brain can be quite abstract. For instance, priming works even when the prime is in lower case (*radio*) and the target in upper case (*RADIO*). Visually, these shapes are radically different. The lowercase *a* looks nothing like the uppercase *A*. Only a cultural convention attaches those two shapes to the same letter. Amazingly, experiments show that, in expert readers, this knowledge has become totally unconscious and is compiled in the early visual system: subliminal priming is just as powerful when the same physical word is repeated (*radio-radio*) as when the case is changed (*radio-RADIO*).[18] Hence unconscious information proceeds all the way up to an abstract representation of letter strings. From the mere glimpse of a word, the brain manages to quickly identify the letters independently of superficial changes in letter shapes.

The next step was to understand where this operation occurs. As my colleagues and I proved, brain imaging is sensitive enough to identify the small activation elicited by an unconscious word.[19] Using functional magnetic resonance imaging (fMRI), we made whole-brain pictures of areas that were affected by subliminal priming. The results showed that a large chunk of the ventral visual cortex could be activated unconsciously. The circuit included a region called the fusiform gyrus, which houses advanced mechanisms of shape recognition and implements the early stages of reading.[20] Here priming did not depend on the shape of the word: this brain area was clearly able to process the abstract identity of a word without caring whether it was in upper or lower case.[21]

Prior to these experiments, some researchers had postulated that the fusiform gyrus always took part in conscious processing. It formed the so-called ventral visual route that allowed us to see shapes. Only the "dorsal route," they thought, linking the occipital visual cortex with the action systems of the parietal cortex, was the seat of unconscious operations.[22] By demonstrating that the ventral route, which cares about the identity of pictures and words, could also operate in an unconscious mode, our experiments and others helped dispel the simplistic idea that the ventral route was conscious while the dorsal route was not.[23] Both circuits, although they are seated high up in the cortex, appeared to be capable of operating below the level of conscious experience.

Binding Without Consciousness

Year after year research on subliminal priming has dispelled many myths about the role of consciousness in our vision. One now-discarded idea was that, although the individual elements of a visual scene could be processed without awareness, consciousness was needed to bind them together. Without conscious attention, features such as motion and color floated freely around and were not bound together into the appropriate objects.[24] The various sites of the brain had to piece the information together into a single "binder" or "object file" before a global percept could arise. Some researchers postulated that this binding process, made possible by neuronal synchrony[25] or reentry,[26] was the hallmark of conscious processing.

We now know that they were wrong: some visual bindings can occur without consciousness. Consider the binding of letters into a word. The letters must clearly be attached together in a precise left-to-right arrangement, so as not to confuse words like *RANGE* and *ANGER*, where the movement of a single letter makes a huge difference. Our experiments demonstrated that such binding is achieved unconsciously.[27] We found that subliminal repetition priming occurred when the word *RANGE* was preceded by *range,* but not when *RANGE* was preceded by *anger*—indicating that subliminal processing is highly sensitive, not just to the presence of letters but also to how they are arranged. In fact, responses to *RANGE* preceded by *anger* were no faster than responses to

RANGE preceded by an unrelated word such as *tulip*. Subliminal perception is not fooled by words that have 80 percent of their letters in common: a single letter can radically alter the pattern of subliminal priming.

In the past ten years, such demonstrations of subliminal perception have been replicated hundreds of times—not just for written words but also for faces, pictures, and drawings.[28] They led to the conclusion that what we experience as a conscious visual scene is a highly processed image, quite different from the raw input that we receive from the eyes. We never see the world as our retina sees it. In fact, it would be a pretty horrible sight: a highly distorted set of light and dark pixels, blown up toward the center of the retina, masked by blood vessels, with a massive hole at the location of the "blind spot" where cables leave for the brain; the image would constantly blur and change as our gaze moved around. What we see, instead, is a three-dimensional scene, corrected for retinal defects, mended at the blind spot, stabilized for our eye and head movements, and massively reinterpreted based on our previous experience of similar visual scenes. All these operations unfold unconsciously—although many of them are so complicated that they resist computer modeling. For instance, our visual system detects the presence of shadows in the image and removes them (figure 10). At a glance, our brain unconsciously infers the sources of lights and deduces the shape, opacity, reflectance, and luminance of the objects.

Whenever we open our eyes, a massively parallel operation takes place in our visual cortex—but we are unaware of it. Uninformed of the inner workings of our vision, we believe that the brain works hard only when we *feel* that we are working hard—for instance, when we're doing math or playing chess. We have no idea how hard it is also working behind the scenes to create this simple impression of a seamless visual world.

Playing Chess Unconsciously

For another demonstration of the power of our unconscious vision, consider chess playing. When grand master Garry Kasparov concentrates on a chess game, does he have to consciously attend to the configuration of pieces in order to notice that, say, a black rook is threatening the white

FIGURE 10. Powerful unconscious computations lie beneath our vision. Glance at this image, and you see a normal-looking checkerboard. You have no doubt that square A is dark and square B is light. But amazingly, they are printed in the same exact shade of gray. (Check this by masking the image with a sheet of paper.) How can we explain this illusion? In a fraction of a second, your brain unconsciously parses the scene into objects, decides that the light comes from the top right, detects that the cylinder casts a shadow on the board, and subtracts this shadow from the image, letting you see what it infers are the true colors of the checkerboard beneath it. Only the final result of these complex computations enters your conscious awareness.

queen? Or can he focus on the master plan, while his visual system automatically processes those relatively trivial relations among pieces?

Our intuition is that in chess experts, the parsing of board games becomes a reflex. Indeed, research proves that a single glance is enough for any grand master to evaluate a chessboard and to remember its configuration in full detail, because he automatically parses it into meaningful chunks.[29] Furthermore, a recent experiment indicates that this segmenting process is truly unconscious: a simplified game can be flashed for 20 milliseconds, sandwiched between masks that make it invisible, and still influence a chess master's decision.[30] The experiment works only on expert chess players, and only if they are solving a meaningful problem, such as determining if the king is under check or not. It implies that the visual system takes into account the identity of the pieces (rook or knight) and their locations, then quickly binds together this information into a meaningful chunk ("black king under check"). These sophisticated operations occur entirely outside conscious awareness.

Seeing Voices

All our examples so far have come from vision. Could consciousness be the glue that binds our distinct sensory modalities into a coherent whole? Do we need to be conscious in order to fuse together visual and auditory signals, as when we enjoy a movie? Again, the surprising answer is no. Even multisensory information can be bound together unconsciously—we become aware only of the result. We owe this conclusion to a remarkable illusion called the "McGurk effect," first described by Harry McGurk and John MacDonald in 1976.[31] The video, which can be found on the Internet,[32] shows a person speaking, and it seems obvious that she is saying *da da da da*. Nothing puzzling—until you close your eyes and realize that the true auditory stimulus is the syllable *ba ba ba*! How does the illusion work? Visually, the mouth of the person moves to say *ga*—but because your ears receive the syllable *ba*, your brain is confronted with a conflict. It solves it, unconsciously, by fusing the two pieces of information. If the two inputs are well synchronized, it binds the information together into a single intermediate percept: the syllable *da*, a compromise between the auditory *ba* and the visual *ga*.

This auditory illusion shows us again how late and reconstructed our conscious experience is. As surprising as it seems, we do not hear the sound waves that reach our ears; nor do we see the photons entering our eyes. What we gain access to is not a raw sensation but an expert reconstruction of the outside world. Behind the scenes, our brain acts as a clever sleuth that ponders all the separate pieces of sensory information we receive, weighs them according to their reliability, and binds them into a coherent whole. Subjectively, it does not feel like any of it is reconstructed. We do not have the impression of *inferring* the identity of the fused sound *da*—we just *hear* it. Nevertheless, during the McGurk effect, what we hear demonstrably arises from sight just as much as from sound.

Where in the brain is this conscious multisensory brew concocted? Brain imaging suggests that it is in the frontal cortex, rather than in the early auditory or visual sensory areas, that the conscious outcome of the McGurk illusion is finally represented.[33] The content of our conscious perception is first distilled within our higher areas, then is sent back to early sensory regions. Clearly, many complex sensory operations unfold sub rosa to assemble the scene that eventually plays out seamlessly in our mind's eye, as if coming straight from our sensory organs.

Can just any information be assembled unconsciously? Probably not. Vision, speech recognition, and expert chess have something in common—they are all extremely automatic and overlearned. This is presumably why their information can be bound without awareness. The neurophysiologist Wolf Singer has suggested that we should perhaps distinguish two types of bindings.[34] Routine bindings would be those that are coded by dedicated neurons committed to specific combinations of sensory inputs. Nonroutine bindings, by contrast, are those that require the de novo creation of unforeseen combinations—and they may be mediated by a more conscious state of brain synchrony.

This more nuanced view of how our cortex synthesizes our perceptions seems much more likely to be correct. From birth on, the brain receives intensive training in what the world looks like. Years of interaction with the environment allow it to compile detailed statistics of which parts of objects tend to frequently co-occur. With intensive experience, visual neurons become dedicated to the specific combination of parts that characterizes a familiar object.[35] After learning, they continue to respond to the appropriate combination even during anesthesia—a clear proof that this form of binding does not require consciousness. Our capacity to recognize written words probably owes much to such unconscious statistical learning: by adulthood, the average reader has seen millions of words, and his or her visual cortex is likely to contain neurons committed to identifying frequent letter strings such as *the, un,* and *tion.*[36] In expert chess players, likewise, a fraction of neurons may become attuned to chessboard configurations. This sort of automatic binding, compiled into dedicated brain circuits, is quite different from, say, the binding of new words into a sentence. When you smile at Groucho Marx's sentence "Time flies like an arrow; fruit flies like a banana," these words bind for the first time in your brain—and part of that combination, at least, seems to require consciousness. Indeed, brain-imaging experiments show that during anesthesia, our brain's capacity to integrate words into sentences is strongly reduced.[37]

Unconscious Meaning?

Our visual system is clever enough to unconsciously assemble several letters into a word—but can the word's meaning also be processed without awareness? Or is consciousness needed to understand even a single word? This deceptively simple question has turned out to be fiendishly difficult to answer. Two generations of scientists have fought over it like mad dogs—each camp persuaded that its answer was obvious.

How could word comprehension *not* require a conscious mind? If one defines consciousness as "the perception of what passes in a man's own mind," as John Locke did in his celebrated *Essay Concerning Human Understanding* (1690), then it is hard to see how the mind could grasp a word's meaning without, at the same time, becoming aware of it. Comprehension (etymologically, "together-catching," the assembling of fragments of meaning in "common sense") and consciousness ("together-knowing") are so closely connected in our mind as to be virtually synonymous.

And yet how could language operate if the elementary process of word comprehension required consciousness? As you read this sentence, do you consciously work out each word's meaning before assembling the words together into a coherent message? No: your conscious mind focuses on the overall gist, the logic of the argument. A glance at each word is enough to place it within the overall structure of discourse. We have no introspection of how a sign evokes a meaning.

So who is right? Thirty years of research in psychology and brain imaging have finally settled the issue. The story of how it was done is interesting, a wild waltz of conjectures and refutations progressively converging toward a stable truth.

It all started in the 1950s with studies of the "cocktail party" effect.[38] Picture yourself at a noisy party. Dozens of conversations around you mix up, but you manage to concentrate on just one of them. Your attention operates as a filter that selects one voice and thwarts all others. Or does it? The British psychologist Donald Broadbent postulated that attention acts as an early filter that interrupts processing at a low level: unattended voices arc blocked at a perceptual level, he surmised, before they can have any influence on comprehension.[39] But this view does not survive scrutiny. Imagine that suddenly one of the party's guests, standing behind you, casually calls your name, even in a low voice. Immediately your attention switches to that speaker. This implies that your brain did indeed process the unattended word, all the way up to a representation of its meaning as a proper name.[40] Careful experimentation confirms this effect and even shows that unattended words can bias a listener's judgment of the conversation that he or she focuses on.[41]

Cocktail party and other divided-attention experiments suggest an unconscious comprehension process, but do they offer watertight evidence? No. In those experiments, listeners deny splitting their attention and swear that they could not hear the unattended stream (that is, before their name was called), but how can wc be sure? Skeptics easily destroy such experiments by denying that the unattended stream is truly unconscious. Perhaps the listener's attention switchcs very quickly from one stream to the other, or perhaps one or two words pass through during a blank period. The cocktail party effect, although impressive in a real-life context, was hard to transform into a laboratory test of unconscious processing.

In the 1970s the Cambridge psychologist Anthony Marcel went one step further. He used the masking technique to flash words below the threshold of conscious perception. With this method, he achieved complete invisibility: every participant, on every trial, denied seeing any word. Even when they were told that a hidden word was present, they could not perceive it. When they were asked to venture a response, they remained unable to say whether the hidden string was an English word or just a random string of consonants. Nevertheless, Marcel was able to show that the participants' brains processed the hidden word unconsciously all the way to its meaning.[42] In a key experiment, he flashed a color word such as *blue* or *red*. Participants denied seeing the word, but when they were subsequently asked to choose a patch of the corresponding color, they were faster by about one-twentieth of a second than when they had been exposed to another, unrelated word. Thus, an unseen color word could prime them to choose the corresponding color. This seemed to imply that their brains had unconsciously registered the meaning of the hidden word.

Marcel's experiments uncovered another striking phenomenon: the brain seemed to unconsciously process all possible meanings of words, even when they were ambiguous or irrelevant.[43] Imagine that I whisper in your ear the word *bank*. A financial institution comes to your mind— but on second thought, perhaps I meant the edge of a river. Consciously, we seem to become aware of only one meaning at a time. Which meaning gets selected is clearly biased by the context: seeing the word *bank* in the context of Robert Redford's beautiful 1992 movie *A River Runs Through It* primes the water-related meaning. In the lab, one can show that even a single word, such as *river*, suffices to make the word *bank* prime the word *water*, while seeing *save* before *bank* primes the word *money*.[44]

Crucially, this adaptation to context seems to occur only at the conscious level. When the prime word was masked down to a subliminal level, Marcel observed a joint activation of both meanings. After flashing the word *bank*, both *money* and *water* were primed—even when a strong context favored the river meaning. Thus our unconscious mind is clever enough to store and retrieve, in parallel, all the possible semantic associations of a word, even when the word is ambiguous and even when only one of its meanings actually fits in the context. The unconscious mind proposes while the conscious mind selects.

The Great Unconscious Wars

Marcel's semantic priming experiments were very creative. They strongly suggested that sophisticated processing of a word's meaning could occur unconsciously. But they were not watertight, and the true skeptics remained unmoved.[45] Their skepticism launched a massive fight between the champions and the detractors of unconscious semantic processing.

Their disbelief was not entirely unjustified. After all, the subliminal influence that Marcel found was so small that it was close to negligible. Flashing a word facilitated processing by a very small amount, sometimes less than one-hundredth of a second. Perhaps this effect arose from a very small fraction of trials on which the hidden word had, in fact, been seen—albeit so briefly as to leave very little or no trace in memory. Marcel's primes were not always unconscious, his detractors argued. In their opinion, the participants' mere verbal report of "I didn't see any words," recorded only at the end of the experiment, failed to provide convincing evidence that they had never seen the prime words. Much greater care was needed to measure prime awareness as objectively as possible, in a separate experiment in which subjects were asked, for instance, to venture a name for the hidden word, or to categorize it according to some criterion. Only random performance on this secondary task, the skeptics contended, would indicate that the primes were truly invisible. And this control task had to be run under exactly the same conditions as in the main experiment. In Marcel's experiments, they argued, either these conditions were not met or, when they were, there was indeed a significant fraction of above-chance responses, suggesting that subjects might have seen a few words.

In response to these critiques, the advocates of unconscious processing tightened up their experimental paradigms. Remarkably, the results still confirmed that words, digits, and even pictures could be unconsciously grasped.[46] In 1996 the Seattle psychologist Anthony Greenwald published in the top-ranking journal *Science* a study that seemed to provide definitive evidence that the emotional meaning of words was processed unconsciously. He had asked participants to classify words as emotionally positive or negative by clicking one of two response keys; unknown to them, each visible target was preceded by a hidden prime. The word pairs were either congruent, reinforcing each other's meaning (both positive or both negative, as when *happy* was followed by *joy*), or incongruent (e.g., *rape* followed by *joy*). When participants responded extremely fast, they performed better on congruent than on incongruent trials. The emotional meanings evoked by the two words seemed to pile up unconsciously, helping the final decision when they shared the same emotion, and hindering it when they did not.

Greenwald's results were strongly replicable. Most subjects not only swore that they could not see the hidden primes but were objectively unable to judge their identity or emotion above chance level. Furthermore, how well they did on such direct guessing tasks was unrelated to how much congruency priming they showed. The priming effect did not seem to be carried by a small set of people who could see the prime words.

Here was, at long last, a genuine demonstration that an emotional meaning could be activated unconsciously.

Or was it? Although the strict referees of *Science* magazine accepted it, Tony Greenwald was a tougher critic of his own work, and a few years later, with his student Richard Abrams, he came up with an alternative interpretation of his own experiment.[47] He pointed out that his experiment had used only a small set of repeated words. Perhaps, he surmised, the participants responded to the same words so often, and under such a tough time pressure, that they ended up associating the letters themselves rather than the meanings with the response categories—thus bypassing meaning. The explanation was not absurd because in the *Science* experiment, subjects repeatedly saw the same words as primes and as targets, and always classified them according to the same rule. After consciously classifying *happy* twenty times as a positive word, Greenwald realized, perhaps their brains wired up a direct nonsemantic route from the meaningless letters *h-a-p-p-y* to the "positive" response.[48]

Alas, this hunch turned out to be correct: in this experiment, priming was indeed subliminal, but it bypassed meaning. First, Greenwald showed that meaningless scrambled primes were just as effective as the real words—*hypap* was just as powerful a prime as *happy*. Second, he carefully manipulated the resemblance of the words that people consciously saw with those that served as hidden primes. In a crucial experiment, two of the conscious words were *tulip* and *humor*, which participants obviously classified as positive. Greenwald then recombined their letters to create a negative word, *tumor*, which he presented only unconsciously.

The fascinating finding was that, unconsciously, the *negative* word *tumor* primed a *positive* response. Subliminally, the participants' brain put *tumor* together with the words *tulip* and *humor* from which it was derived—even though their meaning could not be more different. This was a definite proof that priming depended on a shallow association between specific sets of letters and their corresponding response. Greenwald's experiment involved unconscious perception but not the words' deeper meaning. Under these experimental conditions at least, unconscious processing was not smart at all: instead of caring about a word's meaning, it merely depended on the mapping between letters and responses.

Anthony Greenwald had destroyed the semantic interpretation of his own *Science* paper.

Unconscious Arithmetic

By 1998, although unconscious semantic processing remained as elusive as ever, my colleagues and I realized that Greenwald's experiments were perhaps not the final word. An unusual feature of those experiments is that the participants were asked to respond within a strict deadline of 400 milliseconds. This delay seemed too short to compute the meaning of a rare word such as *tumor*. Given such a tight deadline, the brain had time only to associate letters with responses; perhaps with a more relaxed schedule, it would unconsciously analyze a word's meaning. So Lionel Naccache and I started some experiments that would definitely prove that a word's meaning could be unconsciously activated.[49]

To maximize our chances of obtaining a large unconscious effect, we settled on language's simplest category of meaningful words: numbers. Numbers below ten are special: they are very short words, frequent, extremely familiar, and overlearned since early childhood; their meaning is transparently simple. They can be conveyed in a remarkably compact form—by a single digit. In our experiment, we therefore flashed the numbers 1, 4, 6, and 9, preceded and followed by a string of random letters that made them entirely invisible. Immediately afterward we displayed a second number, this time clearly visible.

We asked our participants to follow the simplest possible instruction: Please tell us, as fast as you can, whether the number that you see is larger or smaller than 5. They had no idea that there was a hidden number; in a separate test, at the end of the experiment, we showed that even when they knew there was one, they could not see it or classify it as large or small. Still, the invisible numbers caused semantic priming. When they were congruent with the target (e.g., both larger than 5), the participants responded more quickly than when they were incongruent (e.g., one smaller and the other larger). For instance, flashing a subliminal digit 9 accelerated the response to 9 and 6, but slowed the response to 4 and 1.

Using brain imaging, we detected a trace of this effect at the cortical level. We observed a very tiny activation in the motor cortex commanding the hand that would have been an appropriate response to the invisible stimulus. Unconscious votes were traversing the brain, from perception to motor control (figure 11). This effect could arise only from

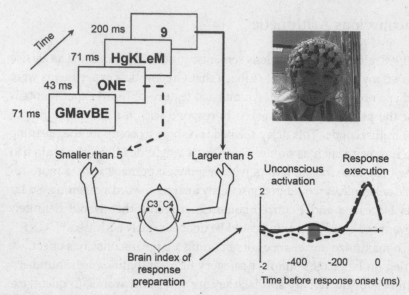

FIGURE 11. Our motor cortex can prepare a response to a stimulus that we do not see. Here, a volunteer was asked to classify numbers as larger or smaller than 5. In this example, the visible target was 9. Just before the target, a hidden number was flashed (the word *one*). Although the hidden number was invisible, it nevertheless sent a small unconscious activation to the motor cortex, commanding the hand that would have been appropriate to respond to it. Thus, an unseen symbol may be identified, processed according to arbitrary instructions, and propagated all the way to the motor cortex.

an unconscious categorization of the meaning of invisible words or digits.

Subsequent work put the final nail in the skeptics' coffin. Our subliminal effect was entirely independent of the notation used for the numbers: *four* primed 4 just as much as an exact repetition of 4 primed 4, suggesting that all the effect arose at the level of abstract meaning. We later showed that priming persisted when the prime was an invisible *visual* number and the target a conscious *spoken* number.[50]

In our initial experiment, the effect might have been caused by a direct association between visual shapes and responses—the same problem that had plagued Greenwald's experiments with emotional words. However, subliminal number priming avoided this criticism. We proved that hidden numbers that had never been consciously seen in the entire experiment still caused semantic priming.[51] By imaging brain activation with functional MRI, we even obtained direct evidence that the "number sense" regions of the brain, in the left and right parietal lobes, were influenced by the unseen number.[52] These regions encode the quantity meaning of numbers[53] and are thought to house neurons tuned to specific quantifies.[54] During subliminal priming, their activity decreased whenever we displayed the same number twice (e.g., *nine* followed by 9). This is a classical phenomenon called "repetition suppression" or "adaptation," which indicates that the neurons recognize that the same item is displayed twice. It seemed that the neurons coding for quantity were habituating to seeing the same number twice, even when the first presentation was unconscious. The evidence had mounted: a higher brain area cared about a specific meaning and could be activated without consciousness.

The final knockout came when our colleagues demonstrated that the number priming effect varies as a direct function of the overlap in number meaning.[55] The strongest priming was obtained by displaying the same quantity twice (e.g., a subliminal *four* preceding 4). The priming decreased slightly for nearby numbers (*three* preceding 4), got even smaller for numbers at a distance of 2 (*two* preceding 4), and so on. Such a semantic distance effect is a hallmark of number meaning. It can arise only if the subject's brain encodes that 4 resembles 3 more than 2 or 1—a definite argument in favor of an unconscious extraction of that number's meaning.

Combining Concepts Without Consciousness

The skeptics' last resort was to accept our demonstration but to assume that numbers were special. Adults have so much experience with this closed set of words, they argued, that it should be no surprise that we can automatically understand them. Other categories of words, however, would be different—surely their meaning would not be represented without consciousness. But even this last line of resistance collapsed when similar priming techniques revealed semantic congruity effects with unseen words outside the number domain.[56] For instance, deciding that the target *piano* is an object rather than an animal can be facilitated by the subliminal presentation of the congruent word *chair*, and hindered by the incongruent word *cat*—even when the primes are never seen throughout the experiment.

Brain-imaging techniques also confirmed the cognitive scientist's conclusions. Recordings of neural activity provided direct evidence that the brain regions involved in semantic processing could be activated without consciousness. In one study, my colleagues and I took advantage of electrodes that had been implanted deep in the brain, in subcortical regions specialized in emotional processing.[57] Naturally, such recordings were performed not in healthy volunteers but in patients with epilepsy. In many hospitals throughout the world, it has become clinical routine to insert electrodes deep inside the patient's skull, in order to identify the source of epileptic discharges and ultimately excise the impaired tissue. In between the seizures, if the patient agrees, we can use the electrodes for a scientific purpose. They grant us access to the average activity of a small brain region or sometimes to the signal emitted by just one neuron.

In our case, the electrodes reached deep into the amygdala, a brain structure involved in emotion. As I explained earlier, the amygdala responds to all sorts of frightening stuff, from snakes and spiders to scary music and strangers' faces—even a subliminal snake or face may trigger it.[58] Our question was, Would this region activate to an unconscious frightening word? So we flashed subliminal words with a disturbing meaning, such as *rape, danger,* or *poison*—and to our great pleasure, an electrical signal appeared, which was absent for neutral words such as *fridge* or *sonata.* The amygdala "saw" words that remained invisible to the patients themselves.

This effect was remarkably slow: it took half a second or more before an invisible word caused an unconscious emotional dip. But the activation was completely unconscious: at the same time that his amygdala fired, a participant denied seeing any word and, when asked to guess, had no idea what it was. Thus a written word could slowly make its way down into the brain, be identified, and even be understood, all without consciousness.

The amygdala is not part of the cortex, so perhaps this makes it special and more automatic. Could the language cortex fire to an unconscious meaning? Further experiments gave a positive answer. They relied on a cortical wave that marks the brain's response to an unexpected meaning. "At breakfast, I like my coffee with cream and socks": as you read such a silly sentence, the bizarre meaning of the final word generates a particular brain wave called the N400. (The *N* refers to its shape, which shows a negative voltage on the top of the head, and the *400* to its peak latency, about 400 milliseconds after the word appears.)

The N400 reflects a sophisticated level of operation, which evaluates how a given word fits within a sentence's context. Its size varies directly with the degree of absurdity: words whose meaning is roughly appropriate cause a very small N400, while utterly unexpected words generate a larger one. Remarkably, this brain event occurs even with words that we do not see—whether they are rendered invisible by masking[59] or by inattention.[60] Networks of neurons in our temporal lobe automatically process not only the various meanings of invisible words but also their compatibility with the past conscious context.

In recent work, Simon van Gaal and I even showed that the N400 wave could reflect an unconscious combination of words.[61] In this experiment, two words appeared in succession, both of them masked below the awareness threshold. They were selected to form unique combinations of positive or negative meanings: "not happy," "very happy," "not sad," and "very sad." Immediately after this subliminal sequence, the subjects saw a positive or negative word (say, *war* or *love*). The N400 wave emitted by this conscious word was modulated by the global unconscious context. Not only did *war* evoke a large N400 when preceded by the incongruous word *happy*, but this effect was strongly modulated, up or down, by the intensifier *very* or the negation *not*. Unconsciously, the brain registered the incongruity of a "very happy war," and judged "not happy war" or "very sad war" as better fits. That experiment is as close as one gets to proving that the brain can unconsciously process the syntax and meaning of a well-formed word phrase.[62]

Perhaps the most remarkable aspect of these experiments is that the N400 wave has exactly the same size whether the words are conscious or invisible. This finding is rife with implications. It means that, in some respects, consciousness is irrelevant to semantics—our brain sometimes performs the same exact operations, all the way up to the meaning level, whether or not we are aware of them. It also means that unconscious stimuli do not always generate minuscule events in the brain. Brain activity can be intense even though the stimulus that causes it remains invisible.

We conclude that an invisible word is fully capable of eliciting a large-scale activation in the brain's meaning networks. However, an important caveat is in order. Accurate reconstruction of the sources of semantic brain waves shows that the unconscious activity is confined to a narrow and specialized brain circuit. During unconscious processing, brain activity remains within the boundaries of the left temporal lobe, which is the primary site of the language networks that process meaning.[63] Later we shall see that conscious words, conversely, gain the upper hand over much larger brain networks that invade the frontal lobes and that underlie the special subjective sense of having the word "in mind." What this means is that, ultimately, unconscious words are not as influential as conscious ones.

Attentive but Unconscious

The discovery that a word or a digit can travel throughout the brain, bias our decisions, and affect our language networks, all the while remaining unseen, was an eye-opener for many cognitive scientists. We had underestimated the power of the unconscious. Our intuitions, it turned out, could not be trusted: we had no way of knowing what cognitive processes could or could not proceed without awareness. The matter was entirely empirical. We had to submit, one by one, each mental faculty to a thorough inspection of its component processes, and decide which of those faculties did or did not appeal to the conscious mind. Only careful experimentation could decide the matter—but with techniques such as masking and attentional blink in our hands, testing the depth and limits of unconscious processing had never been so easy.

The past ten years have now seen a flurry of novel results challenging our picture of the human unconscious. Consider attention. Nothing seems more closely related to consciousness than the capacity to attend to stimuli. Without attention, we may remain totally unaware of external stimuli—as made clear by Dan Simons's gorilla movie and a zillion other effects of inattentional blindness. Whenever there are multiple competing stimuli, attention seems to be a necessary gateway to conscious experience.[64] In such conditions at least, consciousness requires attention. Amazingly, however, the converse statement turns out to be false: several recent experiments demonstrate that our attention can also be deployed unconsciously.[65]

It would be strange indeed if attending required the supervision of awareness. The role of attention, as already noted by William James, is to select "one out of several possible objects of thought." It would be oddly inefficient for our mind to be constantly distracted by dozens or even hundreds of possible thoughts and to examine each of them consciously before deciding which one is worthy of a further look. The determination of which objects are relevant and should be amplified is better left to automatic processes that operate sub rosa, in a massively parallel manner. Unsurprisingly, it turns out that our attentional spotlight is operated by armies of unconscious workers that silently sift through piles of rubble before one of them hits gold and alerts us of its finding.

In recent years, experiment after experiment has revealed the operation of selective attention without consciousness. Suppose we flash a stimulus in the corner of your eye so briefly that you cannot see it. Several experiments have shown that although it remains unconscious, such a flash may still attract your attention: you will become more attentive, and therefore faster and more accurate at attending to other stimuli presented at that same location, although you have no idea that a hidden cue caught your eye.[66] Conversely, a hidden picture may slow you down when its content is irrelevant to the task at hand. Interestingly, this effect works *better* when the distracting stimulus remains unconscious than when it is visible: a conscious distractor can be voluntarily extinguished, while an unconscious one preserves all its nuisance potential because we are unable to learn to control it.[67]

Loud noises, blinking lights, and other unexpected sensory events, as we all know, can irrepressibly attract our attention. However hard we try to ignore them, they invade our mental privacy. Why? They are, in part, an alerting mechanism, keeping us on the watch for potential dangers. As we concentrate on doing our taxes or on playing a favorite video game, it would be unsafe to tune out completely. Unexpected stimuli, such as a scream or the call of our own name, must remain able to break through our current thoughts—and therefore the filter called "selective attention" must continually operate outside our awareness, in order to decide which incoming inputs call for our mental resources. Unconscious attention acts as a constant watchdog.

Psychologists long thought that such automatic and bottom-up processes of the mind were the only ones that operated unconsciously. Psychologists' favorite metaphor for unconscious processing was that of a "spreading activation": a wave that starts from the stimulus and passively spreads through our brain circuits. A hidden prime climbed up the hierarchy of visual areas, progressively contacting processes of recognition, meaning attribution, and motor programming, as it tagged along with, without ever being influenced by, the subject's conscious will, intention, and attention. Thus, the results of subliminal experiments were thought to be independent of the participants' strategies and expectations.[68]

Consider it a major surprise, then, when our experiments shattered this consensus. We proved that subliminal priming is *not* a passive, bottom-up process, operating independent of attention and instructions. In fact, attention determines whether an unconscious stimulus is or is not processed.[69] An unconscious prime that is presented at an unexpected time or place produces virtually no priming onto a subsequent target. Even the mere repetition effect—the accelerated response to *radio* followed by *radio*—varies with how much attention is allocated to these stimuli. The act of attending causes a gain that massively amplifies the brain waves evoked by stimuli presented at the attended time and place. Remarkably, unconscious stimuli benefit from this attentional spotlight just as much as conscious ones do. In other words, attention can amplify a visual stimulus and still leave it too weak to break into our awareness.

Conscious intentions can even affect the orientation of our unconscious attention. Imagine that you are shown a set of shapes and are asked to detect only the squares while ignoring the circles. On a critical trial, a square appears on the right and a circle on the left—but both shapes are masked, so that you fail to detect them. You press randomly, not knowing on which side the square was shown. But a marker of parietal lobe activation called the N2pc reveals an unconscious orientation of your attention toward the appropriate side.[70] Your visual attention is surreptitiously attracted to the correct target, even on totally invisible trials and even if you eventually select the wrong response side. Similarly, during the attentional blink, within an entire stream of letters, the symbol that is arbitrarily designated as a target evokes noticeably more brain activity, even though it remains undetected.[71] On such trials, attention begins to unconsciously sieve the shapes for their relevance, although this process falls short of bringing the target stimulus into participants' conscious awareness.

The Value of an Invisible Coin

How does our attention decide whether a stimulus is relevant? A key component of the selection process is the assignment of a *value* to each potential object of thought. In order to survive, animals must have a very quick way of assigning a positive or negative value to every encounter. Should I stay, or should I go? Should I approach, or should I retreat? Is this a valuable treat or a poisonous trap? Valuation is a specialized process that calls upon evolved neural networks within a set of nuclei called the basal ganglia (because they are located near the base of the brain). And as you may have guessed, they too can operate totally outside our conscious awareness. Even a symbolic value such as money can be unconsciously appraised.

In one experiment, a picture of a penny or a pound sterling coin served as a subliminal incentive (figure 12).[72] The subjects' task was to squeeze a handle, and if they managed to exceed a certain amount of force, they would earn money. At the beginning of each trial, the image of a coin indicated how much money was at stake—and some of these pictures were flashed too fast to be consciously perceived. Although the participants denied having any awareness of either coin image, they exerted a stronger force when their potential gain was a pound than when it was a penny. Furthermore, the expectation of gaining one pound made the subjects' hands sweat in anticipation of this unconscious reward—and the brain's reward circuits were surreptitiously activated. The subjects remained unaware of the reason their behavior varied from trial to trial: they had no idea that their motivation was being unconsciously manipulated.

In another study, the values of the subliminal stimuli were not known in advance but were demonstrably learned during the course of the experiment.[73] The subjects, upon seeing a "signal," had to guess whether to press a button or refrain from pressing it. After each instance, they were told whether they had gained or lost money as a result of pressing or not pressing. Unknown to them, a subliminal shape, flashed within the signal, indicated the correct response; one shape cued the "go" response, another the "withhold" response, and a third was neutral—when it appeared, there was a 50 percent chance that either response would be rewarded.

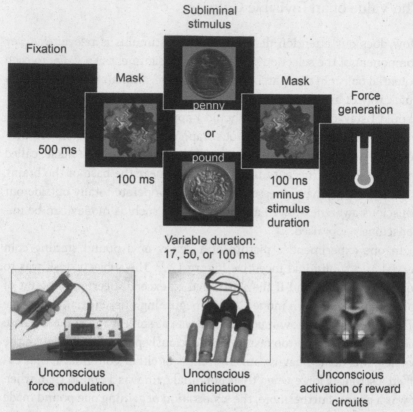

Fixation

Mask

Subliminal
stimulus

Mask

Force
generation

penny

or

pound

500 ms

100 ms

100 ms
minus
stimulus
duration

Variable duration:
17, 50, or 100 ms

Unconscious
force modulation

Unconscious
anticipation

Unconscious
activation of reward
circuits

FIGURE 12. Unconscious incentives can affect our motivations. In this experiment, participants were asked to squeeze a handle as strongly as they could in order to gain money. When a flashed picture specified that the stake was a pound sterling rather than a penny, people exerted a stronger force. They continued to do so even when the image was masked so that they were unaware which coin was presented. The reward circuits of the brain were unconsciously preactivated, and even the hands sweated in anticipation of gain. Thus, an unconscious image can trigger the circuits for motivation, emotion, and reward.

After playing this game for a few minutes, subjects inexplicably got better at this task. They still could not see the shapes that were hidden within the signal, but they had the "hot hand" and began to earn a significant sum of money. Their unconscious value system had kicked in: the positive "go" shape began to trigger key presses, while the negative "withhold" shape evoked systematic withholding. Brain imaging showed that a specific region of the basal ganglia, called the ventral striatum, had attached the relevant values to each shape. In brief, symbols that the subjects had never seen had nevertheless acquired a meaning: one had become repulsive and the other attractive, thus modulating the competition for attention and action.

The outcome of all these experiments is clear: our brain hosts a set of clever unconscious devices that constantly monitor the world around us and assign it values that guide our attention and shape our thinking. Thanks to these subliminal tags, the amorphous stimuli that bombard us become a landscape of opportunities, carefully sorted according to their relevance to our current goals. Only the most relevant events draw our attention and gain a chance to enter our consciousness. Below the level of our awareness, our unconscious brain ceaselessly evaluates dormant opportunities, testifying that our attention largely operates in a subliminal manner.

Unconscious Mathematics

A return from the over-estimation of the property of consciousness is the indispensable preliminary to any genuine insight into the course of psychic events.

—Sigmund Freud, *The Interpretation of Dreams* (1900)

Freud was right: consciousness is overrated. Consider this simple truism: we are conscious only of our conscious thoughts. Because our unconscious operations elude us, we constantly overestimate the role that consciousness plays in our physical and mental lives. By forgetting the amazing power of the unconscious, we overattribute our actions to conscious decisions and therefore mischaracterize our consciousness as a major player in our daily lives. In the words of the Princeton psychologist Julian Jaynes, "Consciousness is a much smaller part of our mental life than we are conscious of, because we cannot be conscious of what we are not conscious of."[74] Paraphrasing Douglas Hofstadter's whimsically circular law of programming ("A project always takes longer than you expect—even when you take into account Hofstadter's Law"), one might elevate this statement to the level of a universal law:

We constantly overestimate our awareness—even when we are aware of the glaring gaps in our awareness.

The corollary is that we dramatically underestimate how much vision, language, and attention can occur outside our awareness. Might some of the mental activities that we consider hallmarks of the conscious mind actually run unconsciously? Consider mathematics. One of the world's greatest mathematicians ever, Henri Poincaré, reported several curious incidents in which his unconscious mind seemed to do all the work:

> I left Caen, where I was living, to go on a geologic excursion under the auspices of the School of Mines. The incidents of the travel made me forget my mathematical work. Having reached Coutances, we entered an omnibus to go some place or other. At the moment when I put my foot on the step, the idea came to me, without anything in my former thoughts seeming to have paved the way for it, that the transformations I had used to define the Fuchsian functions were identical with those of non-Euclidian geometry. I did not verify the idea; I should not have had time, as, upon taking my seat in the omnibus, I went on with a conversation already commenced, but I felt a perfect certainty. On my return to Caen, for conscience sake, I verified the result at my leisure.

And then again:

> I turned my attention to the study of some arithmetical questions apparently without much success and without a suspicion of any connection with my preceding researches. Disgusted with my failure, I went to spend a few days at the seaside and thought of something else. One morning, walking on the bluff, the idea came to me, with just the same characteristics of brevity, suddenness and immediate certainty, that the arithmetic transformations of indefinite ternary quadratic forms were identical with those of non-Euclidian geometry.

These two anecdotes are reported by Jacques Hadamard, a world-class mathematician who dedicated a fascinating book to the mathematician's mind.[75] Hadamard deconstructed the process of mathematical discovery into four successive stages: initiation, incubation, illumination, and verification. *Initiation* covers all the preparatory work, the deliberate conscious exploration of a problem. This frontal attack, unfortunately, often remains fruitless—but all may not be lost, for it launches the unconscious mind on a quest. The *incubation* phase—an invisible brewing period during which the mind remains vaguely preoccupied with the problem but shows no conscious sign of working hard on it—can start. Incubation would remain undetected, were it not for its effects. Suddenly, after a good night's sleep or a relaxing walk, *illumination* occurs: the solution appears in all its glory and invades the mathematician's conscious mind. More often than not, it is correct. However, a slow and effortful process of conscious *verification* is nevertheless required to nail all the details down.

Hadamard's theory is seductive, but does it stand up to scrutiny? Does unconscious incubation truly exist? Or is it just retrospective storytelling glorified by the elation of discovery? Can we truly solve complex problems unconsciously? Cognitive science has only recently begun to bring these questions to the lab. Antoine Bechara, at the University of Iowa, developed a gambling task that studies people's protomathematical intuitions of probability and numerical expectation.[76] In this test, the subjects are given four decks of cards and a loan of $2,000 (in fake bills—psychologists aren't that rich). Turning a card over reveals a positive or negative message (e.g., "you win $100" or "you pay $100"). Participants try to optimize their gains by choosing at will from all four decks. What they do not know is that two of the decks are disadvantageous: they initially provide large earnings but quickly give rise to massive costs, and in the long run the outcome is a net loss. The other two decks lead to moderate ups and downs. In the long run, pulling cards from them yields a small but steady gain.

Initially, the players sample randomly from the four decks. Progressively, however, they develop a conscious hunch, and in the end they can easily report which decks are good and which are bad. But Bechara was interested in the "pre-hunch" period. During this phase, which resembles the mathematician's incubation period, the participants already have a lot of evidence about the four decks but still pull from all of them at random and claim to have no clue as to what they should do. Fascinatingly, just before they choose a card from a bad deck, their hands begin to sweat, thus generating a drop in skin conductance. This physiological marker of the sympathetic nervous system indicates that their brain has already registered the risky decks and is generating a subliminal gut feeling.

The alarm signal probably arises from operations performed in the ventromedial prefrontal cortex—a brain region specializing in unconscious valuation. Brain imaging shows a clear activation of this region, which is predictive of performance, on disadvantageous trials.[77] Patients with lesions to this region do not generate the anticipatory skin conductance in advance of unwittingly choosing from the bad-outcome deck; they do so only later on, once the bad outcome is revealed. The ventromedial and orbifrontal cortex contains a whole array of evaluative processes that constantly monitor our actions and compute their potential value. Bechara's research suggests that these regions often operate outside our conscious awareness. Although we have the impression of making random choices, our behavior may, in fact, be guided by unconscious hunches.

Having a hunch is not exactly the same as resolving a mathematical problem. But an experiment by Ap Dijksterhuis comes closer to Hadamard's taxonomy and suggests that genuine problem solving may indeed benefit from an unconscious incubation period.[78] The Dutch psychologist presented students with a problem in which they were to choose from among four brands of cars, which differed by up to twelve features. The participants read the problem, then half of them were allowed to consciously think about what their choice would be for four minutes; the other half were distracted for the same amount of time (by solving anagrams). Finally, both groups made their choice. Surprisingly, the distracted group picked the best car much more often than the conscious-deliberation group (60 percent versus 22 percent, a remarkably large effect given that choosing at random would result in 25 percent success). The work was replicated in several real-life situations, such as shopping at IKEA: several weeks after a trip there, shoppers who reported putting a lot of conscious effort into their decision were *less* satisfied with their purchases than the buyers who chose impulsively, without much conscious reflection.

Although this experiment does not quite meet the stringent criteria for a fully unconscious experience (because distraction does not fully ensure that the subjects never thought about the problem), it is very suggestive: some aspects of problem solving are better dealt with at the fringes of unconsciousness rather than with a full-blown conscious effort. We are not entirely wrong when we think that sleeping on a problem or letting our mind wander in the shower can produce brilliant insights.

Can the unconscious solve any type of problem? Or, more likely perhaps, are some categories of puzzles especially conducive to being solved by an unconscious hunch? Interestingly, Bechara's and Dijksterhuis's experiments involve similar problems; both require subjects to weigh several parameters. In Bechara's case, they must carefully weigh the gains and losses incurred with each deck of cards. In Dijksterhuis's, they must choose a car based on a weighted average of twelve criteria. When made consciously, such decisions put a heavy load on our working memory: the conscious mind, which typically focuses on one or a few possibilities at a time, is easily overwhelmed. This is probably why the conscious thinkers in Dijksterhuis's experiment did not do so well: they tended to place exaggerated weight on one or two features without seeing the bigger picture. Unconscious processes excel in assigning values to many items and averaging them to reach a decision.

Computing the sum or average of several positive and negative values indeed lies within the normal repertoire of what elementary circuits of neurons can do without consciousness. Even a monkey can learn to make a decision based on the total value brought about by a series of arbitrary shapes, and the firing of parietal neurons keeps track of the sum.[79] In my laboratory, we proved that approximate addition is within grasp of the human unconscious. In one experiment, we flashed a series of five arrows and asked subjects whether more arrows were pointing right or pointing left. When the arrows were made invisible by masking, participants were asked to guess, and indeed they thought that they were responding randomly, but in reality they continued to do much better than chance would predict. Signals from their parietal cortex gave evidence that their brain was unconsciously computing the approximate sum of the overall evidence.[80] The arrows were subjectively invisible, but they still made their way into the brain's weighting and decision systems.

In another experiment, we flashed eight numerals; four of them were visible consciously while the other four were invisible. We asked participants to decide if their mean was larger or smaller than five. The responses were quite accurate on average, but remarkably, the participants considered all eight of the available numbers. Thus, if the conscious numbers were larger than five, but the hidden numbers were smaller than five, the subjects were unconsciously biased to respond "smaller."[81]

The averaging operation that they were asked to perform with the consciously visible numbers extended to the unconscious ones.

Statistics During Sleep

Clearly, then, some elementary mathematical operations, including averaging and comparing, may unfold unconsciously. But what about genuinely creative operations such as Poincaré's insight on the omnibus? Can insight really strike us at any time, even when we least expect it and are thinking of something else? The answer seems to be positive. Our brain acts as a sophisticated statistician that detects meaningful regularities hidden in seemingly random sequences. Such statistical learning is constantly running in the background, even as we sleep.

Ullrich Wagner, Jan Born, and their colleagues tested scientists' claim that they often have a sudden insight upon waking up from a good night's sleep.[82] To bring this idea to the lab, they had subjects participate in a nerdy math experiment: they had to mentally transform a sequence of seven digits into another sequence of seven digits according to an attention-demanding rule. They were asked to name only the last digit of the answer—but finding its value required a long mental calculation. Unknown to them, however, there was a shortcut. The output sequence had a hidden symmetry: the last three digits repeated the previous three in reverse order (e.g., 4 1 4 9 9 4 1), and as a result the last digit was always equal to the second one. Once the participants recognized this shortcut, they could save enormous time and effort by stopping after the second digit. During the initial test, most of the subjects failed to notice the concealed rule. However, a good night's sleep more than doubled the probability that they would have the insight: many participants woke up with the solution in mind! Controls established that the elapsed time was irrelevant; what mattered was sleep. Falling asleep seemed to enable the consolidation of previous knowledge into a more compact form.

We know from animal studies that neurons in the hippocampus and the cortex are active during sleep. Their firing patterns "replay," in fastforward mode, the same sequences of activity that occurred during the previous period of wakefulness.[83] For instance, a rat runs through a maze; then upon falling asleep, his brain reactivates his place-coding neurons so precisely that the pattern can be used to decode the locations where he is mentally traveling—but at a much faster speed, and sometimes even in reverse order. Perhaps this temporal compression offers the possibility of treating a sequence of digits as a near-simultaneous spatial pattern, thus permitting the detection of hidden regularities by classical learning mechanisms. Whatever the neurobiological explanation, sleep is clearly a period of boiling unconscious activity that supports much memory consolidation and insight.

A Subliminal Bag of Tricks

These laboratory demonstrations are a far cry from the type of mathematical thinking that Poincaré had in mind when he was unconsciously exploring Fuchsian functions and non-Euclidean geometry. However, that gap is being reduced as innovative experiments study the greater range of operations that can be performed, at least in part, without awareness.

It was long thought that the mind's "central executive"—a cognitive system that controls our mental operations, avoids automatic responses, switches tasks, and detects our errors—was the sole province of the conscious mind. But recently, sophisticated executive functions have been shown to operate unconsciously, based on invisible stimuli.

One such function is the ability to control ourselves and inhibit our automatic responses. Imagine performing a repetitive task, such as pressing a key whenever a picture appears on screen—except that on rare occasions, the picture depicts a black disk, and then you absolutely have to refrain from clicking. This is called the "stop signal" task, and much research shows that the ability to inhibit a routine response is a marker of the mind's central executive system. The Dutch psychologist Simon van Gaal asked whether refraining from responding requires consciousness: would subjects still manage to avoid clicking if the "stop" signal was subliminal? Amazingly, the answer was yes. When an unconscious "stop" signal was briefly flashed, the participants' hands slowed down, and occasionally, they stopped responding altogether.[84] They did so without understanding why, because the stimulus that triggered this inhibition remained unseen. These findings indicate that *invisible* is not synonymous with *out of control.* Even an invisible stop signal can trigger a wave of activity that spreads deep into the executive networks that allow us to control our actions.[85]

Similarly, we can detect some of our errors without being conscious. In an eye movement task, when the participants' eyes deviate from the plan, the error triggers an activation of the executive control centers in the anterior cingulate cortex—even when participants are unaware of the error and deny that their eyes wandered off the target.[86] Unconscious signals can even cause a partial switch to another task. When subjects are shown a conscious cue that tells them to change from task one to task two, flashing this cue below the threshold for awareness still has the effect of slowing them down and triggering a partial task switch at the cortical level.[87]

In a nutshell, psychology has amply demonstrated not only that subliminal perception exists but that a whole array of mental processes can be launched without consciousness (even though, in most cases, they do not run to full completion). Figure 13 summarizes the various brain regions that, in experiments discussed in this chapter, have been shown to activate in the absence of awareness. The unconscious clearly has a large bag of tricks, from word comprehension to numerical addition, and from error detection to problem solving. Because they operate quickly and in parallel across a broad range of stimuli and responses, these tricks often surpass conscious thought.

Henri Poincaré, in *Science and Hypothesis* (1902), anticipated the superiority of unconscious brute-force processing over slow conscious thinking:

> The subliminal self is in no way inferior to the conscious self; it is not purely automatic; it is capable of discernment; it has tact, delicacy; it knows how to choose, to divine. What do I say? It knows better how to divine than the conscious self, since it succeeds where that has failed. In a word, is not the subliminal self superior to the conscious self?

Contemporary science answers Poincaré's question with a resounding yes. In many respects, our mind's subliminal operations exceed its conscious achievements. Our visual system routinely solves problems of shape perception and invariant recognition that boggle the best computer software. And we tap into this amazing computational power of the unconscious mind whenever we ponder mathematical problems.

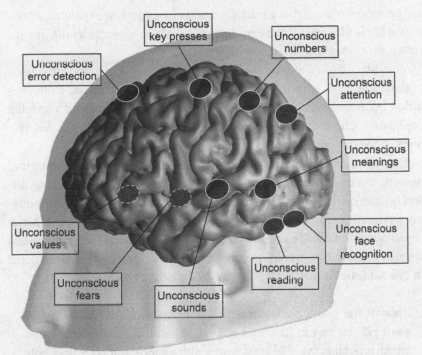

FIGURE 13. An overview of unconscious operations in the human brain. The figure shows only a subset of the many circuits that can activate without awareness. We now believe that virtually any brain processor can operate unconsciously. For greater readability, each computation is pinned to its dominant brain site, but it should be remembered that such neuronal specialization always rests on an entire brain circuit. Some of our unconscious processors are subcortical: they involve groups of neurons located below the surface of the cortex (denoted by dashed ellipses) and often implement functions that appeared early in our evolution, such as the detection of frightening stimuli that warn us of an impending danger. Other computations recruit various sectors of the cortex. Even high-level cortical areas that encode our acquired cultural knowledge, such as reading or arithmetic, may operate outside our awareness.

But we should not get carried away. Some cognitive psychologists go as far as to propose that consciousness is a pure myth, a decorative but powerless feature, like frosting on a cake.[88] All the mental operations that underlie our decisions and behavior, they claim, are accomplished unconsciously. In their view, our awareness is a mere bystander, a backseat driver that contemplates the brain's unconscious accomplishments but lacks effective powers of its own. As in the 1999 movie *The Matrix*, we are prisoners of an elaborate artifice, and our experience of living a conscious life is illusory; all our decisions are made in absentia by the unconscious processes within us.

The next chapter will refute this zombie theory. Consciousness is an evolved function, I argue—a biological property that emerged from evolution because it was useful. Consciousness must therefore fill a specific cognitive niche and address a problem that the specialized parallel systems of the unconscious mind could not.

Ever insightful, Poincaré noted that in spite of the brain's subliminal powers, the mathematician's unconscious cogs did not start clicking unless he had made a massive initial conscious attack on the problem during the initiation phase. And later on, after the "aha" experience, only the conscious mind could carefully verify, step by step, what the unconscious seemed to have discovered. Henry Moore made exactly the same point in *The Sculptor Speaks* (1937):

> Though the non-logical, instinctive, subconscious part of the mind must play its part in [the artist's] work, he also has a conscious mind which is not inactive. The artist works with a concentration of his whole personality, and the conscious part of it resolves conflicts, organizes memories, and prevents him from trying to walk in two directions at the same time.

We are now ready to walk into the unique realm of the conscious mind.

3

WHAT IS CONSCIOUSNESS GOOD FOR?

Why did consciousness evolve? Can some operations be carried out only by a conscious mind? Or is consciousness a mere epiphenomenon, a useless or even illusory feature of our biological makeup? In fact, consciousness supports a number of specific operations that cannot unfold unconsciously. Subliminal information is evanescent, but conscious information is stable—we can hang on to it for as long as we wish. Consciousness also compresses the incoming information, reducing an immense stream of sense data to a small set of carefully selected bite-size symbols. The sampled information can then be routed to another processing stage, allowing us to perform carefully controlled chains of operations, much like a serial computer. This broadcasting function of consciousness is essential. In humans, it is greatly enhanced by language, which lets us distribute our conscious thoughts across the social network.

> The particulars of the distribution of consciousness, so far as we know
> them, point to its being efficacious.
> —William James, *Principles of Psychology* (1890)

In the history of biology, few questions have been debated as heatedly as finalism or teleology—whether it is meaningful to speak of organs as designed or evolved "for" a specific function (a "final cause," or *telos* in Greek). In the pre-Darwinian era, finalism was the norm, as the hand of God was seen as a hidden designer of all things. The great French anatomist Georges Cuvier constantly appealed to teleology when interpreting the functions of the body organs: claws were "for" catching prey, lungs were "for" breathing, and such final causes were the very conditions of existence of the organism as an integrated whole.

Charles Darwin radically altered the picture by pointing to natural selection, rather than design, as an undirected force blindly shaping the biosphere. The Darwinian view of nature has no need for divine intention. Evolved organs are not designed "for" their function; they merely grant their possessor a reproductive advantage. In a dramatic reversal of perspective, antievolutionists seized as counterexamples to Darwin what they viewed as obvious examples of nonadvantageous designs. Why does the peacock carry a huge, visually stunning, but clumsy tail? Why did *Megaloceros*, the extinct Irish elk, carry a gigantic pair of antlers, spanning up to twelve feet, so bulky that it has been blamed for the species' demise? Darwin retorted by pointing to sexual selection: it is advantageous for males, who compete for female attention, to develop elaborate, costly, and symmetrical displays advertising their fitness. The lesson was clear: biological organs did not come labeled with a function, and even clumsy contraptions, tinkered with by evolution, could bring a competitive edge to their possessors.

During the twentieth century, the synthetic theory of evolution further dissolved the teleological picture. The modern vocabulary of evolution and development (evo-devo) now includes an extended toolkit of concepts that collectively account for sophisticated design without a designer:

- Spontaneous pattern generation: The mathematician Alan Turing first described how chemical reactions may lead to the emergence of organized features such as the zebra's stripes or the melon's ribs.[1] On some cone shells, sophisticated pigmentation patterns self-organize under an opaque layer, clearly proving their lack of intrinsic utility—they are a mere offshoot of chemical reactions with their own raison d'être.

- Allometric relations: An increase in the overall size of the organism (which may be advantageous in its own right) may lead to a proportionate change in some of its organs (which may not). The Irish elk's outlandish antlers probably resulted from such an allometric change.[2]

- Spandrels: The late Harvard paleontologist Stephen Jay Gould coined this term to refer to features of the organism that arise as necessary by-products of its architecture but that might later be co-opted (or "exapted") into another role.[3] An example may be the male nipple—an inconsequential but necessary outcome of the organism's *Bauplan* for constructing advantageous female breasts.

Bearing these biological concepts in mind, we can no longer assume that any human biological or psychological trait, including consciousness, necessarily plays a positive functional role in our species' worldwide success. Consciousness could be a happenstance decorative pattern, or the chance outcome of a massive increase in brain size that occurred in our species of the genus *Homo,* or even a mere spandrel, a consequence of other vital changes. This view matches the intuition of the French writer Alexandre Vialatte, who whimsically stated that "consciousness, like the appendix, serves no role but to make us sick." In the 1999 movie *Being John Malkovich,* the puppeteer Craig Schwartz laments the inutility of introspection: "Consciousness is a terrible curse. I think. I feel. I suffer. And all I ask in return is the opportunity to do my work."

Is consciousness a mere epiphenomenon? Should it be likened to the loud roar of a jet engine—a useless and painful but unavoidable consequence of the brain's machinery, inescapably arising from its construction? The British psychologist Max Velmans clearly leans toward this pessimistic conclusion. An impressive array of cognitive functions, he argues, are indifferent to awareness—we may be aware of them, but they would continue to run equally well if we were mere zombies.[4] The popular Danish science writer Tor Nørretranders coined the term "user illusion" to refer to our feeling of being in control, which may well be fallacious; every one of our decisions, he believes, stems from unconscious sources.[5] Many other psychologists agree: consciousness is the proverbial backseat driver, a useless observer of actions that forever lie beyond its control.[6]

In this book, however, I explore a different road—what philosophers call the "functionalist" view of consciousness. Its thesis is that consciousness is useful. Conscious perception transforms incoming information into an internal code that allows it to be processed in unique ways. Consciousness is an elaborate functional property and as such is likely to have been selected, across millions of years of Darwinian evolution, because it fulfills a particular operational role.

Can we determine what that role is? We cannot rewind evolutionary history, but we can use the minimal contrast between seen and unseen images to characterize the uniqueness of conscious operations. Using psychological experiments, we can probe which operations are feasible without consciousness, and which are uniquely deployed when we report awareness. This chapter will show that, far from blacklisting consciousness as a useless feature, these experiments point to consciousness as being highly efficacious.

Unconscious Statistics, Conscious Sampling

My picture of consciousness implies a natural division of labor. In the basement, an army of unconscious workers does the exhausting work, sifting through piles of data. Meanwhile, at the top, a select board of executives, examining only a brief of the situation, slowly makes conscious decisions.

Chapter 2 laid out the powers of our unconscious mind. A great variety of cognitive operations, from perception to language understanding, decision, action, evaluation, and inhibition can unfold, at least partially, in a subliminal mode. Below the conscious stage, myriad unconscious processors, operating in parallel, constantly strive to extract the most detailed and complete interpretation of our environment. They operate as nearly optimal statisticians who exploit every slightest perceptual hint—a faint movement, a shadow, a splotch of light—to calculate the probability that a given property holds true in the outside world. Much as the weather bureau combines dozens of meteorological observations to infer the chance of rain in the next few days, our unconscious perception uses incoming sense data to compute the probability that colors, shapes, animals, or people are present in our surroundings. Our consciousness, on the other hand, offers us only a glimpse of this probabilistic universe—what statisticians call a "sample" from this unconscious distribution. It cuts through all ambiguities and achieves a simplified view, a summary of the best current interpretation of the world, which can then be passed on to our decision-making system.

This division of labor, between an army of unconscious statisticians and a single conscious decision maker, may impose itself on any moving organism by that organism's very necessity of acting upon the world. No one can act on mere probabilities—at some point, a dictatorial process is needed to collapse all uncertainties and decide. *Alea jacta est:* "the die is cast," as Caesar famously said after crossing the Rubicon to seize Rome from the hands of Pompey. Any voluntary action requires tipping the scales to a point of no return. Consciousness may be the brain's scale-tipping device—collapsing all unconscious probabilities into a single conscious sample, so that we can move on to further decisions.

The classical fable of Buridan's ass hints at the usefulness of quickly breaking through complex decisions. In this imaginary tale, a donkey that is thirsty and hungry is placed exactly midway between a pail of water and a stack of hay. Unable to decide between them, the fabled animal dies of both hunger and thirst. The problem seems ridiculous, yet we are constantly confronted with difficult decisions of a similar kind: the world offers us only unlabeled opportunities with uncertain, probabilistic outcomes. Consciousness resolves the issue by bringing to our attention, at any given time, only one out of the thousands of possible interpretations of the incoming world.

The philosopher Charles Sanders Peirce, following in the footsteps of the physicist Hermann von Helmholtz, was among the first to recognize that even our simplest conscious observation results from a bewildering complexity of unconscious probabilistic inferences:

> Looking out my window this lovely spring morning I see an azalea in full bloom. No, no! I do not see that; though that is the only way I can describe what I see. That is a proposition, a sentence, a fact; but what I perceive is not proposition, sentence, fact, but only an image, which I make intelligible in part by means of a statement of fact. This statement is abstract; but what I see is concrete. I perform an abduction when I so much as express in a sentence anything I see. The truth is that the whole fabric of our knowledge is one matted felt of pure hypothesis confirmed and refined by induction. Not the smallest advance can be made in knowledge beyond the stage of vacant staring, without making an abduction at every step.[7]

What Peirce called "abduction" is what a modern cognitive scientist would dub "Bayesian inference," after the Reverend Thomas Bayes (ca. 1701–61), who first explored this domain of mathematics. Bayesian inference consists in using statistical reasoning in a backward manner to infer the hidden causes behind our observations. In classical probability theory, we are typically told what happens (for instance, "someone draws three cards from a deck of fifty-two"); the theory allows us to assign probabilities to specific outcomes (for instance, "What is the probability that all three cards are aces?"). Bayesian theory, however, lets us reason in the converse direction, from outcomes to their unknown origins (for instance, "If someone draws three aces from a deck of fifty-two cards, what is the likelihood that the deck was rigged and comprised more than four aces?"). This is called "reverse inference" or "Bayesian statistics." The hypothesis that the brain acts as a Bayesian statistician is one of the hottest and most debated areas of contemporary neuroscience.

Our brain must perform a kind of reverse inference because all our sensations are ambiguous: many remote objects could have caused them. When I manipulate a plate, for instance, its rim appears to be a perfect circle, but it actually projects on my retina as a distorted ellipse, compatible with myriad other interpretations. Infinitely many potato-shaped objects, of countless orientations in space, could have cast the same projection onto my retina. If I see a circle, it is only because my visual brain, unconsciously pondering the endless possible causes for this sensory input, opts for "circle" as the most probable. Thus, although my perception of the plate as a circle seems immediate, it actually arises from a complex inference that weeds out an inconceivably vast array of other explanations for that particular sensation.

Neuroscience offers much evidence that during the intermediate visual stages, the brain ponders a vast number of alternative interpretations for its sensory inputs. A single neuron, for instance, may perceive only a small segment of an ellipse's overall contour. This information is compatible with a broad variety of shapes and motion patterns. Once visual neurons start talking to one another, however, casting their "votes" for the best percept, the entire population of neurons can converge. When you have eliminated the impossible, Sherlock Holmes famously stated, whatever remains, however improbable, must be the truth.

A strict logic governs the brain's unconscious circuits—they appear ideally organized to perform statistically accurate inferences concerning our sensory inputs. In the middle temporal motion area MT ("area MT"), for instance, neurons perceive the motion of objects only through a narrow peephole (the "receptive field"). At that scale, any motion is ambiguous. If you watch a stick through a peephole, you cannot accurately determine its motion. It could be moving in the direction perpendicular to itself or in countless other directions (figure 14). This basic ambiguity is known as the "aperture problem." Unconsciously, individual neurons

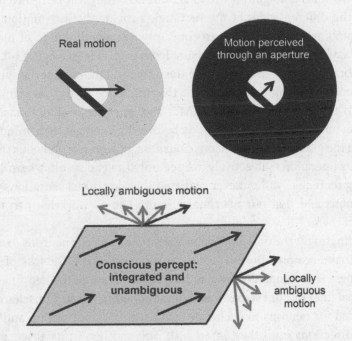

FIGURE 14. Consciousness helps resolve ambiguities. In the region of the cortex that is sensitive to motion, neurons suffer from the "aperture problem." Each of them receives inputs from only a limited aperture, classically called the "receptive field," and thus cannot tell whether the motion is oriented horizontally, perpendicular to the bar, or in any of countless other directions. In our conscious awareness, however, no ambiguity exists: our perceptual system makes a decision and always lets us see the minimal amount of motion, perpendicular to the line. When an entire surface is moving, we perceive the global direction of movement by combining the signals from multiple neurons. Neurons in area MT initially encode each local motion, but they quickly converge to a global interpretation that matches what we consciously perceive. This convergence seems to occur only if the observer is conscious.

in our area MT suffer from it—but at a conscious level, we don't. Even under dire circumstances, we perceive no ambiguity. Our brain makes a decision and lets us see what it considers to be the most likely interpretation, with the minimal amount of motion: the stick always appears to move in the direction perpendicular to itself. An unconscious army of neurons evaluates all the possibilities, but consciousness receives only a stripped-down report.

When we view a more complex moving shape, such as a moving rectangle, the local ambiguities still exist, but now they can be resolved, because the different sides of the rectangle provide distinct motion cues that combine into a unique percept. Only a single direction of motion jointly satisfies the constraints arising from each side (see figure 14). Our visual brain infers it and lets us see the only rigid movement that fits the bill. Neuronal recordings show that this inference takes time: for a full tenth of a second, neurons in area MT "see" only the local motion, and it takes them 120 to 140 milliseconds before they change their mind and encode the global direction.[8] Consciousness, however, is oblivious to this complex operation. Subjectively, we see only the end result, a seamlessly moving rectangle, without ever realizing that our initial sensations were ambiguous and that our neuronal circuits had to work hard to make sense of them.

Fascinatingly, the convergence process that leads our neurons to agree on a single interpretation vanishes under anesthesia.[9] The loss of consciousness is accompanied by a sudden dysfunction of the neuronal circuits that integrate our senses into a single coherent whole. Consciousness is needed for neurons to exchange signals in both bottom-up and top-down directions until they agree with one another. In its absence, the perceptual inference process stops short of generating a single coherent interpretation of the outside world.

The role of consciousness in resolving perceptual ambiguities is nowhere as evident as when we purposely craft an ambiguous visual stimulus. Suppose we present the brain with two superimposed gratings moving in different directions (figure 15). The brain has no way of telling whether the first grating lies in front of the other, or vice versa. Subjectively, however, we do not perceive this basic ambiguity. We never perceive a blend of two possibilities, but our conscious perception decides

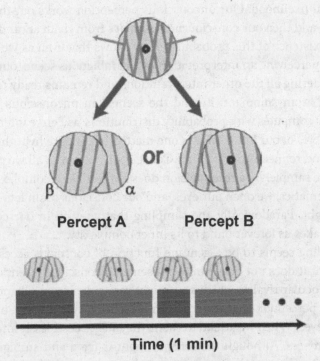

Time (1 min)

FIGURE 15. Consciousness lets us see only one of the plausible interpretations of our sensory inputs. A display consisting of two superimposed gratings is ambiguous: either grating can be perceived to lie in front. But at any given moment, we are aware of only one of those possibilities. Our conscious vision alternates between the two percepts, and the proportion of time spent in one state is a direct reflection of the probability that this interpretation is correct. Thus our unconscious vision computes a landscape of probabilities, and our consciousness samples from it.

and lets us see one of the two gratings in the foreground. The two interpretations alternate: every few seconds, our perception changes and we see the other grating move into the foreground. Alexandre Pouget and his collaborators have shown that, when parameters such as speed and spacing are varied, the time that our conscious vision spends entertaining an interpretation is directly related to its likelihood, given the sensory evidence received.[10] What we see, at any time, tends to be the most likely interpretation, but other possibilities occasionally pop up and stay in our conscious vision for a time duration that is proportional to their statistical likelihood. Our unconscious perception works out the probabilities—and then our consciousness samples from them at random.

The existence of this probabilistic law shows that even as we are consciously perceiving an interpretation of an ambiguous scene, our brain is still pondering all the other interpretations and remains ready to change its mind at any moment. Behind the scenes, an unconscious Sherlock endlessly computes with probability distributions: as Peirce inferred, "the whole fabric of our knowledge is one matted felt of pure hypothesis confirmed and refined by induction." Consciously, however, all we get to see is a single sample. As a result, vision does not feel like a complex exercise in mathematics; we open our eyes, and our conscious brain lets in only a single sight. Paradoxically, the sampling that goes on in our conscious vision makes us forever blind to its inner complexity.

Sampling seems to be a genuine function of conscious access, in the sense that it does not occur in the absence of conscious attention. Consider binocular rivalry, which you might remember from Chapter 1: the unstable perception that results from presenting two distinct images to the two eyes. When we attend to them, the images ceaselessly alternate in our awareness. Although the sensory input is fixed and ambiguous, we perceive it as constantly changing, as we become aware of only one image at a time. Crucially, however, when we orient our attention elsewhere, the rivalry stops.[11] Discrete sampling seems to occur only when we consciously attend. As a consequence, unconscious processes are more objective than conscious ones. Our army of unconscious neurons approximates the true probability distribution of the states of the world, while our consciousness shamelessly reduces it to all-or-none samples.

The whole process bears an intriguing analogy to quantum mechanics (although its neural mechanisms most likely involve only classical physics). Quantum physicists tell us that physical reality consists in a superposition of wave functions that determine the probability of finding a particle in a certain state. Whenever we care to measure, however, these probabilities collapse to a fixed all-or-none state. We never observe strange mixtures such as the famed Schrödinger's cat, half alive and half dead. According to quantum theory, the very act of physical measurement forces the probabilities to collapse into a single discrete measure. In our brain, something similar happens: the very act of consciously attending to an object collapses the probability distribution of its various interpretations and lets us perceive only one of them. Consciousness acts as a discrete measurement device that grants us a single glimpse of the vast underlying sea of unconscious computations.

Still, this seductive analogy may be superficial. Only future research will tell us whether some of the mathematics behind quantum mechanics can be adapted to the cognitive neuroscience of conscious perception. What is certain, though, is that in our brains, such a division of labor is ubiquitous: unconscious processes act as fast and massively parallel statisticians, while consciousness is a slow sampler. We see this not only in vision but also in the domain of language.[12] Whenever we perceive an ambiguous word like *bank*, as we saw in Chapter 2, its two meanings are temporarily primed within our unconscious lexicon, even though we gain conscious awareness of only one of them at a time.[13] The same principle underlies our attention. It feels as if we can attend to only a single location at a time, but the unconscious mechanism by which we select an object is actually probabilistic and considers several hypotheses at once.[14]

An unconscious sleuth even hides in our memory. Try to answer the following question: What percentage of the world's airports is located in the United States? Please venture a guess, even if it feels difficult. Done? Now discard your first guess and give me a second one. Research shows that even your second guess is not random. Furthermore, if you have to bet, you are better off responding with the *average* of your two answers than with either guess alone.[15] Once again, conscious retrieval acts as an invisible hand that draws at random from a hidden distribution of likelihoods. We can take a first sample, a second, and even a third, without exhausting the power of our unconscious mind.

An analogy may be useful: consciousness is like the spokesperson in a large institution. Vast organizations such as the FBI, with their thousands of employees, always possess considerably more knowledge than any single individual can ever grasp. As the sad episode of September 11, 2001, illustrates, it is not always easy to extract the relevant knowledge from the vast arrays of irrelevant beliefs that every employee entertains. In order to avoid drowning in the bottomless sea of facts, the president relies on short briefs compiled by a pyramidal staff, and he lets a single spokesperson express this "common wisdom." Such a hierarchical use of resources is generally rational, even if it implies neglecting the subtle hints that could be the crucial signs that a dramatic event is brewing.

As a large-scale institution with a staff of a hundred billion neurons, the brain must rely on a similar briefing mechanism. The function of consciousness may be to simplify perception by drafting a summary of the current environment before voicing it out loud, in a coherent manner, to all other areas involved in memory, decision, and action.

In order to be useful, the brain's conscious brief must be stable and integrative. During a nationwide crisis, it would be pointless for the FBI to send the president thousands of successive messages, each holding a little bit of truth, and let him figure it out for himself. Similarly, the brain cannot stick to a low-level flux of incoming data: it must assemble the pieces into a coherent story. Like a presidential brief, the brain's conscious summary must contain an interpretation of the environment written in a "language of thought" that is abstract enough to interface with the mechanisms of intention and decision making.

Lasting Thoughts

The improvements we install in our brain when we learn our languages permit us to review, recall, rehearse, redesign our own activities, turning our brains into echo chambers of sorts, in which otherwise evanescent processes can hang around and become objects in their own right. Those that persist the longest, acquiring influence as they persist, we call our conscious thoughts.

—Daniel Dennett, *Kinds of Minds* (1996)

Consciousness is then, as it were, the hyphen which joins what has been to what will be, the bridge which spans the past and the future.

—Henri Bergson, *Huxley Memorial Lecture* (1911)

There may be a very good reason why our consciousness condenses sensory messages into a synthetic code, devoid of gaps and ambiguities: such a code is compact enough to be carried forward in time, entering what we usually call "working memory." Working memory and consciousness seem to be tightly related. One may even argue, with Daniel Dennett, that a main role of consciousness is to create lasting thoughts. Once a piece of information is conscious, it stays fresh in our mind for as long as we care to attend to it and remember it. The conscious brief must be kept stable enough to inform our decisions, even if they take a few minutes to form.

This extended duration, thickening the present moment, is characteristic of our conscious thoughts.

A cellular mechanism of transient memory exists in all mammals, from humans to monkeys, cats, rats, and mice. Its evolutionary advantages are obvious. Organisms that have a memory become detached from pressing environmental contingencies. They are no longer tied to the present but can recall the past and anticipate the future. When an organism's predator hides behind a rock, remembering its invisible presence is a matter of life and death. Many environmental events recur at unspecified time intervals, over vast expanses of space, and indexed by a diversity of cues. The capacity to synthesize information over time, space, and modalities of knowledge, and to rethink it at any time in the future, is a fundamental component of the conscious mind, one that seems likely to have been positively selected for during evolution.

The component of the mind that psychologists call "working memory" is one of the dominant functions of the dorsolateral prefrontal cortex and the areas that it connects with, thus making these areas strong candidates for the depositories of our conscious knowledge.[16] These regions pop up in brain imaging experiments whenever we briefly hold on to a piece of information: a phone number, a color, or the shape of a flashed picture. Prefrontal neurons implement an active memory: long after the picture is gone, they continue to fire throughout the short-term memory task—sometimes as long as dozens of seconds later. And when the prefrontal cortex is impaired or distracted, this memory is lost—it falls into unconscious oblivion.

Patients who suffer from lesions of the prefrontal cortex also exhibit major deficiencies in planning the future. Their remarkable cluster of symptoms suggests a lack of foresight and a stubborn adherence to the present. They seem unable to inhibit unwanted actions and may automatically seize and use tools (utilization behavior) or irrepressibly mimic others (imitation behavior). Their capacities for conscious inhibition, long-term thinking, and planning may be drastically deteriorated. In the most severe cases, apathy and a variety of other symptoms indicate a glaring gap in the quality and contents of mental life. The disorders that relate directly to consciousness include hemineglect (perturbed awareness of one half of space, usually the left), abulia (incapacity to generate voluntary actions), akinetic mutism (inability to generate spontaneous verbal reports, though repetition may be intact), anosognosia (unaware-ness of a major deficit, including paralysis), and impaired autonoetic memory (incapacity to recall and analyze one's own thoughts). Tamper-ing with the prefrontal cortex can even interfere with abilities as basic as perceiving and reflecting upon a brief visual display.[17]

To summarize, the prefrontal cortex seems to play a key role in our ability to maintain information over time, to reflect upon it, and to inte-grate it into our unfolding plans. Is there more direct evidence that such temporally extended reflection necessarily involves consciousness? The cognitive scientists Robert Clark and Larry Squire conducted a wonder-fully simple test of temporal synthesis: time-lapse conditioning of the eyelid reflex.[18] At a precisely timed moment, a pneumatic machine puffs air toward the eye. The reaction is instantaneous: in rabbits and humans alike, the protective membrane of the eyelid immediately closes. Now precede the delivery of air with a brief warning tone. The outcome is called Pavlovian conditioning (in memory of the Russian physiologist Ivan Petrovich Pavlov, who first conditioned dogs to salivate at the sound of a bell, in anticipation of food). After a short training, the eye blinks to the sound itself, in anticipation of the air puff. After a while, an occa-sional presentation of the isolated tone suffices to induce the "eyes wide shut" response.

The eye-closure reflex is fast, but is it conscious or unconscious? The answer, surprisingly, depends on the presence of a temporal gap. In one version of the test, usually termed "delayed conditioning," the tone lasts until the puff arrives. Thus the two stimuli briefly coincide in the ani-mal's brain, making the learning a simple matter of coincidence detec-tion. In the other, called "trace conditioning," the tone is brief, separated from the subsequent air puff by an empty gap. This version, although minimally different, is clearly more challenging. The organism must keep an active memory trace of the past tone in order to discover its system-atic relation to the subsequent air puff. To avoid any confusion, I will call the first version "coincidence-based conditioning" (the first stimulus lasts long enough to coincide with the second, thus removing any need for memory) and the second "memory-trace conditioning" (the subject must keep in mind a memory trace of the sound in order to bridge the tempo-ral gap between it and the obnoxious air puff).

The experimental results are clear: coincidence-based conditioning occurs unconsciously, while for memory-trace conditioning, a conscious mind is required.[19] In fact, coincidence-based conditioning does not require any cortex at all. A decerebrate rabbit, without any cerebral cortex, basal ganglia, limbic system, thalamus, and hypothalamus, still shows eyelid conditioning when the sound and the puff overlap in time. In memory-trace conditioning, however, no learning occurs unless the hippocampus and its connected structures (which include the prefrontal cortex) are intact. In human subjects, memory-trace learning seems to occur if and only if the person reports being aware of the systematic predictive link between the tone and the air puff. Elderly people, amnesiacs, and people who were simply too distracted to notice the temporal relationship show no conditioning at all (whereas these manipulations have no effect whatsoever on coincidence-based conditioning). Brain imaging shows that the subjects who gain awareness are precisely those who activate their prefrontal cortex and hippocampus during the learning.

Overall, the conditioning paradigm suggests that consciousness has a specific evolutionary role: learning over time, rather than simply living in the instant. The system formed by the prefrontal cortex and its interconnected areas, including the hippocampus, may serve the essential role of bridging temporal gaps. Consciousness provides us with a "remembered present," in the words of Gerald Edelman:[20] Thanks to it, a selected subset of our past experiences can be projected into the future and crosslinked with the present sensory data.

What is particularly interesting about the memory-trace conditioning test is that it is simple enough to be administered to all sorts of organisms, from infants to monkeys, rabbits, and mice. When mice take the test, they activate anterior brain regions that are homologous to the human prefrontal cortex.[21] The test may thus be tapping one of the most basic functions of consciousness, an operation so essential that it may also be present in many other species.

If a temporally extended working memory requires consciousness, is it impossible to stretch our unconscious thoughts across time? Empirical measures of the duration of subliminal activity suggest that it is—subliminal thoughts last only for an instant.[22] The lifetime of a subliminal stimulus can be estimated by measuring how long one has to wait before its effect decays to zero. The result is very clear: a visible image may have a long-lasting effect, but an invisible one exerts only a short-lived influence on our thoughts. Whenever we render an image invisible by masking, it nevertheless activates visual, orthographic, lexical, or even semantic representations in the brain, but only for a brief duration. After a second or so, the unconscious activation generally decays to an undetectable level.

Many experiments show that subliminal stimuli undergo a rapid exponential decay in the brain. Summarizing these findings, my colleague Lionel Naccache concluded (contradicting the French psychoanalyst Jacques Lacan) that "the unconscious is not structured as a language but as a decaying exponential."[23] With effort, we may keep subliminal information alive for a slightly longer period—but the quality of this memory is so degraded that our recall, after a few seconds' delay, barely exceeds the level of chance.[24] Only consciousness allows us to entertain lasting thoughts.

The Human Turing Machine

Once information is "in mind," protected from temporal decay, can it enter into specific operations? Do some cognitive operations require consciousness and lie beyond the scope of our unconscious thought processes? The answer seems to be positive: in humans at least, consciousness gives us the power of a sophisticated serial computer.

For instance, try to compute 12 times 13 in your head.

Finished?

Did you feel each of the arithmetic operations churning in your brain, one after the other? Can you faithfully report the successive steps that you took, and the intermediate results that they returned? The answer is usually yes; we are aware of the serial strategies that we deploy to multiply. Personally, I first remembered that 12^2 is 144, then added another 12. Others may multiply the digits one after the other according to the classical multiplication recipe. The point is this: whatever strategy we use, we can consciously report it. And our report is accurate: it can be cross-validated by behavioral measures of response time and eye movements.[25] Such accurate introspection is unusual in psychology. Most mental operations are opaque to the mind's eye; we have no insight into the operations that allow us to recognize a face, plan a step, add two digits, or name a word. Somehow multidigit arithmetic is different: it seems to consist of a series of introspectable steps. I propose that there is a simple reason for it. Complex strategies, formed by stringing together several elementary steps—what computer scientists call "algorithms"—are another of consciousness's uniquely evolved functions.

Would you be able to calculate 12 times 13 unconsciously if the problem was presented to you in a subliminal flash? No, never.[26] A slow dispatching system seems necessary to store intermediate results and pass them on to the next step. The brain must contain a "router" that allows it to flexibly broadcast information to and from its internal routines.[27] This seems to be a major function of consciousness: to collect the information from various processors, synthesize it, and then broadcast the result—a conscious symbol—to other, arbitrarily selected processors. These processors, in turn, apply their unconscious skills to this symbol, and the entire cycle may repeat a number of times. The outcome is a hybrid serial-parallel machine, in which stages of massively parallel computation are interleaved with a serial stage of conscious decision making and information routing.

Together with the physicists Mariano Sigman and Ariel Zylberberg, I have begun to explore the computational properties that such a device would possess.[28] It closely resembles what computer scientists call a "production system," a type of program introduced in the 1960s to implement artificial intelligence tasks. A production system comprises a database, also called "working memory," and a vast array of if-then production rules (e.g., if there is an A in working memory, then change it to the sequence BC). At each step, the system examines whether a rule matches the current state of its working memory. If multiple rules match, then they compete under the aegis of a stochastic prioritizing system. Finally, the winning rule "ignites" and is allowed to change the contents of working memory before the entire process resumes. Thus this sequence of steps amounts to serial cycles of unconscious competition, conscious ignition, and broadcasting.

Remarkably, production systems, although very simple, have the capacity to implement any effective procedure—any thinkable computation. Their power is equivalent to that of the Turing machine, a theoretical device that was invented by the British mathematician Alan Turing in 1936 and that lies at the foundation of the digital computer.[29] Thus our proposal is tantamount to saying that, with its flexible routing capacity, the conscious brain operates as a biological Turing machine. It allows us to slowly churn out series of computations. These computations are very slow because, at each step, the intermediate result must be transiently maintained in consciousness before being passed on to the next stage.

There is an interesting historical twist to this argument. When Alan Turing invented his machine, he was trying to address a challenge posed by the mathematician David Hilbert in 1928: Could a mechanical procedure ever replace the mathematician and, by purely symbolic manipulation, decide whether a given statement of mathematics follows logically from a set of axioms? Turing deliberately designed his machine to mimic "a man in the process of computing a real number" (as he wrote in his seminal 1936 paper). He was not a psychologist, however, and could rely only on his introspection. This is why, I contend, his machine captures only a fraction of the mathematician's mental processes, those that are consciously accessible. The serial and symbolic operations that are captured by a serial Turing machine constitute a reasonably good model of the operations accessible to a conscious human mind.

Don't get me wrong—I do *not* intend to revive the cliché of the brain as a classical computer. With its massively parallel, self-modifiable organization, capable of computing over entire probability distributions rather than discrete symbols, the human brain departs radically from contemporary computers. Neuroscience, indeed, has long rejected the computer metaphor. But the brain's *behavior*, when it engages in long calculations, is roughly captured by a serial production system or a Turing machine.[30] For instance, the time that it takes us to compute a long addition such as 235 + 457 is the sum of the durations of each elementary operation (5 + 7; carry; 3 + 5 + 1; and finally 2 + 4)—as would be expected from the sequential execution of each successive step.[31]

The Turing model is idealized. When we zoom in on human behavior, we see deviations from its predictions. Instead of being neatly separate in time, successive stages slightly overlap and create an undesired cross-talk among operations.[32] During mental arithmetic, the second operation can start before the first one is fully finished. Jérôme Sackur and I studied one of the simplest possible algorithms: take a number n, add 2 to it ($n + 2$), and then decide if the result is larger or smaller than 5 ($n + 2 > 5$?). We observed interference: unconsciously, participants started to compare the initial number n with 5, even before they had obtained the intermediate result $n + 2$.[33] In a computer, such a silly error would never occur; a master clock controls each step, and digital routing ensures that each bit always reaches its intended destination. The brain, however, never evolved for complex arithmetic. Its architecture, selected for survival in a probabilistic world, explains why we make so many errors during mental calculation. We painfully "recycle" our brain networks for serial calculations, using conscious control to exchange information in a slow and serial manner.[34]

If one of the functions of consciousness is to serve as a lingua franca of the brain, a medium for the flexible routing of information across otherwise specialized processors, then a simple prediction ensues: a single routinized operation may unfold unconsciously, but unless the information is conscious, it will be impossible to string together several such steps. In the domain of arithmetic, for instance, our brain might well compute $3 + 2$ unconsciously, but not $(3 + 2)^2$, $(3 + 2) - 1$, or $\frac{1}{3+2}$. Multistep calculations will always require a conscious effort.[35]

Sackur and I set out to test this idea experimentally.[36] We flashed a target digit n and masked it, so that our participants could see it only half the time. We then asked them to perform a variety of operations with it. In three different blocks of trials, they attempted to name it, to add 2 to it (the $n + 2$ task), and to compare it with 5 (the $n > 5$ task). A fourth block required a two-step calculation: add 2, then compare the result with 5 (the $n + 2 > 5$ task). On the first three tasks, people did much better than chance. Even when they swore they hadn't seen anything, we asked them to venture an answer, and they were surprised to discover the extent of their unconscious knowledge. They could name the unseen digit much better than chance alone would predict: nearly half of their verbal responses were correct, whereas with four digits, guessing performance should have been 25 percent. They could even add 2 to it, or decide, above chance level, whether the digit was larger than 5. All these operations, of course, are familiar routines. As we saw in Chapter 2, there is a lot of evidence that they can be partially launched without consciousness. Crucially, however, during the unconscious two-step task ($n + 2 > 5$?), the participants failed: they responded at random. This is strange, because if they had just thought of naming the digit, and used the name to perform the task, they would have reached a very high level of success! Subliminal information was demonstrably present in their brains, since they correctly uttered the hidden number about half of the time—but without consciousness, it could not be channeled through a series of two successive stages.

In Chapter 2, we saw that the brain has no difficulty in unconsciously accumulating information: several successive arrows,[37] digits,[38] and even cues toward buying a car[39] can be added together, and the total evidence can guide our unconscious decisions. Is this a contradiction? No—because the accumulation of multiple pieces of evidence is a single operation for the brain. Once a neuronal accumulator is open, any information, whether conscious or unconscious, can bias it one way or the other. The only step that our unconscious decision-making process does not seem to achieve is a clear decision that can be passed on to the next stage. Although biased by unconscious information, our central accumulator never seems to reach the threshold beyond which it commits to a decision and moves on to the next step. As a consequence, in a complex calculation strategy, our unconscious remains stuck at the level of accumulating evidence for the first operation and never goes on to the second.

A more general consequence is that we cannot reason strategically on an unconscious hunch. Subliminal information cannot enter into our strategic deliberations. This point seems circular, but it isn't. Strategies are, after all, just another type of brain process—so it isn't so trivial that this process cannot be deployed without consciousness. Furthermore, it has genuine empirical consequences. Remember the arrows task, where one views five successive arrows pointing right or left and has to decide where the majority of them point? Any conscious mind quickly realizes that there is a winning strategy: once we have seen three arrows pointing to the same side, the game is over, as no amount of additional information can change the final answer. Participants readily exploit this strategy to get more quickly through the task. However, once again, they can do so only if the information is conscious, not if it is subliminal.[40] When the arrows are masked below the threshold for awareness, all they do is add them up—they cannot unconsciously make the strategic move to the next step.

All together, then, these experiments point to a crucial role for consciousness. We need to be conscious in order to rationally think through a problem. The mighty unconscious generates sophisticated hunches, but only a conscious mind can follow a rational strategy, step after step. By acting as a router, feeding information through any arbitrary string of successive processes, consciousness seems to give us access to a whole new mode of operation—the brain's Turing machine.

A Social Sharing Device

Consciousness is properly only a connecting network between man and man; it is only as such that it has had to develop: the recluse and wild-beast species of men would not have needed it.

—Friedrich Nietzsche, *The Gay Science* (1882)

In *Homo sapiens,* conscious information does not propagate solely within one individual's head. Thanks to language, it can also jump from mind to mind. During human evolution, social information sharing may have been one of the essential functions of consciousness. Nietzsche's "wild-beast species" probably relied on consciousness as a nonverbal buffer and router for millions of years—but only in the genus *Homo* did a sophisticated capacity to communicate those conscious states emerge. Thanks to human language, as well as to nonverbal pointing and gesturing, the conscious synthesis that emerges in one mind can be rapidly transferred to others. This active social transmission of a conscious symbol offers new computational abilities. Humans can create "multicore" social algorithms that do not draw solely on the knowledge available to a single mind but rather allow the confrontation of multiple points of view, variable levels of expertise, and a diversity of sources of knowledge.

It is no accident that verbal reportability—the capacity to put a thought into words—is considered a key criterion for conscious perception. We do not usually conclude that someone is aware of a piece of information unless he or she can, at least in part, formulate it with language (assuming, of course, that he is not paralyzed, aphasic, or too young to speak). In humans, the "verbal formulator" that allows us to express the contents of our mind is an essential component that can be deployed only when we are conscious.[41]

I do not mean, of course, that we can always accurately express our conscious thoughts with Proustian accuracy. Consciousness overflows language: we perceive vastly more than we can describe. The fullness of our experience of a Caravaggio painting, a gorgeous sunset over the Grand Canyon, or the changing expressions on a baby's face eludes exhaustive verbal description—which probably contributes in no small part to the fascination they exert. Nevertheless, and virtually by definition, whatever we are aware of can be at least partially framed in a linguistic format. Language provides a categorical and syntactic formulation of conscious thoughts that jointly lets us structure our mental world and share it with other human minds.

Sharing information with others is a second reason our brain finds it advantageous to abstract from the details of our present sensations and create a conscious "brief." Words and gestures provide us with only a slow communication channel—only 40 to 60 bits per second,[42] or about 300 times slower than the (now antiquated) 14,400-baud faxes that revolutionized our offices in the 1990s. Hence our brain drastically compresses the information to a condensed set of symbols that are assembled into short strings, which are then sent over the social network. It would actually be pointless to transmit to others a precise mental image of what I see from my own point of view; what others want is not a detailed description of the world as I see it, but a summary of the aspects that are likely to also be true from my interlocutor's viewpoint: a multisensory, viewer-invariant, and durable synthesis of the environment. In humans, at least, consciousness seems to condense information into exactly the kind of précis that other minds are likely to find useful.

The reader may object that language often serves trivial goals, such as exchanging the latest gossip about which Hollywood actress slept with whom. According to the Oxford anthropologist Robin Dunbar, close to two-thirds of our conversations may concern such social topics; he even proposed the "grooming and gossip" theory of language evolution, according to which language emerged solely as a bonding device.[43]

Can we prove that our conversations are more than tabloids? Can we show that they pass on to others precisely the sort of condensed information that is needed to make collective decisions? The Iranian psychologist Bahador Bahrami recently proved this idea using a clever experiment.[44] He had pairs of subjects perform a simple perceptual task. They were shown two displays, and their goal was to decide, on each trial, whether the first or the second contained a near-threshold target image. The two participants were first asked to give independent responses. The computer then revealed their choices, and if they disagreed, the subjects were asked to re-solve the conflict through a brief discussion.

What is particularly smart about this experiment is that, in the end, on each trial, the pair of subjects behaved as a single participant: they al-ways provided a single answer, whose accuracy could be gauged using exactly the same good old methods of psychophysics that are classically used to evaluate a single person's behavior. And the results were clear: as long as the two participants' abilities were reasonably similar, pairing them yielded a significant improvement in accuracy. The group systematically outperformed the best of its individual members—giving substance to the familiar saying "Two heads are better than one."

A great advantage of Bahrami's setup is that it can be modeled math-ematically. Assuming that each person perceives the world with his or her personal noise level, it is easy to compute how their sensations should be combined: the strength of the signals that each player perceived on a given trial should be inversely weighted by the player's average noise level, then averaged together to yield a single compound sensation. This optimal rule for multibrain decisions is, in fact, exactly identical to the law governing multisensory integration *within* a single brain. It can be approximated by a very simple rule of thumb: in most cases, people need to communicate not the nuances of what they saw (which would be im-possible) but simply a categorical answer (in this case, the first or the sec-ond display) accompanied by a judgment of confidence (or lack thereof).

It turned out that the successful pairs of participants spontaneously adopted this strategy. They talked about their confidence level using words such as *certain, very unsure,* or *just guessing.* Some of them even designed a numerical scale to precisely gauge their degree of certainty. Using such confidence-sharing schemes, their paired performance shot up to a very high level, essentially indistinguishable from the theoretical optimum.

Bahrami's experiment readily explains why judgments of confidence occupy such a central location in our conscious minds. In order to be useful to us and to others, each of our conscious thoughts must be ear-marked with a confidence label. Not only do we know that we know, or that we don't, but whenever we are conscious of a piece of information, we can ascribe to it a precise degree of certainty or uncertainty. Further-more, socially, we constantly endeavor to monitor the reliability of our sources, keeping in mind who said what to whom, and whether they were right or wrong (which is precisely what makes gossip a central feature of our conversations). These evolutions, largely unique to the human brain, point to the evaluation of uncertainty as an indispensable component of our social decision-making algorithm.

Bayesian decision theory tells us that the very same decision-making rules should apply to our own thoughts and to those that we receive from others. In both cases, optimal decision making demands that each source of information, whether internal or external, should be weighted, as accurately as possible, by an estimate of its reliability, before all the information is brought together into a single decision space. Prior to hominization, the primate prefrontal cortex already provided a workspace where past and present sources of information, duly weighted by their reliability, could be compiled to guide decisions. From there, a key evolutionary step, perhaps unique to humans, seems to have opened this workspace to social inputs from other minds. The development of this social interface allowed us to reap the benefits of a collective decision-making algorithm: by comparing our knowledge with that of others, we achieve better decisions.

Thanks to brain imaging, we are beginning to elucidate which brain networks support information sharing and reliability estimation. Whenever we deploy our social competence, the most anterior sectors of the prefrontal cortex, in the frontal pole and along the midline of the brain (within the ventromedial prefrontal cortex), are systematically activated. Posterior activations often occur as well, in a region lying at the junction of the temporal and parietal lobes, as well as along the brain's midline (the precuneus). These distributed areas form a brain-scale network, tightly interconnected by powerful long-distance fiber tracks, involving the prefrontal cortex as a central node. This network figures prominently among the circuits that turn on during rest, whenever we have a few seconds to ourselves: we spontaneously return to this "default mode" system of social tracking in our free time.[45]

Most remarkably, as would be expected from the social decision-making hypothesis, many of these regions activate both when we think about ourselves—for instance, when we introspect about our level of confidence in our own decisions[46]—and when we reflect upon the thoughts of others.[47] The frontal pole and ventromedial prefrontal cortex, in particular, show very similar response profiles during judgments about ourselves and about others[48]—to such an extent that thinking hard about one may prime the other.[49] Thus this network appears ideally suited to evaluate the reliability of our own knowledge and compare it with the information we receive from others.

In brief, within the human brain lies a set of neural structures uniquely adapted to the representation of our social knowledge. We use the same database to encode our self-knowledge and to accumulate information about others. These brain networks build a mental image of our own self as a peculiar character sitting next to others in a mental database of our social acquaintances. Each of us represents "oneself as another," as the French philosopher Paul Ricoeur puts it.[50]

If this view of the self is correct, then the neural underpinnings of our own identity are built up in a rather indirect manner. We spend our life monitoring our behavior as well as that of others, and our statistical brain constantly draws inferences about what it observes, literally "making up its mind" as it proceeds.[51] Learning who we are is a statistical deduction from observation. Having spent a lifetime with ourselves, we reach a view of our own character, knowledge, and confidence that is only a bit more refined than our view of other people's personalities. Furthermore, our brain does enjoy privileged access to some of its inner workings.[52] Introspection makes our conscious motives and strategies transparent to us, while we have no sure means of deciphering them in others. Yet we never genuinely know our true selves. We remain largely ignorant of the actual unconscious determinants of our behavior, and therefore we cannot accurately predict what our behavior will be in circumstances beyond the safety zone of our past experience. The Greek motto "Know thyself," when applied to the minute details of our behavior, remains an inaccessible ideal. Our "self" is just a database that gets filled in through our social experiences, in the same format with which we attempt to understand other minds, and therefore it is just as likely to include glaring gaps, misunderstandings, and delusions.

Needless to say, these limits of the human condition have not escaped novelists. In his introspective novel *Thinks . . .* , the British contemporary writer David Lodge depicts his two main characters, the English teacher Helen and the artificial intelligence mogul Ralph, exchanging thoughtful reflections upon the self, while lightly flirting at night in an outdoor Jacuzzi:

> *Helen:* I suppose it must have a thermostat. Does that make it conscious?
> *Ralph:* Not self-conscious. It doesn't know it's having a good time—unlike you and me.
> *Helen:* I thought there was no such thing as the self.
> *Ralph:* No such *thing*, no, if you mean a fixed discrete entity. But of course there are selves. We make them up all the time. Like you make up stories.
> *Helen:* Are you saying our lives are just fictions?
> *Ralph:* In a way. It's one of the things we do with our spare brain capacity. We make up stories about ourselves.

Partially deluding ourselves may be the price we pay for a uniquely human evolution of consciousness: the ability to communicate our conscious knowledge with others, in rudimentary form, but with exactly the sort of confidence evaluation that is mathematically needed to reach a useful collective decision. Imperfect as it is, our human ability for introspecting and social sharing has created alphabets, cathedrals, jet planes, and lobster Thermidor. For the first time in evolution, it has also allowed us to voluntarily create fictive worlds: we can tweak the social decision-making algorithm to our advantage by faking, forging, counterfeiting, fibbing, lying, perjuring, denying, forswearing, arguing, refuting, and rebuffing. Vladimir Nabokov, in his *Lectures on Literature* (1980), saw it all:

> Literature was not born the day when a boy crying "wolf, wolf" came running out of the Neanderthal valley with a big gray wolf at his heels; literature was born on the day when a boy came crying "wolf, wolf" and there was no wolf behind him.

Consciousness is the mind's virtual-reality simulator. But how does the brain make up the mind?

4

THE SIGNATURES OF A CONSCIOUS THOUGHT

Brain-imaging techniques have led to a breakthrough in consciousness research. They have revealed how brain activity unfolds as a piece of information gains access to consciousness, and how this activity differs during unconscious processing. Comparing these two states reveals what I call a "signature of consciousness": a reliable marker that the stimulus was consciously perceived. In this chapter, I describe four signatures of consciousness. First, although a subliminal stimulus can propagate deeply into the cortex, this brain activity is strongly amplified when the threshold for awareness is crossed. It then invades many additional regions, leading to a sudden ignition of parietal and prefrontal circuits (signature 1). In the electroencephalogram, conscious access appears as a late slow wave called the P3 wave (signature 2). This event emerges as late as one-third of a second after the stimulus; our consciousness lags behind the external world. By tracking brain activity with electrodes placed deep inside the brain, two more signatures can be observed: a late and sudden burst of high-frequency oscillations (signature 3), and a synchronization of information exchanges across distant brain regions (signature 4). All these events provide reliable indexes of conscious processing.

A person . . . is a shadow which we can never penetrate, of which there can be no such thing as direct knowledge.

—Marcel Proust, *The Guermantes Way* (1921)

Marcel Proust's metaphor renews a worn-out cliché: the mind as a fortress. Withdrawn behind our mental walls, hidden from others' inquisitive gaze, we may freely think whatever we want. Our consciousness is an impenetrable sanctuary where our minds go free-wheeling, while our colleagues, friends, and spouses think we are attending to their words. Julian Jaynes pictures it as "a secret theater of speechless monologue and prevenient counsel, an invisible mansion of all moods, musings, and mysteries, an infinite resort of disappointments and discoveries." How could scientists ever infiltrate this inner bastion?

And yet in the space of only twenty years, the unthinkable happened. In 1990 the skull became transparent: the Japanese researcher Seiji Ogawa and his colleagues invented functional magnetic resonance imaging (fMRI), a powerful and harmless technique that, without the use of any injection, allows us to visualize the activity of the whole brain.[1] Functional MRI capitalizes on the coupling of brain cells with blood vessels. Whenever a neuronal circuit increases its activity, the glial cells that surround these neurons sense the surge in synaptic activity. To quickly compensate for this heightened energy consumption, they open up the local arteries. Two or three seconds later the blood flow increases, bringing in more oxygen and glucose. Red blood cells abound, carrying hemoglobin molecules that convey the oxygen. The great feat of fMRI consists in detecting the physical properties of the hemoglobin molecule at a distance: the hemoglobin without oxygen acts as a small magnet, while the hemoglobin with oxygen does not. Magnetic resonance machines are giant magnets that are tuned to pick up these very small distortions in magnetic fields, thus indirectly reflecting the recent neuronal activity in every piece of brain tissue.

Functional MRI easily visualizes the state of activity of the living human brain at millimeter resolution, up to several times per second. Unfortunately, it cannot track the time course of neuronal firing, but other techniques are now available to precisely time the electrical currents at synapses, again without opening the skull. Electroencephalography (EEG for short), the good old-fashioned recording of brain waves invented in the 1930s, has been perfected into a high-powered technique, with up to 256 electrodes providing high-quality digital recordings of brain activity with millisecond resolution over the whole head. In the 1960s an even better technology emerged: magnetoencephalography (MEG), the ultra-precise recording of the minuscule magnetic waves that accompany the discharge of currents in cortical neurons. Both EEG and MEG can be recorded very simply, either by placing small electrical leads on the head

(EEG) or by placing very sensitive detectors of magnetic fields around it (MEG).

With fMRI, EEG, and MEG in hand, we can now track the entire sequence of brain activation as a visual stimulus travels from the retina to the highest reaches of the frontal cortex. In combination with the techniques of cognitive psychology, these tools offer a new window into the conscious mind. As we discussed in Chapter 1, many experimental stimuli provide optimal contrasts between conscious and unconscious states. Through masking or inattention, we can take any visible image and make it vanish from sight. We can even place it at threshold, so that it is perceived only half the time and therefore varies only in its subjective awareness. In the best experiments, stimulus, task, and performance are tightly equalized. As a result, consciousness is the only variable that is experimentally manipulated: the subject reports seeing in one case and not seeing in the other.

All that remains, then, is to examine what difference consciousness makes at the brain level. What specific circuits, if any, activate only on conscious trials? Does conscious perception elicit unique brain events, specific waves, or oscillations? Such markers, if they could be found, would serve as signatures of consciousness. The presence of these patterns of neural activity, like a signature on a document, would reliably index conscious perception.

In this chapter, we will see that several signatures of consciousness can be found. Thanks to brain imaging, the mystery of consciousness has finally been cracked open.

The Avalanche of Consciousness

In 2000 the Israeli scientist Kalanit Grill-Spector, then at the Weizmann Institute of Science in Tel Aviv, performed a simple masking experiment.[2] She flashed pictures for a very brief duration, which varied between one-fiftieth and one-eighth of a second, and followed them with a scrambled image. As a result, some images remained detectable while others became downright invisible—they fell above or below the threshold for conscious perception. The participants' reports traced a beautiful curve: images presented below 50 milliseconds were very hard to see, while those shown for 100 milliseconds or more were visible.

Grill-Spector then scanned the participants' visual cortex (at that time, it was not easy to scan the whole brain). What she observed was a clear dissociation. In early visual areas, activity was present irrespective of consciousness. The primary visual cortex and surrounding regions were basically activated by all images, regardless of the amount of masking. In the higher visual centers of the cortex, however, within the fusiform gyrus and the lateral occipitotemporal region, a tight correlation emerged between brain activation and conscious reports. These regions are involved in sorting out categories of pictures such as faces, objects, words, and places, and in creating an invariant representation of their appearance. It seemed that, whenever brain activation reached this level, the image was likely to become conscious.

At just about the same time, I was doing similar experiments on the perception of masked words.[3] My scanner provided whole-brain images of the areas that activated whenever subjects watched words that were flashed just above or just below the threshold for conscious perception. And the results were clear: even the higher visual areas of the fusiform gyrus could be activated in the absence of any consciousness. In fact, quite abstract cerebral operations, involving advanced regions of the temporal and parietal lobes, could be performed subliminally—for instance, the recognition that *piano* and *PIANO* are the same word, or that the digit *3* and the word *three* mean the same quantity.[4]

Nevertheless, when the threshold for conscious perception was crossed, I too saw massive changes in those higher visual centers. Their activity was strongly amplified. In the key region for letter recognition, the "visual word form area," brain activation was multiplied by twelve! Furthermore, an entire set of additional regions appeared that had simply been absent when the word was masked and remained unconscious. These regions were broadly distributed in the parietal and frontal lobes, even reaching into the depths of the anterior cingulate gyrus in the midline of the two hemispheres (figure 16).

By measuring the amplitude of this activity, we discovered that the amplification factor, which distinguishes conscious from unconscious processing, varies across the successive regions of the visual input pathway. At the first cortical stage, the primary visual cortex, the activation evoked by an unseen flashed word is strong enough to be easily detectable. However, as it progresses forward into the cortex, masking makes it

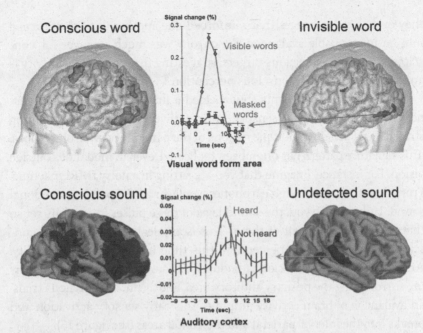

FIGURE 16. The first signature of conscious perception is an intense ignition of distributed brain regions, including bilateral prefrontal and parietal regions. A word made subliminal by masking (above) activates specialized reading circuits, but the very same word, when seen, causes an enormous amplification of activity that invades the parietal and prefrontal lobes. Similarly, auditory areas can be activated by an unconscious chord (below), but the very same sound, when consciously detected, invades extensive sectors of inferior parietal and prefrontal cortex.

lose strength. Subliminal perception can thus be compared to a surf wave that looms large on the horizon but merely licks your feet when it reaches the shore.[5] By comparison, conscious perception is a tsunami— or perhaps an avalanche is a better metaphor, because conscious activation seems to pick up strength as it progresses, much as a minuscule snowball gathers snow and ultimately triggers a landslide.

To bring this point home, in my experiments I flashed words for only 43 milliseconds, thereby injecting minimal evidence into the retina. Nevertheless, activation progressed forward and, on conscious trials, ceaselessly amplified itself until it caused a major activation in many regions. Distant brain regions also became tightly correlated: the incoming wave peaked and receded simultaneously in all areas, suggesting that they exchanged messages that reinforced one another until they turned into an unstoppable avalanche. Synchrony was much stronger for conscious than for unconscious targets, suggesting that correlated activity is an important factor in conscious perception.[6]

These simple experiments thus yielded a first signature of consciousness: an amplification of sensory brain activity, progressively gathering strength and invading multiple regions of the parietal and prefrontal lobes. This signature pattern has often been replicated, even in modalities outside vision. For instance, imagine that you are sitting in a noisy fMRI machine. From time to time, through earphones, you hear a brief pulse of additional sound. Unknown to you, the sound level of these pulses is carefully set so that you detect only half of them. This is an ideal way to compare conscious and unconscious perception, this time in the auditory modality. And the result is equally clear: unconscious sounds activate only the cortex surrounding the primary auditory area, and again, on conscious trials, an avalanche of brain activity amplifies this early sensory activation and breaks into the inferior parietal and prefrontal areas (see figure 16).[7]

For a third example, consider motor action. Suppose that you are told to move whenever you see a target, but to refrain from responding if you see a "no-go" cue just before the target.[8] This is a typical task of response inhibition: you need to exert conscious control in order to inhibit the strong tendency to respond with the dominant "go" response on "no-go" trials. Now imagine that the "no-go" cue, on half the trials, is presented just below the threshold for conscious perception. How can you possibly follow an order that you do not perceive? Fascinatingly, your brain stands up to this impossible challenge. Even on subliminal trials, participants' responses slow down ever so slightly, suggesting that the brain partially deploys its inhibition powers unconsciously (as we saw in Chapter 2). Brain imaging shows that this subliminal inhibition relies on two regions associated with the control of motor commands: the presupplementary motor area and the anterior insula. However, once again conscious perception causes a massive change: when the "no-go" cue is visible, activation nearly doubles in these two control regions, and it invades a massively larger network of areas in the parietal and prefrontal lobes (figure 17). By now, this parietal and prefrontal circuit should be familiar: its sudden activation systematically appears as a reproducible signature of conscious awareness.[9]

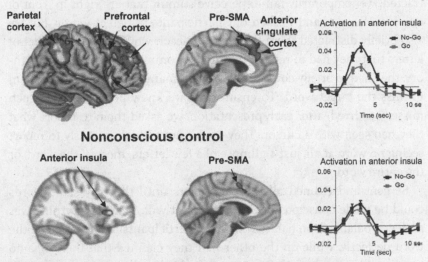

FIGURE 17. Actions that are controlled consciously or unconsciously rely on partially distinct brain circuits. An invisible "no-go" signal reaches a few specialized brain regions, such as the anterior insula and the presupplementary motor area (pre-SMA), that monitor our motor actions and keep them in check (right column). The same signal, when made visible, activates many more regions of the parietal and prefrontal lobes that are associated with voluntary control.

Timing the Conscious Avalanche

Although functional magnetic resonance imaging is a wonderful tool for localizing *where* in the brain the activation occurs, it is unable to tell us precisely *when*. We cannot really use it to measure how fast, and in which order, the successive brain areas light up when we become aware of a stimulus. To accurately time the conscious avalanche, the more precise methods of electro- and magnetoencephalography (EEG and MEG) are the perfect tools. A few electrodes pasted onto the skin or magnetic sensors surrounding the head let us track brain activity with millisecond precision.

In 1995 Claire Sergent and I designed a careful EEG study that, for the first time, isolated the time course of conscious access.[10] We tracked the cortical fate of *identical* images that sometimes were consciously perceived and sometimes went totally undetected (figure 18). We capitalized on the attentional blink phenomenon—the fact that, when briefly distracted, we temporarily fail to perceive stimuli that are right in front of our eyes. Sergent and I asked our participants to detect words, but we also briefly distracted them by preceding each word with another set of letters that they had to report. In order to commit these letters to memory, they had to briefly concentrate, and on many trials, this caused them to miss the target word. To ensure that we knew precisely when such misses occurred, after each presentation, we asked them to report what they had seen with a cursor. They could move it continuously to report seeing no word at all, just a glimpse of a few letters, most of the word, or the entire word.

Sergent and I tuned all the parameters until the very same words could be made conscious or unconscious at will. When everything was perfectly balanced, on half the trials the participants reported seeing the word perfectly, while on the other half they claimed that there was no word at all. Their conscious reports varied in an all-or-none manner: either they perceived the word or they entirely missed it, but they rarely reported a partial perception of the letters.[11]

Simultaneously, our recordings showed that the brain was also undergoing a sudden change of mind, discontinuously jumping from the invisible to the perceived state. Initially, within the early visual system, the visible and invisible words evoked no difference in activity at all. Conscious and unconscious words, like any visual stimulation, evoked an indistinguishable stream of brain waves over the posterior part of the visual cortex. These waves are called the P1 and the N1, to indicate that the first one is positive and peaks around 100 milliseconds, while the second is negative and reaches its maximum at about 170 milliseconds. Both waves reflected the progression of visual information through a hierarchy of visual areas—and this initial progression seemed totally unaffected by consciousness. Activation was very strong and exactly as intense when the word could be reported as when it remained totally invisible. Clearly, the word was entering the visual cortex normally, whether the viewer would later report seeing it or not.

Just a few hundredths of a second later, however, the pattern of activation changed radically. Suddenly, between 200 and 300 milliseconds after the word onset, brain activity faded on unconscious trials, whereas

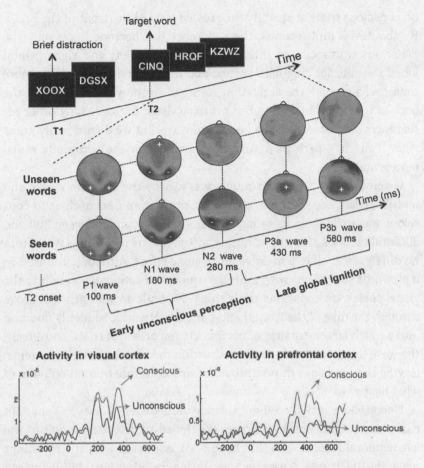

FIGURE 18. Slow positive waves over the top and back of the head provide a second signature of conscious perception. In this experiment, words were flashed during the attentional blink, at the very moment when viewers were distracted by another task. As a result, viewers missed half of the words: they frequently reported that they could not see them. Brain waves recording on the surface of the head tracked the fate of words they saw and those they didn't see. Initially, both elicited identical activations of the visual cortex. But conscious and unconscious trials suddenly diverged at around 200 milliseconds. For conscious words only, the wave of activity is amplified and flows into the prefrontal cortex and many other associative regions, then back to visual areas. This global ignition causes a large positive voltage on the top of the head—the P3 wave.

on conscious trials, it steadily progressed toward the front of the brain. By about 400 milliseconds, the difference had become huge: only the conscious words caused intense activity in the left and right frontal lobes, the anterior cingulate cortex, and the parietal cortex. After more than half a second, the activation returned to the visual regions at the back of the brain, including the primary visual cortex. Many other researchers have observed this backward wave, but we do not really know what it means—perhaps a sustained memory of the conscious visual representation.[12]

Given that our original stimulus was *exactly* the same on visible and invisible trials, the swiftness of the transition from unconscious to conscious was striking. In less than one-tenth of a second, between 200 and 300 milliseconds after the stimulus appeared, our recordings went from no difference at all to a massive all-or-none effect. Although it looked as if all words started out with a similar amount of activity flowing into the visual cortex, on conscious trials this wave built up strength and broke through the dike of the frontal and parietal networks, suddenly flooding into a much larger expanse of cortex. On unconscious trials, conversely, the wave remained safely contained within the brain's posterior systems, leaving the conscious mind untouched and therefore in total oblivion of what happened.

Unconscious activity did not subside immediately, however. For about half a second, unconscious waves continued to reverberate within the left temporal lobe, at sites that have been associated with the meanings of words. In Chapter 2, we saw how, during the attentional blink, unseen words continue to activate their meanings.[13] This unconscious interpretation occurs within the confines of the temporal lobe. Only its overflow into the broader ranges of the frontal and parietal lobes signals conscious perception.

The conscious avalanche produces a simple marker that is easily picked up by electrodes glued to the top of the head. On conscious trials only, an ample voltage wave sweeps through this region. It starts around 270 milliseconds and peaks anywhere between 350 and 500 milliseconds. This slow and massive event has been called the P3 wave (because it is the third large positive peak after a stimulus appears) or the P300 wave (because it often starts around 300 milliseconds).[14] It is only a few microvolts in size, a million times smaller than an AA battery. However, such a surge of electrical activity is easily measurable with modern amplifiers. The P3 wave is our second signature of consciousness. A variety of paradigms have now shown that it can be easily recorded whenever we suddenly gain access to a conscious percept.[15]

By looking closer at our recordings, we discover that the evolution of the P3 wave also explains *why* our participants failed to see the target word. In our experiment, there were in fact *two* P3 waves. The first P3 was evoked by the initial string of letters, which served to distract attention and was always consciously perceived. The second was elicited by the target word when it was seen. Fascinatingly, there was a systematic trade-off between these two events. Whenever the first P3 wave was large and long, the second one was much more likely to be absent—and those were precisely the trials on which the target was likely to be missed. Conscious access thus operated as a push-and-pull system: whenever the brain was occupied for a long time by the first string, as indexed by a long P3 wave, it could not simultaneously attend to the second word. Consciousness of one seemed to exclude consciousness of the other.

René Descartes would have been delighted; he was the first to note that "we cannot be very attentive to several things at once," a limitation of consciousness that he attributed to the simple mechanical fact that the pineal gland could lean to only one side at a time. Leaving aside this discredited brain localization, Descartes was right: our conscious brain cannot experience two ignitions at once and lets us perceive only a single conscious "chunk" at a given time. Whenever the prefrontal and parietal lobes are jointly engaged in processing a first stimulus, they cannot simultaneously reengage toward a second one. The very act of concentrating on the first item often prevents us from perceiving the second. Sometimes we do end up perceiving it—but then its P3 wave is sharply delayed.[16] This is the "refractory period" phenomenon that we met in Chapter 1: before a second target enters consciousness, it must wait until the conscious mind is done with the first one.

Consciousness Lags Behind the World

An important consequence of these observations is that our consciousness of unexpected events lags considerably behind the real world. Not only do we consciously perceive only a very small proportion of the sensory signals that bombard us, but when we do, it is with a time lag of at least one-third of a second. In this respect, our brain is like an astronomer who watches for supernovae. Because the speed of light is finite, the news from distant stars takes millions of years to reach us. Likewise, because our brain accumulates evidence at a sluggish speed, the information that we attribute to the conscious "present" is outdated by at least one-third of a second. The duration of this blind period may even exceed half a second when the input is so faint that it calls for a slow accumulation of evidence before crossing the threshold for conscious perception. (This is analogous to the astronomer's long-exposure shooting, which lets the light from faint stars accumulate on a sensitive photographic plate.)[17] As we just saw, consciousness can be delayed even further when the mind is occupied elsewhere. This is why you should not use your cell phone while driving—even a seemingly reflex response, such as hitting the brakes when you see the taillights of the car in front of you, slows down when your conscious mind is distracted.[18]

We are all blind to the limits of our attention and do not realize that our subjective perception lags behind the objective events in the outside world. But most of the time it does not matter. We can enjoy a beautiful sunset or listen to a symphony orchestra concert without realizing that the colors we see and the music we hear date from half a second ago. When we are passively listening, we do not really care exactly when the sounds were emitted. And even when we need to act, the world is often slow enough for our delayed conscious responses to remain roughly appropriate. It is only when we try to act "in real time" that we realize how slow our awareness is. Any pianist who rushes through an allegro knows better than to attempt to control each of his flying fingers—conscious control is way too slow to tramp into this fast dance. To appreciate the slowness of your consciousness, try to photograph a fast and unpredictable event, such as a lizard sticking its tongue out: by the time your finger presses the shutter, the event that you hoped to capture on film is long gone.

Fortunately, our brain also contains exquisite mechanisms that compensate for these delays. First, we often rely on an unconscious "autopilot." As René Descartes observed long ago, a burnt finger retracts from the fire way before we become aware of the pain. Our eyes and hands often react appropriately because they are guided by a whole array of fast sensory-motor loops that operate outside our conscious awareness. These motor circuits may certainly be set up according to our conscious intentions, as when we cautiously reach toward a candle's flame. But then the action itself unfolds unconsciously, and our fingers adjust to a sudden shift in the target's location with an amazingly swift move, way before we consciously detect any change.[19]

Anticipation is a second mechanism that compensates for the sluggishness of our consciousness. Virtually all our sensory and motor areas contain temporal learning mechanisms that anticipate events in the outside world. When such events unfold in a predictable manner, these brain mechanisms generate accurate anticipations, which let us perceive the events closer to the time when they actually occur. An unfortunate consequence is that, when an unanticipated event occurs—for instance, a brief flash of light—we misperceive its onset. Relative to a dot moving at predictable speed, a flash of light appears to lag behind its true position.[20] This "flash lag" effect, whereby we always perceive a predictable stimulus sooner than an unpredictable one, is a living testimony to the long and winding paths that lead to the fortress of the conscious mind.

Only when our brain's anticipation mechanisms fail do we become acutely aware of the long delay that our consciousness imposes. If you accidentally drop a glass of milk, you experience this phenomenon firsthand: for a split second, you become acutely aware that your consciousness gropes hopelessly behind the event, and you can only lament your own slowness.

Error perception actually operates in two steps, much like the perception of any other physical attribute: unconscious appraisal followed by conscious ignition. Suppose that you are asked to move your eyes in a countermanding fashion: whenever a flash of light appears, you are to gaze away from it. More often than not, however, when the flash appears, your eyes will not move directly away from it; they will first be magnetically attracted toward it and only later turn away. What is fascinating is that you may not be aware of your initial error. On some trials, you may have the feeling that your eyes move away directly, even though they do not. Electroencephalography can be used to monitor how such an unconscious error is encoded in the brain.[21] Initially, during the first one-fifth of a second, the cortex reacts virtually identically to conscious and unconscious errors. An autopilot system in the cingulate gyrus notices that the motor plan does not unfold according to instructions and fires vigorously to signal the error—even when it remains unconscious.[22] Like other sensory responses, this initial brain response is fully unconscious and often remains undetected. When we gain full awareness of our erroneous action, however, a late brain response ensues, a strong positive response that can be recorded from the top of the scalp. Although it has been given a different name, "error-related positivity" (or Pe for short), this response is virtually indistinguishable from the familiar P3 wave that accompanies our conscious perception of sensory events. Thus actions and sensations seem to be consciously perceived in a very similar manner. Once again the P3 wave appears to be a reliable signature of the brain's conscious appraisal—and this signature arises quite late after the event that triggered it.[23]

Isolating the Conscious Moment

The critical reader may remain skeptical: Have we truly identified a unique signature of conscious access? Could the observed ignition of parietal and prefrontal networks, and the accompanying P3 wave, have other explanations? In the past decade, neuroscientists have endeavored to refine their experiments in order to control for all possible confounding factors. Although the jury is still out, some of these ingenious experiments convincingly isolate conscious perception from other sensory and motor events. Let us look at how they work.

Conscious perception entails many consequences. Whenever we become aware of an event, myriad possibilities open up. We can report it, either verbally or with gestures. We can store it in memory and later recall it. We can evaluate it or act upon it. All these processes are deployed only after we become aware—and thus they might be confused with conscious access. Does the brain activity that we observe on conscious trials have anything specific to do with conscious access?

To address this difficult issue, my fellow researchers and I have tried hard to match conscious and unconscious trials. By design, our initial experiments asked our participants to act similarly in both cases. In our attentional blink study, for instance, participants first had to remember the target letters, then decide whether they had also seen a word or not.[24] Arguably, deciding that one has *not* seen a word is just as difficult as, if not more difficult than, deciding that one has seen one. Furthermore, participants made both "seen" and "unseen" responses using the same kind of movement, a left- or right-hand key press. None of these factors could explain our finding of a large P3 wave, with strong parietal and prefrontal activation, on seen-word trials but not on unseen-word ones.

However, the devil's advocate might argue that seeing a word triggers a series of brain processes at a precise moment in time, whereas "not seeing" clearly cannot be associated with such a sharp onset; one has to wait until the end of a trial in order to decide that one hasn't seen anything. Could such a temporal dilution explain the differences in brain activation?

Using a clever trick, Hakwan Lau and Richard Passingham rejected this possibility.[25] They relied on the surprising phenomenon of blindsight. As we saw in Chapter 2, subliminal images that are flashed for a short duration, although invisible, may still induce cortical activations that sometimes reach the motor cortex. As a result, participants accurately respond to a target that they deny seeing—hence the term *blindsight*. Lau and Passingham cleverly used this effect to equalize objective motor performance on conscious and unconscious trials: participants did *exactly* the same thing in both cases. Even with this subtle control, greater conscious visibility was again associated with a stronger activation of the left prefrontal cortex. These results were obtained in healthy volunteers, but also in the classical blindsight patient G.Y., this time with a full-blown pattern of distributed parietal and prefrontal activation on conscious trials.[26]

Great, says the devil's advocate; you have equalized the responses, but now the conscious and unconscious stimuli differ. Can you equalize *both* the stimuli *and* the responses, keeping *everything* identical except the subjective feeling of conscious vision? Only then will I be truly convinced that you have nailed down the signatures of consciousness.

Does it sound impossible? It is not. During his Ph.D. research, the Israeli psychologist Moti Salti, with his mentor Dominique Lamy, accomplished this remarkable feat and thereby confirmed that the P3 wave is a signature of conscious access.[27] The simple experimental trick was to sort out the trials on the basis of the participant's response. Salti flashed an array of lines at one of four locations and asked each participant to give two immediate responses: (1) Where was the flash? (2) Did you see it, or did you just guess? On the basis of this information, he could easily separate different types of trials. Many were "aware, correct" trials in which the participants reported seeing the target and, of course, responded correctly. However, due to blindsight, there were also a large number of "unaware, correct" trials in which the participants denied seeing anything yet responded correctly.

So here was the perfect control: same stimulus, same response, but different awareness. EEG recordings showed that all the early brain activations, up to about 250 milliseconds, were strictly identical. The two types of trials differed in only one feature: the P3 wave, which after 270 milliseconds grew to a massively larger size on conscious trials than on unconscious trials. Not only its amplitude but also its topography was distinctive: while unconscious stimuli evoked a small positive wave over the posterior parietal cortex, presumably reflecting the unconscious processing chain that led to the correct response, only conscious perception elicited an expansion of this activation into the left and right frontal lobes.

Playing the devil's advocate himself, Salti considered whether his results could be explained by a mixture of unconscious trials, some with random responding and others with a normal-size P3. His analyses squarely rejected this alternative model. A small posterior P3 did occur on unconscious trials, but it was too small, too short, and too posterior to match the one seen on conscious trials. It merely indicated that, on unseen trials, the avalanche of brain activity started but quickly fizzled and stopped short of triggering a global P3 event. Only the full-size P3, when it extended bilaterally over the prefrontal cortex, genuinely indexed a neural process that was unique to conscious perception.

Igniting the Conscious Brain

Whenever we become aware of an unexpected piece of information, the brain suddenly seems to burst into a large-scale activity pattern. My colleagues and I have called this property "global ignition."[28] We were inspired by the Canadian neurophysiologist Donald Hebb, who first analyzed the behavior of collective assemblies of neurons in his 1949 best seller *The Organization of Behavior*.[29] Hebb explained, in very intuitive terms, how a network of neurons that excite one another can quickly fall into a global pattern of synchronized activity—much as an audience, after the first few handclaps, suddenly bursts into broad applause. Like the enthusiastic spectators who stand up after a concert and contagiously spread the applause, the large pyramidal neurons in the upper layers of cortex broadcast their excitation to a large audience of receiving neurons. Global ignition, my colleagues and I have suggested, occurs when this broadcast excitation exceeds a threshold and becomes self-reinforcing: some neurons excite others that, in turn, return the excitation.[30] The net result is an explosion of activity: the neurons that are strongly interconnected burst into a self-sustained state of high-level activity, a reverberating "cell assembly," as Hebb called it.

This collective phenomenon resembles what physicists call a "phase transition," or mathematicians a "bifurcation": a sudden, nearly discontinuous change in the state of a physical system. Water that freezes into an ice cube epitomizes the phase transition from liquid to solid. Early on in our thinking about consciousness, my colleagues and I noted that the concept of phase transition captures many properties of conscious perception.[31] Like freezing, consciousness exhibits a threshold: a brief stimulus remains subliminal, while an incrementally longer one becomes fully visible. Most physical self-amplifying systems possess a tipping point where global change happens or fails depending on minute impurities or noise. The brain, we reasoned, may be no exception.

Does a conscious message trigger a brain-scale phase transition in our cortical activity, freezing brain areas together into a coherent state? If so, how could we prove it? To find out, Antoine Del Cul and I designed a simple experiment.[32] We continuously varied one physical parameter of a display, similar to slowly dropping the temperature of a water vial. We then examined whether subjective reports, as well as objective markers of brain activity, behaved in a discontinuous manner and suddenly burst, as if they were undergoing a drastic phase transition.

In our experiment, we flashed a digit for just a single frame of our video screen (16 milliseconds), then a blank, and finally a mask made of random letters. We varied the duration of the blank in small steps of 16 milliseconds. What did viewers report? Did their perception change continuously? No—it followed the all-or-none pattern of a phase transition. At long delays, they could see the digit—but at short delays, they saw only the letters: the digit was masked. Crucially, these two states were separated by a clear threshold. Perception was nonlinear: as the delay increased, visibility did not improve smoothly (participants did not report seeing more and more of the digit) but showed a sudden step (now I see it, now I don't). A delay of about 50 milliseconds separated the perceived and unperceived trials.[33]

With this finding at hand, we then turned to EEG recordings and investigated which brain events also occurred in a steplike response to the masked digits. Once again the results pointed to the P3 waveform. All the preceding events either did not vary with the stimulus at all or, when they did, evolved in a manner that did not resemble the participants' subjective reports.

We found, for instance, that the initial response of the visual cortex, indexed by P1 and N1 waves, was essentially unaffected by the digit-letters delay. This should not be surprising: after all, the very same digit was presented on all trials, for the same duration, so we were witnessing the first stages of its entry into the brain, which were essentially constant, whether the digit was ultimately seen or not.

The following waves, in the left and right visual areas, still behaved in a continuous manner. The size of these visual activations grew in direct proportion to the duration of the digit's presence on screen, prior to its interruption by the mask. The flashed digit was able to progress into the brain to the point where its activity was cut short by the letter mask. As a result, the brain waves increased in duration and size, in strict proportion to the digit-to-letters time lag. This proportionality to the stimulus did not correspond to the nonlinear all-or-none experience that the participants reported. It implied that these waves too did not relate to the participants' consciousness. At this stage, activity was still strong on trials in which people strongly denied seeing any digit.

Starting at 270 milliseconds after the digit's onset, however, our recordings suddenly exhibited the global ignition pattern (figure 19). The brain waves showed a sudden divergence, with an avalanche of activation that built up quickly and strongly on trials where the participant

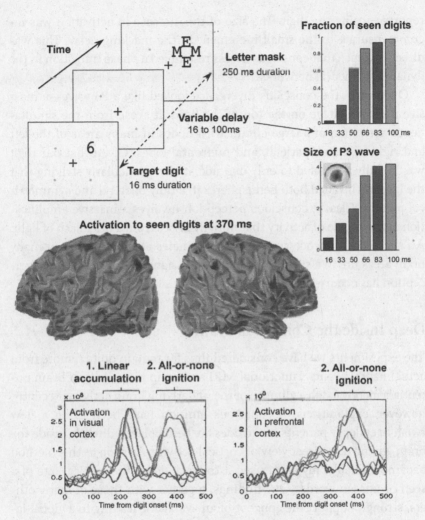

FIGURE 19. Conscious perception triggers a sudden change in late brain activity—what physicists call a "nonlinear phase transition." In this experiment, a digit was flashed, and after a variable delay, a set of letters masked it. Activation of the visual cortex increased smoothly as the delay increased. Conscious perception, however, was discontinuous: the digit suddenly became visible when the delay passed a threshold of about 50 milliseconds. Once again, the late P3 wave appeared as a signature of conscious perception. Starting around 300 milliseconds after the digit, several regions of the cortex, including the frontal lobes, ignited suddenly, in an all-or-none manner, only when participants reported seeing the digit.

reported seeing the digit. The size of the increase in activation was incommensurate to the small increment in the masking delay. This was direct evidence that conscious access resembled a phase transition in the dynamics of neural networks.

Once again the conscious divergence looked like a P3 wave—a massive positive voltage on the top of the head. It arose from the simultaneous activation of a large circuit with nodes in many areas of the left and right occipital, parietal, and prefrontal lobes. Given that our digit was initially presented to only one side, it was particularly striking that the ignition invaded both hemispheres in a fully bilateral and symmetrical pattern. Clearly, conscious perception involves a massive amplification of the trickle of activity that initially arises from a brief flash of light. An avalanche of processing stages culminates at the point when many brain areas fire in a synchronized manner, signaling that conscious perception has occurred.

Deep Inside the Conscious Brain

The experiments we have considered thus far remain quite remote from actual neural events. Functional MRI and scalp recordings of brain potentials merely catch a glimpse of the underlying brain activity. Recently, however, explorations of conscious ignition have been given a new twist: in epilepsy patients, electrodes are being placed directly inside the brain, giving us a direct view of cortical activity. As soon as this method became available, my team used it to track down the cortical fate of a seen or unseen word.[34] Our findings, together with those of many others, strongly support the concept of an avalanche leading to a global ignition.[35]

In one study, we combined data from ten patients to paint a picture of the step-by-step progression of a word into the cortex.[36] Through electrodes placed all along the visual pathway, we could monitor the progression of our stimulus through successive stages and sort them out as a function of whether the patient reported seeing it or not seeing it. The initial activation was very similar, but the two traces quickly diverged for seen and unseen trials. After about 300 milliseconds, the difference became massive. On unseen trials, activity died out so quickly that frontal activation was virtually absent. On seen trials, however, it was massively amplified. In one-third of a second, the brain shifted from a very small difference to a massive all-or-none ignition.

With our focal electrodes, we could evaluate how far a conscious message was broadcast. Remember that we were recording from electrode sites chosen solely for the monitoring of epilepsy. Thus their location bore no specific relation to the goal of our study. Nevertheless, nearly 70 percent of them showed a significant influence of the consciously perceived words—as opposed to only 25 percent for unconsciously perceived words. The simple conclusion: unconscious information remains confined to a narrow brain circuit, while consciously perceived information is globally distributed to the vast majority of the cortex for an extended time.

Intracranial recordings also provided a unique window into the temporal pattern of cortical activity. Electrophysiologists distinguish many different rhythms in the EEG signal. The awake brain emits a variety of electrical fluctuations that are coarsely defined by their frequency bands, conventionally labeled with Greek letters. The bestiary of brain oscillations includes the alpha band (8 to 13 hertz), the beta band (13 to 30 hertz), and the gamma band (30 hertz and higher). When a stimulus enters the brain, it perturbs the ongoing fluctuations by reducing or shifting them, as well as by imposing new frequencies of its own. Analyzing these rhythmic effects in our data led us to a new view of the signatures of conscious ignition.

Whenever we presented a subject with a word, whether it was seen or unseen, we saw a wave of enhanced gamma-band activity in the brain. The brain emitted enhanced electrical fluctuations in this high-frequency band, which typically reflects neuronal discharges, within the first 200 milliseconds after the word appeared. However, this burst of gamma rhythms later died out for the unseen words, while it remained sustained for the seen words. By 300 milliseconds, an all-or-none difference was in place. The very same pattern was observed by Rafi Malach and his colleagues at the Weizmann Institute (figure 20).[37] A massive increase in gamma-band power, starting around 300 milliseconds after the stimulus, thus constitutes our third signature of conscious perception.

These results shed new light on an old hypothesis concerning the role of 40-hertz oscillations in conscious perception. As early as the 1990s, the late Nobel Prize winner Francis Crick, together with Christof Koch, speculated that consciousness might be reflected in a brain oscillation at

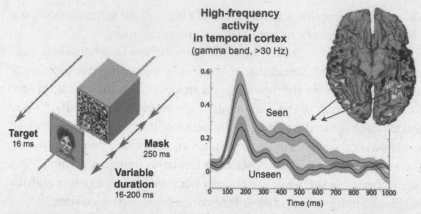

FIGURE 20. A long burst of high-frequency activity accompanies the conscious perception of a flashed picture and therefore constitutes a third signature of consciousness. In rare cases of epilepsy, electrodes can be placed atop the cortex, where they pick up the avalanche of activity evoked by a flashed picture. When viewers failed to see the picture, only a brief burst of high-frequency activity traversed the ventral visual cortex. When they saw the picture, however, the avalanche self-amplified until it caused a full-blown all-or-none ignition. Conscious perception was characterized by a lasting burst of high-frequency electrical activity, which indicates a strong activation of local neuronal circuits.

around 40 hertz (25 pulses per second), reflecting the circulation of information between the cortex and the thalamus. We now know that this hypothesis was too strong: even an unconscious stimulus can induce high-frequency activity, not only at 40 hertz but over the entire gamma band.[38] Indeed, we should not be surprised that high-frequency activity accompanies both conscious and unconscious processing: such activity is present in virtually any group of active cortical neurons, whenever inhibition is present to sculpt the neuronal discharges into high-frequency rhythmic patterns.[39] But what our experiments show is that such activity is strongly enhanced during the ignited conscious state. It is the late amplification of gamma-band activity, rather than its mere presence, that constitutes a signature of conscious perception.

The Brain Web

Why does the brain generate synchronized neuronal oscillations? Probably because synchrony facilitates the transmission of information.[40]

Within the vast neuronal forests of the cortex, with their millions of cells discharging at random, it would be easy to lose track of a small assembly of active neurons. If they shout in unison, however, their voice is much more likely to be heard and relayed. Excitatory neurons often orchestrate their discharges in order to broadcast a significant message. In essence, synchrony opens up a channel of communication between distant neurons.[41] Neurons that oscillate together share windows of opportunity during which they are all ready to receive signals from one another. The synchrony that we researchers observe in our macroscopic recordings may indicate that, at the microscopic scale, thousands of neurons are exchanging information. What may be particularly significant for conscious experience are instances when such exchanges occur not only between two local regions but across many distant regions of the cortex, thus forming a coherent brain-scale assembly.

In agreement with this idea, several teams have observed that the massive synchronization of electromagnetic signals across the cortex constitutes a fourth signature of conscious perception.[42] Once again the effect occurs primarily within a late time window: about 300 milliseconds after an image appears, many distant electrodes start to synchronize—but only if the image is consciously perceived (figure 21). Invisible images create only a temporary synchrony, spatially restricted to the back of the brain, where operations unfold without awareness. Conscious perception, by contrast, involves long-distance communication and a massive exchange of reciprocal signals that has been termed a "brain web."[43] The frequency at which this brain web is established varies across studies, but it typically occurs in the lower frequencies of the beta band (13–30 hertz) or the theta band (3–8 hertz). Presumably these slow carrier frequencies are the most convenient for bridging over the significant delays that are involved in transmitting information across distances of several centimeters.

We still do not understand exactly how millions of neuronal discharges, distributed across time and space, encode a conscious representation. Evidence is mounting that frequency analysis, although a useful mathematical technique, cannot be the whole answer. Most of the time the brain does not truly oscillate at a precise frequency. Rather, neuronal activity fluctuates in broadband patterns that wax and wane, span many frequencies, yet somehow remain synchronized across the vast distances

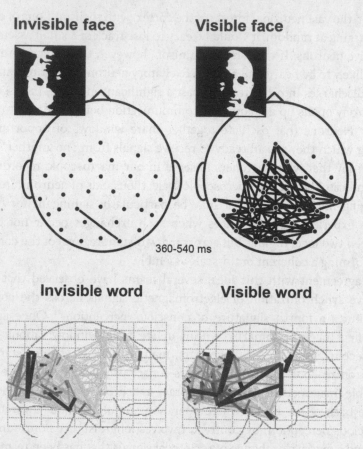

FIGURE 21. The synchronization of many distant brain regions, forming a global "brain web," provides a fourth signature of consciousness. About a third of a second after seeing a face (above), electrical brain signals synchronize. (Each line represents a highly synchronized pair of electrodes.) High-frequency oscillations in the gamma band (greater than 30 hertz) fluctuate in sync, suggesting that the underlying regions exchange messages at a high rate through a web of connections. Similarly, during conscious word perception (below), causal relations show a massive bidirectional increase between distant cortical regions, particularly with the frontal lobe. Only a modest and local synchronization occurs when the participants fail to perceive the face or the word.

of the brain. Furthermore, frequencies tend to be "nested" inside one another: high-frequency bursts fall at predictable moments relative to lower-frequency fluctuations.[44] We need new mathematical instruments in order to understand these complicated patterns.

One interesting tool that my colleagues and I have applied to our brain recordings is "Granger causality analysis." Back in 1969 the British economist Clive Granger invented this method to determine when two time series—for instance, two economic indicators—are related in such a manner that one may be said to "cause" the other. Recently, the method has been extended to neuroscience. The brain is so tightly interconnected that causality is an essential but challenging issue to determine. Does activation progress in a bottom-up fashion, from sensory receptors to higher-order integrative centers in the cortex? Or is there also a significant top-down component, in which the higher regions send descending prediction signals that shape what we consciously perceive? Anatomically, bottom-up and top-down pathways are both present throughout the cortex. Most long-distance connections are bidirectional, and the descending top-down projections often vastly outnumber the ascending ones. We are still largely ignorant of the reason for this arrangement, and whether it plays a role in consciousness.

Granger causality analysis has allowed us to shed some light on this issue. Given two temporal signals, the method asks whether one signal precedes the other and predicts its future values. According to this mathematical tool, signal A is said to "cause" signal B if the past states of A predict the present state of signal B better than the past states of signal B alone does. Note that nothing, in this definition, precludes a causal relation in both directions: A may influence B at the same time as B influences A.

When my colleagues and I applied Granger causality analysis to our intracranial recordings, we found that it clarified the dynamics of conscious ignition.[45] Specifically during the consciously perceived trials, we observed a massive increase in *bidirectional* causality throughout the brain. Once again this "causal explosion" emerged all of a sudden around 300 milliseconds. By that time, the vast majority of our recording sites had become integrated into a massive web of tangled relations, running primarily in the forward direction, from the visual cortex to the frontal lobe, but also in the converse top-down direction.

The forward-moving wave is consistent with an obvious intuition: sensory information must climb up the hierarchy of cortical areas, from primary visual cortex to increasingly abstract representations of the stimulus. But what are we to make of the opposite descending wave? We can interpret it either as an attention signal, which amplifies the incoming activity, or as a confirmation signal, a simple check that the input is consistent with the current interpretation at a higher level. The most encompassing description is that the brain falls into a "distributed attractor"—a large-scale pattern of ignited brain regions that, for a short while, produces a sustained state of reverberating activity.

No such thing happened on unconscious trials; the brain web never ignited. There was only a transient period of causal interrelations in the ventral visual cortex, but it did not last much beyond 300 milliseconds. Quite interestingly, this period was dominated by descending top-down causal signals. It looked as if the anterior regions were desperately interrogating sensory areas. Their failure to respond with a consistent signal resulted in the absence of conscious perception.

The Tipping Point and Its Precursors

Let me summarize our conclusions so far. Conscious perception results from a wave of neuronal activity that tips the cortex over its ignition threshold. A conscious stimulus triggers a self-amplifying avalanche of neural activity that ultimately ignites many regions into a tangled state. During that conscious state, which starts approximately 300 milliseconds after stimulus onset, the frontal regions of the brain are being informed of sensory inputs in a bottom-up manner, but these regions also send massive projections in the converse direction, top-down, and to many distributed areas. The end result is a brain web of synchronized areas whose various facets provide us with many signatures of consciousness: distributed activation, particularly in the frontal and parietal lobes, a P3 wave, gamma-band amplification, and massive long-distance synchrony.

The avalanche metaphor, with its tipping point, helps resolve some of the controversies surrounding the issue of exactly *when* conscious perception arises in the brain. My own data, as well as that of many colleagues, point to a late onset, close to one-third of a second after the visual stimulation started, but other laboratories have found much earlier differences between conscious and unconscious trials—sometimes as early as 100 milliseconds.[46] Are they wrong? No. With enough sensitivity, one can often detect small changes in brain activity that predate full-blown ignition. But do these differences already index a conscious brain? No. First of all, they are not always detected—there are now a fair number of excellent experiments, using the same exact stimulation on seen and unseen trials, in which the only correlate of conscious perception is the late ignition.[47] Second, the shape of the early changes does not fit with conscious reports—during masking, for instance, early events increase linearly with stimulus duration, while subjective perception is nonlinear. Finally, early events typically exhibit only a small amplification on conscious trials, on top of a large subliminal activation.[48] Again, such a small change does not fit the bill: it means that a large activation remains present on trials where the person reports no awareness at all.

So why does early visual activity predict consciousness in some experiments? Most likely, random fluctuations in ascending activity increase the chances that the brain will later burst into a state of global ignition. On average, positive fluctuations tip the scales toward conscious perception—much as a single snowball can trigger a full-blown avalanche, or the famed butterfly, a catastrophic hurricane. Just as an avalanche is a probabilistic event, not a certain one, the cascade of brain activity that eventually leads to conscious perception is not fully deterministic: the very same stimulus may at times be perceived and at others remain undetected. What makes the difference? Unpredictable fluctuations in neuronal firing sometimes fit with the incoming stimulus, and sometimes fight against it. When we average thousands of trials in which conscious perception does or does not occur, these small biases emerge from the noise as a statistically significant effect. Everything else being equal, the initial visual activation is a tad larger on a seen trial than on an unseen one. Concluding that, at that stage, the brain is already conscious would be just as wrong as saying that the first snowball *is* already the avalanche.

Some experiments even detect a correlate of conscious perception in brain signals that are recorded *before* a visual stimulus is presented.[49] Now that seems even stranger: How can brain activity already contain a marker of conscious perception for a stimulus that will be presented a few seconds later? Is this a case of precognition? Clearly not. What we are witnessing is simply the preconditions that are, *on average*, more likely to cause a full-blown avalanche of conscious perception.

Remember that brain activity is in constant flux. Some of these fluctuations help us perceive the desired target stimuli, while others hinder our ability to concentrate on the task. Brain imaging is now sensitive enough to pick up the signals that, prior to a stimulus, already index the readiness of the cortex to perceive it. As a result, when we average backward in time, starting from the knowledge that conscious perception did occur, we find that these early events act as partial predictors of later awareness. However, they are not yet constitutive of a conscious state. Conscious perception seems to arise later on, when preexisting biases and incoming evidence combine into a full-blown ignition.

These observations point to an all-important conclusion: we must learn to distinguish the mere *correlates of consciousness* from the genuine *signatures of consciousness*. Although the quest for the brain mechanisms of conscious experience is often described as a search for neural correlates of consciousness, this phrase is inadequate. Correlation is not causation, and a mere correlate is therefore insufficient. Too many brain events correlate with conscious perception—including, as we just saw, fluctuations that precede the stimulus itself and thus cannot logically be considered as coding for it. What we are looking for is not just any statistical relation between brain activity and conscious perception, but a systematic signature of consciousness, which is present whenever conscious perception occurs and absent whenever it does not, and which encodes the full subjective experience that a person reports.

Decoding a Conscious Thought

Let us play the devil's advocate again. Might global ignition act as a mere alert tone, a siren that plays whenever we become aware of something? Might it bear no specific relation to the details of our conscious thoughts? Might it just be a surge of global excitation, unrelated to the actual *contents* of subjective experience?

Many general-purpose nuclei in the brain stem and thalamus do indeed seem to label the moments that call for our attention. The locus coeruleus, for instance, is a cluster of neurons located down in the brain stem that deliver a particular neurotransmitter, norepinephrine, to a large expanse of the cortex whenever a stressful attention-demanding event occurs. A burst of norepinephrine may well accompany the exciting event of becoming aware of a visual percept, and some have suggested that this is exactly what is reflected by the massive P3 wave that we observe on the scalp during conscious access.[50] The discharge of nor-epinephrine neurons would bear no unique relation to consciousness; it would constitute a nonspecific signal, essential for our overall vigilance but devoid of the fine-grained distinctions that form the fabric of our conscious mental life.[51] Calling such a brain event the medium of consciousness would be like confusing the thump of the Sunday newspaper on our doorstep with the actual text that conveys the news.

So how could we separate the genuine conscious code from its accompanying unconscious bells and whistles? In principle, the answer is easy. We need to search the brain for a decodable neural representation whose content correlates 100 percent with our subjective awareness.[52] The conscious code that we are looking for should contain a full record of the subject's experience, replete with exactly the same level of detail as the person perceives. It should be insensitive to features that she misses, even if they are physically present in the input. Conversely, it should encode the subjective content of conscious perception, even if that perception is an illusion or a hallucination. It should also preserve our subjective sense of perceived similarity: when we see a diamond and a square as two distinct shapes, rather than rotated versions of each other, so should the brain's conscious representation.

The conscious code should also be highly invariant: it should stay put whenever we feel that the world is stable, but change as soon as we see it moving. This criterion strongly constrains the search for signatures of consciousness, because it almost certainly excludes all our early sensory areas. As we walk down a corridor, the walls project a constantly changing image on our retinas—but we are oblivious to this visual motion and perceive a stable room. Motion is omnipresent in our early visual areas but not in our awareness. Three or four times per second, our eyes jiggle around. As a result, on the retina as well as in most of our visual areas, the entire image of the world slips back and forth. Fortunately, we remain oblivious to this nauseating swirling: our perception remains steady. Even when we gaze at a moving target, we do not perceive the background scenery gliding in the opposite direction. In the cortex, our conscious code must therefore be similarly stabilized. Somehow, thanks to the motion sensors in our inner ear and to predictions arising from our motor commands, we manage to subtract out our own motion and perceive our environment as an invariant entity. Only when these predictive motor signals are bypassed—for instance, when you move your eye by gently poking it with a finger—does the whole world seem moving.

The visual slippage induced by our own motion is just one of many cues that our brain edits out of our conscious brief. Many other features set our conscious world apart from the blurry signals that reach our senses. When we watch TV, for instance, the image flickers 50 to 60 times per second, and recordings show that this hidden rhythm enters our primary visual cortex, where neurons flicker at the same frequency.[53] Fortunately we do not perceive those rhythmic flashes; the fine-grained temporal information that is present in our visual areas is filtered out before it reaches our awareness. Likewise, a very fine mesh of lines is encoded by our primary visual cortex, even though it cannot be seen.[54]

But our consciousness is not just nearly blind: it is an active observer that dramatically enhances and transforms the incoming image. On the retina and at the earliest stages of cortical processing, the center of our vision is massively expanded relative to the periphery: many more neurons care about the center of our gaze than about the surroundings. Yet we do not perceive the world as through a giant magnifying lens; nor do we experience a sudden expansion of whichever face or word we decide to look at. Consciousness ceaselessly stabilizes our perception.

As a final example of the massive discrepancy between the initial sense data and our conscious perception of them, consider color. Outside the center of our gaze, the retina contains very few color-sensitive cones—and yet we are not color-blind in the periphery of our visual field. We do not walk into a black and white world, marveling at how color appears whenever we gaze at something. Rather, our conscious world appears in full color. Each of our retinas even has a huge gap called the "blind spot" at the place where our optic nerve leaves—and yet fortunately, we do not perceive a black hole in our inner picture of the world.

All these arguments prove that early visual responses cannot contain the conscious code. Much processing is needed before our brain solves the perceptual jigsaw puzzle and pieces together a stable view of the world. This is probably why the signatures of consciousness occur so late in time: a third of a second may be the minimum time needed for our cortex to see through all the haphazard jigsaw puzzle pieces and put together a stable representation of the world.

If this view is correct, then this late brain activity should embrace a full record of our conscious experience—a complete code of our thoughts. If we could read this code, we would gain full access to any person's inner world, subjectivity and illusions included.

Is this prospect science fiction? Not quite. By selectively recording from single neurons in the human brain, the neuroscientist Quian Quiroga and his Israeli colleagues Itzhak Fried and Rafi Malach have opened up the doors of conscious perception.[55] They discovered neurons that react only to specific pictures, places, or people—and ignite only when conscious perception occurs. Their finding provides decisive evidence against the nonspecific interpretation. During global ignition, the brain is not globally excited. Rather, a very precise set of neurons is active, and its contours sharply delineate the subjective contents of consciousness.

How can neurons be recorded from deep inside the human brain? I have already explained that neurosurgeons now monitor epileptic fits by placing an array of electrodes inside the skull. Usually these electrodes are large and record indiscriminately from thousands of cells. However, building upon earlier pioneering work,[56] the neurosurgeon Itzhak Fried developed a delicate system of very fine electrodes that are specifically designed to record from individual neurons.[57] In the human brain, as in that of most other animals, cortical neurons exchange discrete electrical signals; they are called "spikes" because they appear as very sharp deviations of the electrical potential on an oscilloscope. Excitatory neurons typically emit a few spikes per second, and each of them quickly propagates along the axon to reach both local and distant targets. Thanks to Fried's intrepid experiments, it has become possible to record, for hours or even days, *all* the spikes that a given neuron emits, while the patient, fully awake, lives a normal life.

When Fried and his collaborators placed electrodes in the anterior temporal lobe, they immediately came up with a remarkable finding. They discovered that individual human neurons can be extraordinarily selective to a picture, a name, and even a concept. By bombarding a patient with hundreds of pictures of faces, places, objects, and words, they usually found that just one or two pictures triggered a given cell. One neuron, for instance, discharged to pictures of Bill Clinton and to no other person![58] Over the years, human neurons have been reported to respond selectively to a cornucopia of photos, including members of the patient's family, famous locations such as the Sydney Opera House or the White House, and even to television celebrities such as Jennifer Aniston and Homer Simpson. Remarkably, the written word often sufficed to activate them: the same neuron would discharge to the words *Sydney Opera* and to the sight of this famous landmark.

It is fascinating that, by blindly inserting an electrode and listening in on a random neuron, we can find a Bill Clinton cell. This implies that, at any given time, millions of such cells must be discharging in response to the scenes we see. Together, anterior temporal lobe neurons are thought to form a distributed internal code for people, locations, and other memorable concepts. Each specific picture, such as Clinton's face, induces a particular pattern of active and inactive neurons. The code is so accurate that, by looking at which neurons fire and which remain silent, we can train a computer to guess, with very high accuracy, what picture the person is seeing.[59]

Clearly, then, these neurons are highly specific to the current visual scene, yet highly invariant. What their discharges index is neither a global arousal signal nor myriad changing details, but the gist of the current picture—just the right sort of stable representation that we would expect to encode for our conscious thoughts. So do these neurons bear any relation to their owner's conscious experience? Yes. Crucially, in the anterior temporal region, many neurons fire *only* if a certain picture is consciously seen. In one experiment, the pictures were masked by nonsense images and were flashed so briefly that many of them could not be seen.[60] On each trial, the patient reported whether he had recognized the picture. The majority of cells emitted spikes only when the patient reported seeing the picture. The visual display was exactly the same on conscious and unconscious trials, yet the cell's firing reflected the person's subjective perception rather than the objective stimulus.

Figure 22 shows a cell whose firing was triggered by a picture of the World Trade Center. The neuron discharged only on conscious trials. Whenever the patient reported seeing nothing, because the picture was masked beyond recognition, the cell remained absolutely silent. Even for a fixed amount of objective physical stimulation, when the very same picture was presented for a fixed amount of time, subjectivity mattered. With picture duration set precisely at the threshold of awareness, the person

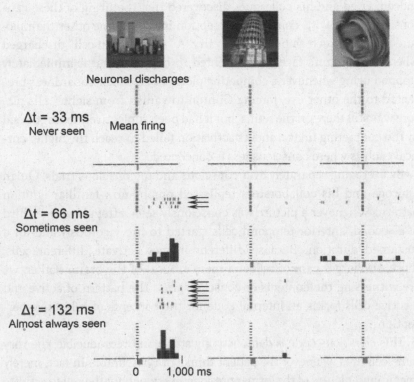

FIGURE 22. Individual neurons track our conscious percepts: they fire only when we consciously perceive a specific picture. In this example, a neuron in the human anterior temporal lobe fired selectively to a picture of the World Trade Center, but virtually only when that picture was consciously seen. As the duration of presentation increased, conscious perception became more frequent. Neuronal discharges occurred only when the person reported seeing the picture (trials marked by an arrow). The neuron was selective and did not discharge much to other pictures, such as a face or the Leaning Tower of Pisa. Its late and sustained firing indexed a specific content of awareness. Millions of such neurons, firing together, code for what we see.

reported seeing the picture about half the time—and the cell's spikes tracked just the trials with conscious perception. The cell's firing was so reproducible that it was possible to draw a line and separate the seen and unseen trials from the number of observed spikes. In a nutshell, a subjective state of mind could be decoded from an objective state of the brain.

If anterior temporal cells encode conscious perception, then their discharges should be unrelated to *how* consciousness is manipulated.

Indeed, Fried and his colleagues discovered that the firing of these neurons correlates with conscious perception in paradigms other than picture masking, such as binocular rivalry. A "Bill Clinton cell" discharged whenever Clinton's face was presented to one eye—but it immediately stopped firing whenever a competing picture of a checkerboard was presented to the other eye, forcing Clinton to vanish from sight.[61] His picture was still there on the retina, but it had been subjectively extinguished by the competing image, and its activation failed to reach the higher cortical centers where consciousness is concocted.

By averaging separately over conscious and unconscious trials, Quian Quiroga and his collaborators replicated our by-now-familiar ignition pattern. Whenever a picture was consciously seen, after about one-third of a second, anterior temporal cells started to fire vigorously and for a sustained duration. Because different images activate different cells, those discharges cannot reflect a mere arousal of the brain. Rather, we are witnessing the contents of consciousness. The pattern of active and inactive cells forms an internal code for the contents of subjective perception.

This conscious code is demonstrably stable and reproducible: the very same cell fires whenever the patient thinks of Bill Clinton. In fact, merely imagining a picture of the former president is enough for the cell to activate in the absence of any objective external stimulation. The majority of anterior temporal neurons exhibit the same selectivity for actual and imagined pictures.[62] Memory recall also activates them. One cell, which fired when the patient was viewing a video of *The Simpsons*, discharged again every time the patient, in full darkness, recalled seeing that movie clip.

Although individual neurons track what we imagine and perceive, it would be wrong to conclude that a single cell suffices to induce a conscious thought. Conscious information is probably distributed within myriad cells. Imagine several million neurons, spread throughout the associative areas of the cortex, each coding for a fragment of the visual scene. Their synchronous discharges form macroscopic brain potentials, strong enough to be picked up by classical electrodes located in or even outside the skull. The firing of a single cell is undetectable at a distance, but because conscious perception mobilizes huge cell assemblies, we can, to some extent, determine whether a person is seeing a face or a building, just from the topography of the large electrical potentials emitted by her visual cortex.[63] Likewise, the location and even the number of items that a person keeps in his short-term memory can be determined from the pattern of slow brain waves over the parietal cortex.[64]

Because the conscious code is stable and present for quite some time, even fMRI, a rather coarse method that averages across millions of neurons, may decipher it. In one recent experiment, after a patient saw a face or a house, a distinct pattern of activity appeared in the anterior part of the ventral temporal lobe, and it sufficed to determine what the person had seen.[65] That pattern remained stable over many trials, while no such reproducible activity occurred on unconscious trials.

So imagine that you have been shrunk down to submillimeter size and are sent into the cortex. There you are surrounded by thousands of neuronal discharges. How can you recognize which of these spikes encode a conscious percept? You would have to search for sets of spikes with three distinctive features: *stability* over time, *reproducibility* across trials, and *invariance* over superficial changes that leave the content intact. These criteria are met, for instance, in the posterior cingulate cortex, a high-level integration area located in the midline parietal cortex. There, the neural activity evoked by a visual stimulus remains stable as long as the object itself stays put, even when the eyes move.[66] Furthermore, neurons in this region are tuned to the location of objects in the outside world: even as we look around, they maintain an invariant level of firing. This point is far from trivial because during eye movements, the entire visual image slips over our primary visual cortex—yet somehow, by the time it gets to the posterior cingulate, the image has been stabilized.

The posterior cingulate region, where invariant-location cells dwell, is closely connected to a site called the parahippocampal gyrus (next to the hippocampus), where "place cells" are found.[67] These neurons fire whenever an animal occupies a certain location in space—for instance, the northwest corner of a familiar room. Place cells too are highly invariant over a variety of sensory cues, and they even maintain their space-selective firing as the animal wanders around in full darkness. Fascinatingly, these neurons demonstrably encode where the animal *thinks* it is. If a rat is "teleported" by suddenly switching the colors of the floor, walls, and ceiling so that they resemble another familiar room, place cells in the hippocampus briefly oscillate between the two interpretations, then settle into a firing pattern appropriate to the illusory room.[68] The decoding of neural signals in this region is so advanced that it has become possible to tell where the animal is (or thinks it is) from the collective firing pattern of nerve cells—and even to do so during sleep, when the spatial trajectory is merely imagined. In a few years, it does not seem so far-fetched to think that similar abstract codes, encrypting the very fabric of our thoughts, will become decodable in the human brain.

In summary, neurophysiology has now cracked wide open the mystery box of conscious experience. During conscious perception, patterns of neuronal activity unique to a given picture or concept can be recorded at various sites in the brain. Such cells fire strongly if and only if the person reports perceiving a picture—whether it is real or imaginary. Each conscious visual scene appears to be encoded by a reproducible pattern of neuronal activity that remains stable for half a second or more, as long as the person sees it.

Inducing a Hallucination

Is this it? Has our search for the neural signatures of consciousness reached a happy end? Not quite. One more criterion must be met. To qualify as a genuine signature of consciousness, brain activity should not only occur whenever the corresponding conscious content does; it must also demonstrably *cause* this content to pop into our awareness.

The prediction is simple: if we managed to induce a certain state of brain activity, we should evoke the corresponding state of mind. If a *Matrix*-like stimulator could re-create, in our brain, the precise state of neuronal firing that our circuits were in the last time we saw a sunset, we should visualize it with full clarity—a full-blown hallucination indistinguishable from the original experience.

Such a re-creation of brain states may sound far-fetched, but it is not; it happens every night. During dreams, we lie motionless, but our mind flies, simply because our brain fires organized trains of spikes that evoke precise mental content. In rats, neuronal recordings during sleep show a replay of neuronal patterns in the cortex and hippocampus that directly correlate with the content of the animals' experience during the previous day.[69] And in humans, the cortical areas that are active just seconds prior to wakening can predict the content of the reported dream.[70] For instance, whenever the activity concentrates in a region that is known to be specialized for faces, the dreamer predictably reports the presence of other people in her dream.

These fascinating findings demonstrate a correspondence between neural states and mental states—but they still do not establish causality. Proving that a pattern of brain activity causes a mental state is one of the hardest problems facing neuroscientists. Virtually all our noninvasive brain-imaging methods are correlative rather than causative—they involve the passive observation of a correlation between brain activation and mental states. Two special methods, however, allow us to safely stimulate the human brain, with techniques that are both harmless and reversible.

In healthy participants, we can activate the brain from outside with a technique called transcranial magnetic stimulation (TMS). Pioneered at the beginning of the twentieth century,[71] and later revived by modern technologies,[72] this technique has now come into widespread use (figure 23). Here is how it works. A battery of accumulators suddenly delivers a strong electrical current to a coil placed atop the head. This current induces a magnetic field that penetrates the head and generates a discharge at a precise "sweet spot" in the underlying cortex. Safety guidelines

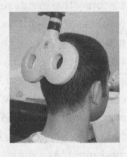

FIGURE 23. Transcranial magnetic stimulation can be used to interfere with human brain activity and induce changes in conscious experience. Pioneered by S. P. Thompson (1910; left) and by C. E. Magnusson and H. C. Stevens (1911; middle), the technique has now become much simpler and cheaper (right). The application of a transient magnetic field induces a pulse of current inside the cortex, which may disrupt an ongoing perception or even cause an illusory experience, such as seeing a flash of light. Such experiments prove the existence of a causal link between brain activity and conscious experience.

ensure that the technique is harmless: only an audible click and, occasionally, an unpleasant muscle twitch occur. In this manner, any normal brain can be stimulated within virtually any region of the cortex, with precise timing.

For greater spatial precision, an alternative is to stimulate the neurons directly with electrodes placed inside the brain. This option is of course available only for patients with epilepsy, Parkinson's, or movement disorders, who are increasingly explored with intracranial electrodes. With the patient's agreement, small currents can be injected into these wires, in sync with an external stimulus. An electrical discharge can even be applied during surgery. Because the brain is devoid of pain receptors, such electrical stimulation is harmless and can be very informative in order to identify regions of crucial importance that the scalpel must spare, such as the language circuits. Many hospitals throughout the world routinely carry on such uncanny intraoperative experiments. Lying on the operating table, skull half open but fully awake, a patient carefully describes his experience as an electrode injects a small amount of current at a precise spot in his brain.

The results of these investigations are extremely rewarding. Many stimulation studies, both in humans and in nonhuman primates, have demonstrated a direct causal mapping between neural states and conscious perception. The mere stimulation of neuronal circuits, in the absence of any objective event, suffices to cause a conscious subjective feeling whose content varies with the stimulated circuit. For instance, transcranial magnetic stimulation of the visual cortex, in full darkness, creates an impression of light, technically known as a phosphene: just after the current is applied, a faint spot of light appears at a location that varies with the site of cortical stimulation. Move the stimulation coil over to the side of the brain, over an area called MT/V5, which responds to motion, and the percept suddenly changes: the brain's owner now reports an impression of fleeting movement. At a different site, color sensations can also be evoked.

Neuronal recordings have long established that each parameter of the visual scene maps onto a distinct site of the visual cortex. In different sectors of the occipital cortex, a mosaic of neurons responds to shape, motion, or color. Stimulation studies now show that the relation between these neurons' firing and the corresponding perception is causal. A focal discharge at any of these sites, even in the absence of an image, can evoke the corresponding smidgen of consciousness, with appropriate qualities of luminance or color.

With intracranial electrodes, the effects of stimulation can be even more specific.[73] Sparking off an electrode atop the face region of the ventral visual cortex can immediately induce the subjective perception of a face. Moving the stimulation forward into the anterior temporal lobe can awaken complex memories drawn from the patient's past experience. One patient smelled burnt toast. Another saw and heard a full orchestra playing, with all its instruments. Others experienced even more complex and dramatically vivid dreamlike states: they saw themselves giving birth, lived through a horror movie, or were projected back into a Proustian episode of their childhood. Wilder Penfield, the Canadian neurosurgeon who pioneered these experiments, concluded that our cortical microcircuits contain a dormant record of the major and minor events of our lives, ready to be awakened by brain stimulation.

A systematic exploration suggests that every cortical site holds its own specialized piece of knowledge. Consider the insula, a deep sheath of cortex that is buried beneath the frontal and temporal lobes. Stimulating it can have a diversity of unpleasant effects, including a sensation of suffocation, burning, stinging, tingling, warmth, nausea, or falling.[74] Move the electrode to a location farther below the surface of the cortex, the subthalamic nucleus, and the same electrical pulse may induce an immediate state of depression, complete with crying and sobbing, monotone voice, miserable body posture, and glum thoughts. Stimulating parts of the parietal lobe may cause a feeling of vertigo and even the bizarre out-of-body experience of levitating to the ceiling and looking down at one's own body.[75]

If you had any lingering doubts that your mental life arises entirely from the activity of the brain, these examples should lift them. Brain stimulation seems capable of bringing about virtually any experience, from orgasm to déjà vu. But this fact by itself does not directly speak to the issue of the causal mechanisms of consciousness. Neural activity, after arising at the stimulation site, immediately spreads to other circuits, blurring the causal story. Indeed, recent research suggests that the initial bit of induced activity is unconscious: only if the activation spreads to distant regions of parietal and prefrontal cortex does conscious experience occur.

Consider, for instance, the striking dissociation recently reported by the French neuroscientist Michel Desmurget.[76] When he stimulated the premotor cortex at a relatively low threshold, during surgery, the patient's arm moved, but the person denied that anything had happened (she could not see her limbs). Conversely, when Desmurget stimulated the inferior parietal cortex, the patient reported a conscious urge to move, and with higher current, she swore that she had moved her hand—but in reality her body had remained perfectly still.

These results have a major implication: not all brain circuits are equally important for conscious experience. Peripheral sensory and motor circuits can be activated without necessarily generating a conscious experience. Higher-order regions of the temporal, parietal, and prefrontal cortexes, on the other hand, are more intimately associated with a reportable conscious experience, since their stimulation can induce purely subjective hallucinations that have no foundation in objective reality.

The logical next step is to create perceived and unperceived brain stimulation with a minimal difference and examine how the results differ. Like many scientists before them, the London neuroscientists Paul Taylor, Vincent Walsh, and Martin Eimer used transcranial magnetic stimulation of the primary visual cortex to induce visual phosphenes— hallucinations of light created solely by cortical activity.[77] But very cleverly, they tuned the intensity of the injected current until the patient reported seeing a spot of light about half the time. They also managed to track the activity induced by this threshold-level pulse throughout the brain, by recording the subject's EEG, millisecond by millisecond, at various times after the onset of the stimulation.

The results were illuminating. The initial part of the injected pulse bore no relation at all to consciousness. For a full 160 milliseconds, brain activity unfolded identically on visible and invisible trials. Only after this long period did our good old friend the P3 wave appear on the surface of the head, with a much stronger intensity on perceived than on unperceived trials. Only its onset was earlier than usual (about 200 milliseconds): the magnetic pulse, unlike an external light, bypassed the initial processing stages of vision, thus shortening the duration of conscious access by one-tenth of a second.

Brain stimulation thus demonstrates a causal relation between cortical activity and conscious experience. Even in full darkness, a pulse of stimulation to the visual cortex can induce a visual experience. However, this relation is indirect: local activity is insufficient to create a conscious perception; before it gains access to consciousness, the induced activity must first be dispatched to distant brain sites. Once again the late part of the firing train, when activation diffuses to higher cortical centers and creates a distributed brain web, seems to be what causes conscious perception. During the formation of this conscious brain web, neural activity circulates broadly in the cortex and often returns to sensory areas, thus tying together the neuronal fragments of a perceived picture. Only then do we experience "seeing."

Destroying Consciousness

If we can create a conscious percept, can we also destroy it? Assuming that the late activation of a global brain web causes all our conscious experiences, then tampering with it should eradicate conscious perception. The experiment is, again, conceptually simple. First present the subject with a visible stimulus, well above the normal threshold for conscious perception, and then use a pulse of current to zap the late long-distance network that supports consciousness. The subject should report that there was no stimulus at all—that he was unaware of seeing anything. Or, imagine that the pulse does not simply destroy the global state of neuronal activity but replaces it with a different one. Then the subject should report becoming conscious of the content attached to the substituted neuronal state—a subjective experience that may have nothing to do with the true state of the world.

Although this may sound like science fiction, several variants of this experiment have already been performed, with considerable success. One version used a dual transcranial magnetic stimulator, which can induce currents in two distinct brain regions at two arbitrary moments. The recipe is simple: first, excite the motion area MT/V5 with a pulse of electrical current; check that, by itself, this discharge evokes a conscious feeling of visual movement; then apply a second pulse of current, for instance to the primary visual cortex. Amazingly, it works: the second pulse eradicates the conscious feeling of seeing that the first pulse was able to induce. This result proves that the initial pulse, by itself, fails to cause a conscious experience: the induced activation must loop back to the primary visual cortex before being consciously perceived.[78] Consciousness lives in the loops: reverberating neuronal activity, circulating in the web of our cortical connections, causes our conscious experiences.

Even more fascinating, cortical stimulation can be combined with genuine visual images to create novel illusions. For instance, stimulating the visual cortex one-fifth of a second after briefly flashing a picture can induce its replay in consciousness: the participant reports seeing the picture a second time, confirming that a trace of it was still lingering in the visual cortex 200 milliseconds after its first appearance.[79] The effect is particularly strong when the person is told to keep the picture in memory. Those results suggest that, when we hold an image in mind, our brain literally keeps it alive in the firing of neurons in the visual cortex, at a subthreshold level, ready to be reenacted by a pulse of stimulation.[80]

How global is the brain web that creates our conscious world? According to the Dutch neurophysiologist Viktor Lamme, whenever two areas form a local loop, such that area A speaks to area B, and then B talks back to A, this is already sufficient to induce a form of consciousness.[81] Such a loop makes the activation reverberate, causing "recurrent processing," the reinjection of information into the same circuit that originated it. "We could even define consciousness as recurrent processing," Lamme writes.[82] For him, any neuronal loop holds a little piece of awareness. However, I doubt that this view is correct. Our cortex is full of closed loops: neurons communicate reciprocally at all scales, from millimeter-size local microcircuits to global highways spanning centimeters. It would be really surprising if each of these loops, however tiny, sufficed to bring about a fragment of consciousness.[83] Much more plausible, in my opinion, is the view that reverberating activity is a necessary but not sufficient condition for conscious experience. Only the long-distance loops, bringing in prefrontal and parietal regions, would create a conscious code.

What would be the role of the short local loops? They are probably indispensable for early *unconscious* visual operations, during which we piece together the multiple fragments of a scene.[84] With their very small receptive fields, visual neurons cannot immediately apprehend the global properties of the image, such as the presence of a large shadow (as in the shadow illusion shown in figure 10). Interactions among many neurons are needed before such global properties are established.[85]

So is it the local loops or the global loops that induce consciousness? Some scientists argue for local loops because they tend to disappear under anesthesia,[86] but such evidence is inconclusive: reverberating activity may be one of the first features to go when the brain is bathed in anesthetics, a consequence rather than a cause of the loss of consciousness.

Tampering with brain activity using the finer technique of brain stimulation tells another story. Zapping the short-distance loops within primary visual cortex, about 60 milliseconds after flashing a visual image, does affect conscious perception, but crucially the very same stimulation also disrupts *unconscious* processing.[87] Blindsight, the capacity to make above-chance judgments on subliminal visual information, is destroyed together with conscious sight. This observation implies that the initial stages of local cortical processing, when activity circulates in local loops, are not exclusively associated with conscious perception. They correspond to unconscious operations and merely set the brain on the appropriate course that, much later, will result in conscious perception.

If my view is correct, then conscious appraisal arises from the later activation of multiple synchronized regions of parietal and prefrontal cortex—and thus zapping those regions should have a major effect. Indeed, a great variety of studies in normal subjects, using TMS to interfere with brain activity, have now demonstrated that parietal or frontal stimulation creates a transient invisibility. Virtually all the visual conditions of stimulation that make pictures temporarily invisible, such as masking and inattentional blindness, can be strongly enhanced by briefly disrupting the left or right parietal region.[88] For instance, a faint but otherwise visible patch of color vanishes from sight when a parietal region is zapped.[89]

Most remarkable is a study performed by Hakwan Lau and his team, then at the University of Oxford, in which the left and right prefrontal regions were both temporarily obliterated.[90] Each dorsolateral prefrontal lobe was bombarded by 600 pulses, grouped into short bouts of 20 seconds, first left, then right. The paradigm is called "theta-burst" because the current pulses are arranged to specifically disrupt the theta rhythm (5 cycles per second), one of the preferred frequencies at which the cortex passes messages over long distances. Bilateral theta-burst stimulation has a long-lasting effect that amounts to a virtual lobotomy: for about twenty minutes, the frontal lobes are inhibited, leaving the experimenters ample time to evaluate the impact on perception.

The results were subtle. Objectively, nothing was changed: the stoned participants continued to perform equally well in judging which shape had been shown (a diamond or a square, presented close to the threshold for conscious perception). Their subjective reports, however, told another story. For several minutes, they lost confidence in their judgments. They became unable to rate how well they perceived the stimuli, and they had a subjective feeling that their vision had become unreliable. Like the philosopher's zombie, they perceived and acted well, but without a normal sense of how well they were doing.

Before the participants were zapped, their ratings of stimulus visibility correlated well with their objective performance: like any of us, whenever they felt that they could see the stimulus, they could indeed identify its shape with near-perfect accuracy, and whenever they felt that the shapes were invisible, their responses were essentially random. During the temporary lobotomy, however, this correlation was lost. Quite surprisingly, the participants' subjective reports became unrelated to their actual behavior. This is the exact definition of *blindsight*—a dissociation between subjective perception and objective behavior. This condition, which is usually associated with a major brain lesion, could now be reproduced in any normal brain by interfering with the operation of the left and right frontal lobes. Clearly, these regions play a causal role in the cortical loops of consciousness.

A Thing Which Thinks

But what then am I? A thing which thinks. What is a thing which thinks? It is a thing which doubts, understands, affirms, desires, wills, refuses, which also imagines and feels.

—René Descartes, *Meditation II* (1641)

Putting together all the evidence inescapably leads us to a reductionist conclusion. All our conscious experiences, from the sound of an orchestra to the smell of burnt toast, result from a similar source: the activity of massive cerebral circuits that have reproducible neuronal signatures. During conscious perception, groups of neurons begin to fire in a coordinated manner, first in local specialized regions, then in the vast expanses of our cortex. Ultimately, they invade much of the prefrontal and parietal lobes, while remaining tightly synchronized with earlier sensory regions.

It is at this point, where a coherent brain web suddenly ignites, that conscious awareness seems to be established.

In this chapter, we discovered no fewer than four reliable signatures of consciousness—physiological markers that index whether the participant experienced a conscious percept. First, a conscious stimulus causes an intense neuronal activation that leads to a sudden ignition of parietal and prefrontal circuits. Second, in the EEG, conscious access is accompanied by a slow wave called the P3 wave, which emerges as late as one-third of a second after the stimulus. Third, conscious ignition also triggers a late and sudden burst of high-frequency oscillations. Finally, many regions exchange bidirectional and synchronized messages over long distances in the cortex, thus forming a global brain web.

One or more of these events could still be an epiphenomenon to consciousness, much like the steam whistle in a locomotive—systematically accompanying it but contributing nothing to it. Causality remains hard to assess using neuroscience methods. Nevertheless, several pioneering experiments have begun to demonstrate that interfering with high-level cortical circuitry can disrupt subjective perception while leaving unconscious processing intact. Other stimulation experiments have induced hallucinations such as illusory points of light or an anomalous sense of body motion. While these studies are too rudimentary to paint a detailed picture of the conscious state, they leave no doubt that the electrical activity of neurons can cause a state of mind, or equally easily, destroy an existing one.

In principle, we neuroscientists believe in the philosopher's fantasy of a "brain in a vat," powerfully illustrated by the movie *The Matrix*. By stimulating the appropriate neurons and silencing others, we should be able to re-create, at any given time, hallucinations of any of the myriad subjective states that people routinely entertain. Neural avalanches should cause mental symphonies.

At present, technology remains far behind the Wachowski brothers' fantasy. We cannot yet control the billions of neurons that would be needed to accurately paint, on the surface of the cortex, the neural equivalent of a busy Chicago street or a Bahamas sunset. But are such fantasies forever beyond our reach? I wouldn't bet on it. In the hands of contemporary bioengineers, motivated by the need to restore functions in blind, paralyzed, or Parkinsonian patients, neurotechnologies are quickly progressing. Silicon chips with thousands of electrodes can now be implanted in the cortex of experimental animals, dramatically increasing the bandwidth of brain-computer interfaces.

Even more exciting are the recent breakthroughs in optogenetics, a fascinating technique that drives neurons by light rather than by electrical current. The crux of the technique is the discovery, in algae and bacteria, of light-sensitive molecules, called "opsins," that convert light's photons into electrical signals, the neuron's basic currency. The genes for opsins are known, and their properties can be genetically engineered. Injecting a virus carrying these genes into an animal's brain, and restricting their expression to a precise subset of neurons, has made it possible to add new photoreceptors to the brain's toolkit. Deep inside the cortex, in dark places normally insensitive to light, shining a laser suddenly triggers a flood of neuronal spikes with millisecond precision.

Using optogenetics, neuroscientists can selectively activate or inhibit any brain circuit.[91] The technique has even been used to awaken a sleeping mouse by stimulating its hypothalamus.[92] Soon we should be able to induce even more differentiated states of brain activity—and therefore re-create, de novo, a specific conscious percept. Stay tuned, as the next ten years are likely to yield major new insights into the neuronal code that supports our mental life.

5

THEORIZING CONSCIOUSNESS

We have discovered signatures of conscious processing, but what do they mean? Why do they occur? We have reached the point where we need a theory to explain how subjective introspection relates to objective measurements. In this chapter, I introduce the "global neuronal workspace" hypothesis, my laboratory's fifteen-year effort to make sense of consciousness. The proposal is simple: consciousness is brain-wide information sharing. The human brain has developed efficient long-distance networks, particularly in the prefrontal cortex, to select relevant information and disseminate It throughout the brain. Consciousness is an evolved device that allows us to attend to a piece of information and keep it active within this broadcasting system. Once the information is conscious, it can be flexibly routed to other areas according to our current goals. Thus we can name it, evaluate it, memorize it, or use it to plan the future. Computer simulations of neural networks show that the global neuronal workspace hypothesis generates precisely the signatures that we see in experimental brain recordings. It can also explain why vast amounts of knowledge remain inaccessible to our consciousness.

> I shall consider human actions and desires . . . as though I were
> concerned with lines, planes, and solids.
> —Baruch Spinoza, *Ethics* (1677)

The discovery of signatures of consciousness is a major advance, but these brain waves and neuronal spikes still do not explain what consciousness *is* or why it occurs. Why should late neuronal firing, cortical ignition, and brain-scale synchrony ever create a subjective state of mind? How do these brain events, however complex, elicit a mental experience? Why should the firing of neurons in brain area V4 elicit a perception of color, and those in area V5 a sense of motion? Although neuroscience has identified many empirical correspondences between brain activity and mental life, the conceptual chasm between brain and mind seems as broad as it ever was.

In the absence of an explicit theory, the contemporary search for the neural correlates of consciousness may seem as vain as Descartes's ancient proposal that the pineal gland is the seat of the soul. This hypothesis seems deficient because it upholds the very division that a theory of consciousness is supposed to resolve: the intuitive idea that the neural and the mental belong to entirely different realms. The mere observation of a systematic relationship between these two domains cannot suffice. What is required is an overarching theoretical framework, a set of bridging laws that thoroughly explain how mental events relate to brain activity patterns.

The enigmas that baffle contemporary neuroscientists are not so different from the ones that physicists resolved in the nineteenth and twentieth centuries. How, they wondered, do the macroscopic properties of ordinary matter arise from a mere arrangement of atoms? Whence the solidity of a table, if it consists almost entirely of a void, sparsely populated by a few atoms of carbon, oxygen, and hydrogen? What is a liquid? A solid? A crystal? A gas? A burning flame? How do their shapes and other tangible features arise from a loose cloth of atoms? Answering these questions required an acute dissection of the components of matter, but this bottom-up analysis was not enough: a synthetic mathematical theory was needed. The kinetic theory of gases, first established by James Clerk Maxwell and Ludwig Boltzmann, famously explained how macroscopic variables of pressure and temperature emerged from the motion of atoms in a gas. It was the first in a long line of mathematical models of matter—a reductionist chain that now accounts for substances as diverse as our glues and soap bubbles, the water percolating in our coffeepots and the plasma in our distant sun.

A similar theoretical effort is now needed to close the gap between brain and mind. No experiment will ever show how the hundred billion neurons in the human brain fire at the moment of conscious perception. Only mathematical theory can explain how the mental reduces to the neural. Neuroscience needs a series of bridging laws, analogous to the Maxwell-Boltzmann theory of gases, that connect one domain with the other. This is no easy task: the "condensed matter" of the brain is perhaps the most complex object on earth. Unlike the simple structure of a gas, a model of the brain will require many nested levels of explanation. In a dizzying arrangement of Russian dolls, cognition arises from a sophisticated arrangement of mental routines or processors, each implemented by circuits distributed across the brain, themselves made up of dozens of cell types. Even a single neuron, with its tens of thousands of synapses, is a universe of trafficking molecules that will provide modeling work for centuries.

In spite of these difficulties, in the past fifteen years, my colleagues Jean-Pierre Changeux, Lionel Naccache, and I have started to bridge the gap. We have sketched a specific theory of consciousness, the "global neuronal workspace," that is the condensed synthesis of sixty years of psychological modeling. In this chapter, I hope to convince you that, although precise mathematical laws are still far on the horizon, we now have a few glimpses into the nature of consciousness, how it arises from coordinated brain activity, and why it exhibits the signatures that we see in our experiments.

Consciousness Is Global Information Sharing

What kind of information-processing architecture underlies the conscious mind? What is its raison d'être, its functional role in the information-based economy of the brain? My proposal can be stated succinctly.[1] When we say that we are aware of a certain piece of information, what we mean is just this: the information has entered into a specific storage area that makes it available to the rest of the brain. Among the millions of mental representations that constantly crisscross our brains in an unconscious manner, one is selected because of its relevance to our present goals. Consciousness makes it globally available to all our high-level decision systems. We possess a mental router, an evolved architecture for extracting relevant information and dispatching it. The psychologist Bernard Baars calls it a "global workspace": an internal system, detached

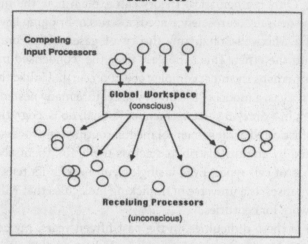

Baars 1989

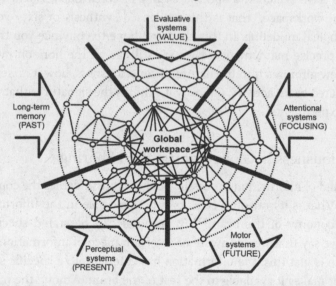

Dehaene and Changeux, 1998

FIGURE 24. Global neuronal workspace theory proposes that what we experience as consciousness is the global sharing of information. The brain contains dozens of local processors (represented by circles), each specialized for one type of operation. A specific communication system, the "global workspace," allows them to flexibly share information. At any given moment, the workspace selects a subset of processors, establishes a coherent representation of the information they encode, holds it in mind for an arbitrary duration, and disseminates it back to virtually any of the other processors. Whenever a piece of information accesses the workspace, it becomes conscious.

from the outside world, that allows us to freely entertain our private mental images and to spread them across the mind's vast array of specialized processors (figure 24).

According to this theory, consciousness is just brain-wide information sharing. Whatever we become conscious of, we can hold it in our mind long after the corresponding stimulation has disappeared from the outside world. That's because our brain has brought it into the workspace, which maintains it independently of the time and place at which we first perceived it. As a result, we may use it in whatever way we please. In particular, we can dispatch it to our language processors and name it; this is why the capacity to report is a key feature of a conscious state. But we can also store it in long-term memory or use it for our future plans, whatever they are. The flexible dissemination of information, I argue, is a characteristic property of the conscious state.

The workspace idea represents a synthesis of many earlier proposals in the psychology of attention and consciousness. As early as 1870, the French philosopher Hippolyte Taine introduced the metaphor of a "theater of consciousness."[2] The conscious mind, he explained, is like a narrow stage that lets us hear only a single actor:

> You may compare the mind of a man to the stage of a theatre, very narrow at the footlights but constantly broadening as it goes back. At the footlights, there is hardly room for more than one actor. . . . As one goes further and further away from the footlights, there are other figures less and less distinct as they are more distant from the lights. And beyond these groups, in the wings and altogether in the background, are innumerable obscure shapes that a sudden call may bring forward and even within direct range of the footlights. Undefined evolutions constantly take place throughout this seething mass of actors of all kinds, to furnish the chorus leaders who in turn, as in a magic lantern picture, pass before our eyes.

Decades before Freud, Taine's metaphor implied that while only a single item makes it into our awareness, our mind must comprise an enormous variety of unconscious processors. What a massive support staff for a one-man show! At any given moment, the content of our consciousness arises from myriad covert operations, a backstage ballet that remains hidden from sight.

The philosopher Daniel Dennett reminds us that we must be wary of the theater allegory, for it can lead to a great sin: the "homunculus fallacy."[3] If consciousness is a stage, who is the audience? Do "they" too have little brains, with a ministage and all? And who, in turn, watches it? One must constantly resist the absurd Disney-like fantasy of a homunculus standing in our brains, peering at our screens and commanding our acts. There is no "I" who looks inside us. The stage itself is the "I." There is nothing wrong with the stage metaphor, provided that we eliminate the intelligence of the audience and replace it with explicit operations of an algorithmic nature. As Dennett whimsically states, "One discharges fancy homunculi from one's scheme by organizing armies of idiots to do the work."[4]

Bernard Baars's version of the workspace model eliminates the homunculus. The audience of the global workspace is not a little man in the head but a collection of other unconscious processors that receive a broadcast message and act upon it, each according to its own competence. Collective intelligence arises from the broad exchange of messages selected for their pertinence. This idea is not new—it dates back to the inception of artificial intelligence, when researchers proposed that subprograms would exchange data via a shared "blackboard," a common data structure similar to the "clipboard" in a personal computer. The conscious workspace is the clipboard of the mind.

Taine's narrow stage, too tiny to let more than a single actor perform at a time, vividly exemplifies another idea with a long history: that consciousness arises from a limited-capacity system that deals with only one thought at a time. During World War II, the British psychologist Donald Broadbent developed a better metaphor, borrowed from the newborn theory of information and computing.[5] Studying airplane pilots, he realized that, even with training, they could not easily attend to two simultaneous trains of speech, one in each ear. Conscious perception, he surmised, must involve a "limited-capacity channel"—a slow bottleneck that processes only one item at a time. The subsequent discovery of the attentional blink and the psychological refractory period, as we saw in Chapter 2, strongly supported this notion: while our attention is attracted by a first item, we become utterly blind to others. Modern cognitive psychologists have developed a variety of essentially equivalent metaphors, picturing conscious access as a "central bottleneck"[6] or a "second processing stage,"[7] a VIP lounge to which only the happy few are admitted.

A third metaphor emerged in the 1960s and 1970s: it depicted consciousness as a high-level "supervision system," a high-powered central executive that controls the flow of information in the rest of the nervous system.[8] As William James had noted in his 1890 masterpiece *The Principles of Psychology*, consciousness looks like "an organ added for the sake of steering a nervous system grown too complex to regulate itself."[9] Taken literally, this statement smacks of dualism: consciousness is not an outsider added to the nervous system but a full in-house participant. In this sense, our nervous system does achieve the remarkable feat of "regulating itself," but it does so in a hierarchical manner. The higher centers of the prefrontal cortex, which are more recent in evolution, take the lead over the lower-level systems hosted in posterior cortical areas and subcortical nuclei—often to inhibit them.[10]

The neuropsychologists Michael Posner and Tim Shallice proposed that information becomes conscious whenever it is represented within this high-level regulatory system. We now know that this view cannot be quite right; as we saw in Chapter 2, even a subliminal stimulus, without being seen, may partially trigger some of the inhibitory and regulatory functions of the supervisory executive system.[11] However, conversely, any information that reaches the conscious workspace immediately becomes capable of regulating, in an extremely deep and extensive manner, all our thoughts. Executive attention is just one of the many systems that receive inputs from the global workspace. As a result, whatever we are aware of becomes available to drive our decisions and our intentional actions, giving rise to the feeling that they are "under control." Language, long-term memory, attention, and intention systems are all part of this inner circle of intercommunicating devices that exchange conscious information. Thanks to this workspace architecture, whatever we are aware of can be arbitrarily rerouted and become the subject of a sentence, the crux of a memory, the focus of our attention, or the core of our next voluntary act.

Beyond Modularity

Like the psychologist Bernard Baars, I believe that consciousness reduces to what the workspace does: it makes relevant information globally accessible and flexibly broadcasts it to a variety of brain systems. In principle, nothing prevents the reproduction of these functions in nonbiological hardware such as a silicon-based computer. In practice, however, the relevant operations are far from trivial. We do not yet know exactly how the brain implements them, or how we could endow a machine with them. Computer software tends to be organized in a rigidly modular fashion: each routine receives specific inputs and transforms them according to precise rules in order to generate well-defined outputs. A word processor may hold a piece of information (say, a block of text) for a while, but the computer as a whole has no means of deciding whether this piece of information is globally relevant, or of making it broadly accessible to other programs. As a result, our computers remain despairingly narrow-minded. They carry out their tasks to perfection, but what is known inside a module, however intelligent, cannot be shared with others. Only a rudimentary mechanism, the clipboard, allows computer programs to share their knowledge—but only under the supervision of an intelligent deus ex machina: the human user.

Our cortex, unlike the computer, seems to have resolved this problem by simultaneously embracing a modular set of processors and a flexible routing system. Many sectors of the cortex are dedicated to a specific process. Entire patches are composed solely of face-specific neurons that react only when a face appears on the retina.[12] Regions of the parietal and motor cortexes are dedicated to specific motor acts or to the particular body parts that perform them. Even more abstract sectors encode our knowledge of numbers, animals, objects, and verbs. If workspace theory is right, consciousness may have evolved to mitigate this modularity. Thanks to the global neuronal workspace, information can be shared freely across the modular processors of our brain. This global availability of information is precisely what we subjectively experience as a conscious state.[13]

The evolutionary advantages of this arrangement are obvious. Modularity is useful because different domains of knowledge require different tunings of the cortex: the circuits for navigating in space perform different operations than those that recognize a landscape or store a past event in memory. But decisions must often be based on the pooling of multiple sources of knowledge. Picture a thirsty elephant, alone in the savannah. Its survival depends on finding the next waterhole. Its decision to walk toward a distant and invisible location must be based on the most efficient use of available information, including a mental map of space; the visual recognition of landmarks, trees, and paths; and a recall of past successes and failures at finding water. Long-term decisions of such a vital nature, leading the animal through an exhausting journey under the African sun, must make use of all existing sources of data. Consciousness may have evolved, aeons ago, in order to flexibly tap into all the sources of knowledge that might be relevant to our current needs.[14]

An Evolved Communication Network

According to this evolutionary argument, consciousness implies connectivity. Flexible information sharing requires a specific neuronal architecture to link the many distant and specialized regions of the cortex into a coherent role. Can we identify such a structure inside our brains? As early as the late nineteenth century, the Spanish neuroanatomist Santiago Ramón y Cajal noted a peculiar aspect of brain tissue. Unlike the dense mosaic of cells that make up our skin, the brain comprises enormously elongated cells: neurons. With their long axon, neurons possess the property, unique among cells, of measuring up to meters in size. A single neuron in the motor cortex may send its axon to extraordinarily distant regions of the spinal cord, in order to command specific muscles. Most interestingly, Cajal discovered that long-distance projection cells are quite dense in the cortex (figure 25), the thin mantle that forms the surface of our two hemispheres. From their locations in the cortex, nerve cells shaped like pyramids often send their axons all the way to the back of the brain or to the other hemisphere. Their axons group together into dense bundles of fibers that form cables of several millimeters in diameter and up to several centimeters in length. Using magnetic resonance imaging, we can now easily detect these crisscrossing fiber bundles in the living human brain.

Importantly, not all brain areas are equally well connected. Sensory

FIGURE 25. Long-distance neuronal connections may support the global neuronal workspace. The famous neuroanatomist Santiago Ramón y Cajal, who dissected the human brain in the nineteenth century, already noted how large cortical neurons, shaped like pyramids, sent their axons to very distant regions (left). We now know that these long-distance projections convey sensory information to a densely connected network of parietal, temporal, and prefrontal regions (right). A lesion in these long-distance projections may cause spatial neglect, a selective loss of visual awareness of one side of space.

regions, such as the primary visual area V1, tend to be choosy and to establish only a small set of connections, primarily with their neighbors. Early visual regions are arranged in a coarse hierarchy: area V1 speaks primarily to V2, which in turns speaks to V3 and V4, and so on. As a result, early visual operations are functionally encapsulated: visual neurons initially receive only a small fraction of the retinal input and process it in relative isolation, without any "awareness" of the overall picture.

In the higher association areas of the cortex, however, connectivity loses its local nearest-neighbor or point-to-point character, thus breaking the modularity of cognitive operations. Neurons with long-distance axons are most abundant in the prefrontal cortex, the anterior part of the brain. This region connects to many other sites in the inferior parietal lobe, the middle and anterior temporal lobe, and the anterior and posterior cingulate areas that lie on the brain's midline. These regions have been identified as major hubs—the brain's main interconnection centers.[15] All are heavily connected by reciprocal projections: if area A projects to area B, then almost invariably B also sends a projection back to A (figure 25). Furthermore, long-distance connections tend to form triangles: if area A projects jointly to areas B and C, then they, in turn, are very likely to be interconnected.[16]

These cortical regions are strongly connected to additional players, such as the central lateral and intralaminar nuclei of the thalamus (involved in attention, vigilance, and synchronization), the basal ganglia (crucial for decision making and action), and the hippocampus (essential for memorizing the episodes of our lives and for recalling them). Pathways linking the cortex with the thalamus are especially important. The thalamus is a collection of nuclei, each of which enters into a tight loop with at least one region of the cortex and often many of them at once. Virtually all regions of the cortex that are directly interconnected also share information via a parallel information route through a deep thalamic relay.[17] Inputs from the thalamus to the cortex also play a fundamental role in exciting the cortex and maintaining it in an "up" state of sustained activity.[18] As we shall see, the reduced activity of the thalamus and its interconnections play a key role in coma and vegetative states, when the brain loses its mind.

The workspace thus rests on a dense network of interconnected brain regions—a decentralized organization without a single physical meeting site. At the top of the cortical hierarchy, an elitist board of executives, distributed in distant territories, stays in sync by exchanging a plethora of messages. Strikingly, this anatomical network of interconnected high-level areas, involving primarily the prefrontal and parietal lobes, coincides with the one that I described in Chapter 4 and whose sudden activation constituted our first signature of conscious processing. We are now in a position to understand why these associative areas systematically ignite whenever a piece of information enters our awareness: those regions possess precisely the long-distance connectivity needed to broadcast messages across the long distances of the brain.

The pyramidal neurons of the cortex that participate in this long-distance network are well adapted to the task (figure 26). To harbor the complex molecular machinery needed to sustain their immense axons, they possess giant cell bodies. Remember that the cell's nucleus is where the genetic information is encoded in DNA—and yet the receptor molecules that are transcribed there must somehow make their way to synapses centimeters away. The large nerve cells capable of performing this spectacular feat tend to concentrate in specific layers of the cortex—the layers II and III, which are especially responsible for the callosal connections that distribute information across the two hemispheres.

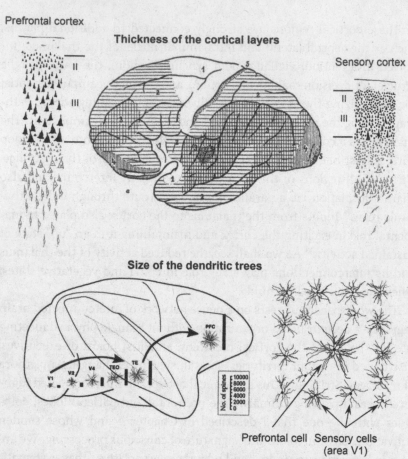

FIGURE 26. Large pyramidal neurons are adapted to the global broadcasting of conscious information, particularly in the prefrontal cortex. The whole cortex is organized in layers, and layers II and III contain the large pyramidal neurons whose long axons project to distant regions. These layers are much thicker in the prefrontal cortex than in sensory areas (above). The thickness of layers II and III roughly delineates the regions that are maximally active during conscious perception. These neurons also exhibit adaptations to the reception of global messages. Their dendritic trees (below), which receive projections from other regions, are much larger in the prefrontal cortex than in other regions. These adaptations to long-distance communication are more prominent in the human brain than in the brains of other primate species.

As early as the 1920s, the Austrian neuroanatomist Constantin von Economo observed that these layers were not equally distributed. They were much thicker in prefrontal and cingulate cortex, as well as in associative areas of the parietal and temporal lobes—precisely the tightly interconnected regions that activate during conscious perception and processing.

More recently, Guy Elston, in Queensland, Australia, and Javier DeFelipe, in Spain, have observed that these giant workspace neurons also possess immense dendrites, the neuron's receiving antennas, making them particularly suited to the gathering of messages arising from many distant regions.[19] Pyramidal neurons collect information from other neurons through their dendrites (the word comes from the Greek root for "tree"), the dense arborescence that collects incoming signals. At the place where an incoming neuron makes a synapse, the receiver neuron grows a microscopic anatomical structure called a spine—a mushroom-shaped protuberance. Vast numbers of spines densely cover the dendritic tree. Crucially for the workspace hypothesis, Elston and DeFelipe showed that the dendrites are much larger, and the spines much more numerous, in the prefrontal cortex than in posterior regions of the brain (see figure 26).

Furthermore, those adaptations to long-distance communication are particularly obvious in the human brain.[20] Relative to our primate cousins, our prefrontal neurons are more branched and contain more spines. Their dense jungle of dendrites is controlled by a family of genes that are uniquely mutated in humans.[21] The list includes FoxP2, the famous gene with two mutations specific to the *Homo* lineage,[22] which modulates our language networks,[23] and whose disruption creates a massive impairment in articulation and speech.[24] The FoxP2 family includes several genes responsible for building neurons, dendrites, axons, and synapses. In an amazing feat of genomic technology, scientists created mutant mice carrying the two human FoxP2 mutations—and sure enough, they grew pyramidal neurons with much larger, humanlike dendrites and a greater facility to learn (although they still didn't speak).[25]

Because of FoxP2 and its associated gene family, each human prefrontal neuron may host fifteen thousand spines or more. This implies that it is talking to just about as many other neurons, most of them located very far away in the cortex and thalamus. This anatomical arrangement looks like the perfect adaptation to meet the challenge of collecting information anywhere in the brain and, once it has been deemed relevant enough to enter the global workspace, broadcast it back to thousands of sites.

Suppose we could track all the connections that are activated as we consciously recognize a face—much as the FBI traces a phone call through successive telecom hubs. What kind of network would we see? Initially, very short connections, located inside our retinas, clean up the incoming image. The compressed image is then sent, via the massive cable of the optic nerve, to the visual thalamus, then on to the primary visual area in the occipital lobe. Via local U-shaped fibers, it gets progressively transmitted to several clusters of neurons in the right fusiform gyrus, where researchers have discovered "face clusters"—patches of neurons tuned to faces. All this activity remains unconscious. What happens next? Where do the fibers go? The Swiss anatomist Stéphanie Clarke found the surprising answer:[26] all of a sudden, long-distance axons allow the visual information to be dispatched to virtually any corner of the brain. From the right inferior temporal lobe, massive and direct connections project, in a single synaptic step, to distant areas of the associative cortex, including those in the opposite hemisphere. The projections concentrate in the inferior frontal cortex (Broca's area) and in the temporal association cortex (Wernicke's area). Both regions are key nodes of the human language network—and at this stage, therefore, words begin to be attached to the incoming visual information.

Because these regions themselves participate in a broader network of workspace areas, the information can now be further disseminated to the entire inner circle of higher-level executive systems; it can circulate in a reverberating assembly of active neurons. According to my theory, access to this dense network is all that is needed for the incoming information to become conscious.

Sculpting a Conscious Thought

Try to evaluate the sheer number of conscious thoughts that you can entertain: all the faces, objects, and scenes that you recognize; every single shade of emotion that you have experienced, from brutish anger to subtle schadenfreude; every piece of geographical trivia, historical information, mathematical knowledge, or mere gossip, true or false, that you have ever heard or may hear; the pronunciation and meaning of every word you know or could know, in any of the world's languages. . . . Isn't the list endless? And yet any of them could, in the next minute, become the subject of your conscious thoughts. How can such a cornucopia of states be encoded in the neuronal workspace? What is the neural code for consciousness, and how does it support a near-infinite repertoire of ideas?

The neuroscientist Giulio Tononi points out that the sheer size of our repertoire of ideas sharply constrains the neural code for conscious thoughts.[27] Its primary characteristic must be an enormous degree of differentiation: the combinations of active and inactive neurons in our global workspace must be able to form billions of different activity patterns. Each of our potential conscious mental states must be assigned to a different state of neuronal activity, well delineated from all the others. As a result, our conscious states must exhibit sharp boundaries: either it's a bird, or it's a plane, or it's Superman, but not all at the same time. A clear mind, with myriad potential thoughts, requires a brain with myriad potential states.

In his book *The Organization of Behavior* (1949), Donald Hebb had already proposed a visionary theory of how the brain might encode thoughts. He introduced the concept of "cell assemblies"—sets of neurons that are interconnected by excitatory synapses and that therefore tend to remain activated long after any external stimulus is gone. "Any frequently repeated, particular stimulation," he surmised, "will lead to the slow development of a 'cell-assembly,' a diffuse structure comprising cells in the cortex and diencephalon (and also, perhaps, in the basal ganglia of the cerebrum), capable of acting briefly as a closed system."[28]

All the neurons in a cell assembly support one another by sending excitatory pulses. As a result, they form a delimited "hill" of activity in neural space. And because many such local assemblies can activate independently at different places in the brain, the outcome is a combinatorial code capable of representing billions of states. For instance, any visual object can be represented by a combination of color, size, and fragments of shapes. Recordings from the visual cortex support this idea: a fire extinguisher, for instance, seems to be encoded by a combination of active "patches" of neurons, each comprising a few hundred active neurons and each representing a particular part (handle, body, hose, etc.).[29]

In 1959, the artificial intelligence pioneer John Selfridge introduced another useful metaphor: the "pandemonium."[30] He envisioned the brain as a hierarchy of specialized "daemons," each of which proposes a tentative interpretation of the incoming image. Thirty years of neurophysiological research, including the spectacular discovery of visual cells tuned to lines, colors, eyes, faces, and even U.S. presidents and Hollywood stars, have brought strong support to this idea. In Selfridge's model, the daemons yelled their preferred interpretation at one another, in direct proportion to how well the incoming image favored their own interpretation. Waves of shouting were propagated through a hierarchy of increasingly abstract units, allowing neurons to respond to increasingly abstract features of the image—for instance, three daemons shouting for the presence of eyes, nose, and hair would together conspire to excite a fourth daemon coding for the presence of a face. By listening to the most vocal daemons, a decision system could form an opinion of the incoming image—a conscious percept.

Selfridge's pandemonium model received one important improvement. Originally, it was organized according to a strict feed-forward hierarchy: the daemons bellowed only at their hierarchical superiors, but a high-ranking daemon never yelled back at a low-ranking one or even at another daemon of the same rank. In reality, however, neural systems do not merely report to their superiors; they also chat among themselves. The cortex is full of loops and bidirectional projections.[31] Even individual neurons dialogue with each other: if neuron α projects to neuron β, then β probably projects back to α.[32] At any level, interconnected neurons support each other, and those at the top of the hierarchy can talk back to their subordinates, so that messages propagate downward at least as much as upward.

Simulation and mathematical modeling of realistic "connectionist" models with many such loops show that they possess a very useful property. When a subset of neurons is excited, the entire group self-organizes into "attractor states": groups of neurons form reproducible patterns of activity that remain stable for a long duration.[33] As anticipated by Hebb, interconnected neurons tend to form stable cell assemblies.

As a coding scheme, these recurrent networks possess an additional advantage—they often converge to a consensus. In neuronal networks that are endowed with recurrent connections, unlike Selfridge's daemons, the neurons do not simply yell stubbornly at one another: they progressively come to an intelligent agreement, a unified interpretation of the perceived scene. The neurons that receive the greatest amount of activation mutually support one another and progressively suppress any alternative interpretation. As a result, missing parts of the image can be restored and noisy bits can be removed. After several iterations, the neuronal representation encodes a cleaned-up, interpreted version of the perceived image. It also becomes more stable, resistant to noise, internally coherent, and distinct from other attractor states. Francis Crick and Christof Koch describe this representation as a winning "neural coalition" and suggest that it is the perfect vehicle for a conscious representation.[34]

The term "coalition" points to another essential aspect of the conscious neuronal code: it must be tightly integrated.[35] Each of our conscious moments coheres as one single piece. When contemplating Leonardo da Vinci's *Mona Lisa*, we do not perceive a disemboweled Picasso with detached hands, Cheshire cat smile, and floating eyes. We retrieve all these sensory elements and many others (a name, a meaning, a connection to our memories of Leonardo's genius)—and they are somehow bound together into a coherent whole. Yet each of them is initially processed by a distinct group of neurons, spread centimeters apart on the surface of the ventral visual cortex. How do they get attached to one another?

One solution is the formation of a global assembly, thanks to the hubs provided by the higher sectors of cortex. These hubs, which the neurologist Antonio Damasio calls "convergence zones,"[36] are particularly predominant in the prefrontal cortex but also in other sectors of the anterior temporal lobe, inferior parietal lobe, and a midline region called the precuneus. All send and receive numerous projections to and from a broad variety of distant brain regions, allowing the neurons there to integrate information over space and time. Multiple sensory modules can therefore converge onto a single coherent interpretation ("a seductive Italian woman"). This global interpretation may, in turn, be broadcast back to the areas from which the sensory signals originally arose. The outcome is an integrated whole. Because of neurons with long-distance top-down axons, projecting back from the prefrontal cortex and its associated high-level network of areas onto the lower-level sensory areas, global broadcasting creates the conditions for the emergence of a single state of consciousness, at once differentiated and integrated.

This permanent back-and-forth communication is called "reentry" by the Nobel Prize winner Gerald Edelman.[37] Model neuronal networks suggest that reentry allows for a sophisticated computation of the best possible statistical interpretation of the visual scene.[38] Each group of neurons acts as an expert statistician, and multiple groups collaborate to explain the features of the input.[39] For instance, a "shadow" expert decides that it can account for the dark zone of the image—but only if the light comes from the top left. A "lighting" expert agrees and, using this hypothesis, explains why the top parts of the objects are illuminated. A third expert then decides that, once these two effects are accounted for, the remaining image looks like a face. These exchanges continue until every bit of the image has received a tentative interpretation.

The Shape of an Idea

Cell assemblies, a pandemonium, competing coalitions, attractors, convergence zones with reentry . . . each of these hypotheses seems to hold a grain of truth, and my own theory of a global neuronal workspace draws heavily from them.[40] It proposes that a conscious state is encoded by the stable activation, for a few tenths of a second, of a subset of active workspace neurons. These neurons are distributed in many brain areas, and they all code for different facets of the same mental representation. Becoming aware of the *Mona Lisa* involves the joint activation of millions of neurons that care about objects, fragments of meaning, and memories.

During conscious access, thanks to the workspace neurons' long axons, all these neurons exchange reciprocal messages, in a massively parallel attempt to achieve a coherent and synchronous interpretation. Conscious perception is complete when they converge. The cell assembly that encodes this conscious content is spread throughout the brain: fragments of relevant information, each distilled by a distinct brain region, cohere because all the neurons are kept in sync, in a top-down manner, by neurons with long-distance axons.

Neuronal synchrony may be a key ingredient. There is growing evidence that distant neurons form giant assemblies by synchronizing their spikes with ongoing background electrical oscillations.[41] If this picture is correct, the brain web that encodes each of our thoughts resembles a swarm of fireflies that harmonize their discharges according to the overall rhythm of the group's pattern. In the absence of consciousness, moderate-size cell assemblies may still synchronize locally—for instance, when we unconsciously encode a word's meaning inside the language networks of our left temporal lobe. However, because the prefrontal cortex does not gain access to the corresponding message, it cannot be broadly shared and therefore remains unconscious.

Let us conjure one more mental image of this neuronal code for consciousness. Picture the sixteen billion cortical neurons in your cortex. Each of them cares about a small range of stimuli. Their sheer diversity is flabbergasting: in the visual cortex alone, one finds neurons that care about faces, hands, objects, perspective, shape, lines, curves, colors, 3-D depth . . . Each cell conveys only a few bits of information about the perceived scene. Collectively, though, they are capable of representing an immense repertoire of thoughts. The global workspace model claims that, at any given moment, out of this enormous potential set, a single object of thought gets selected and becomes the focus of our consciousness. At this moment, all the relevant neurons activate in partial synchrony, under the aegis of a subset of prefrontal cortex neurons.

It is crucial to understand that, in this sort of coding scheme, the silent neurons, which do *not* fire, also encode information. Their muteness implicitly signals to others that their preferred feature is not present or is irrelevant to the current mental scene. A conscious content is defined just as much by its silent neurons as by its active ones.

In the final analysis, conscious perception may be likened to the sculpting of a statue. Starting with a raw block of marble and chipping away most of it, the artist progressively exposes his vision. Likewise, starting with hundreds of millions of workspace neurons, initially uncommitted and firing at their baseline rate, our brain lets us perceive the world by silencing most of them, keeping only a small fraction of them active. The active set of neurons delineates, quite literally, the contours of a conscious thought.

The landscape of active and inactive neurons can explain our second signature of consciousness: the P3 wave that I described in Chapter 4, a large positive voltage that peaks at the top of the scalp. During conscious perception, a small subset of workspace neurons becomes active and defines the current content of our thoughts, while the rest are inhibited. The active neurons broadcast their message throughout the cortex by sending spikes down their long axons. At most places, however, these signals land on inhibitory neurons. They act as a silencer that hushes entire groups of neurons: "Please remain silent, your features are irrelevant." A conscious idea is encoded by small patches of active and synchronized cells, together with a massive crown of inhibited neurons.

Now, the geometrical layout of the cells is such that, in the active ones, synaptic currents travel from the superficial dendrites toward the cells' bodies. Because all these neurons are parallel to one another, their electrical currents add up, and, on the surface of the head, they create a slow negative wave over the regions that encode the conscious stimulus.[42] The inhibited neurons, however, dominate the picture—and their activity adds up to form a *positive* electrical potential. Because many more neurons are inhibited than are activated, all these positive voltages end up forming a large wave on the head—the P3 wave that we easily detect whenever conscious access occurs.[43] We have explained our second signature of consciousness.

The theory readily explains why the P3 wave is so strong, generic, and reproducible: it mostly indicates what the current thought is *not* about. It is the focal negativities that define the contents of consciousness, not the diffuse positivity. In agreement with this idea, Edward Vogel and his colleagues at the University of Oregon have published beautiful demonstrations of negative voltages over the parietal cortex that track the current contents of our working memory for spatial patterns.[44] Whenever we memorize an array of objects, slow negative voltages indicate exactly how many objects we saw and where they were. These voltages last for as long as we keep the objects in mind; they increase when we add objects to our memory, saturate when we cannot keep up, collapse when we forget, and faithfully track the number of items that we remember. In Edward Vogel's work, negative voltages directly delineate a conscious representation—exactly as our theory predicts.

Simulating a Conscious Ignition

The science of reality is no longer content with the phenomenological how, the mathematical how is what it seeks.

—Gaston Bachelard, *The Formation of the Scientific Mind* (1938)

Conscious access carves a thought in us by sculpting a pattern of active and inactive neurons into our global workspace network. Although this metaphorical vision may suffice to boost our intuition of what consciousness is, it should ultimately be replaced by a more sophisticated mathematical theory of how neural networks operate, and why they generate the neurophysiological signatures that we can observe in our macroscopic recordings. In an effort in this direction, Jean-Pierre Changeux and I have started to develop computer simulations of neural networks that capture some of the basic properties of conscious access.[45]

Our modest goal was to probe how neurons would behave once they were connected according to the precepts of global workspace theory (figure 27). To re-create, in the computer, the dynamics of a small coalition of neurons, we started with "integrate and fire" neurons—simplified equations that mimic the spiking of nerve cells. Each neuron possessed realistic synapses, with parameters that captured several major types of receptors for neurotransmitters in the living brain.

We then wired these virtual neurons into local cortical columns, mimicking the subdivision of the cortex into interconnected layers of cells. The concept of a neuronal "column" comes from the fact that the neurons that lie on top of one another, perpendicular to the surface of the cortex, tend to be tightly interconnected, to share similar responses, and to originate from divisions of the same founder cell during development. Our model respected this biological arrangement: the neurons in our simulated columns tended to support one another and to respond to similar inputs.

We also included a small thalamus—a structure consisting of multiple nuclei, each strongly connected with a sector of cortex or with a broad array of cortical locations. We hooked it up with realistic connection strengths and timing delays, taking into account the distances that the spikes had to travel along the axons. The result was a coarse model of the basic computational unit in the primate brain: the thalamocortical column. We made sure that this model operated in a realistic manner— even in the absence of inputs, the virtual neurons fired spontaneously and generated an electroencephalogram somewhat like the one generated by the human cortex.

Once we had a good model of the thalamocortical column, we inter-

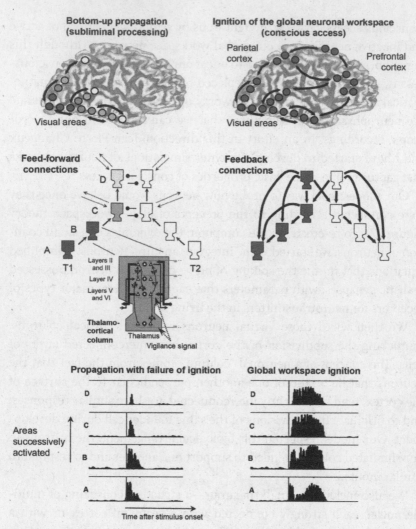

FIGURE 27. A computer simulation mimics the signatures of unconscious and conscious perception. Jean-Pierre Changeux and I simulated, in the computer, a subset of the many visual, parietal, and prefrontal areas that contribute to sub-liminal and conscious processing (above). Four hierarchical regions were linked by feed-forward and long-distance feedback connections (middle). Each simu-lated area comprised cortical cells that were organized in layers and connected to neurons in the thalamus. When we stimulated the network with a brief input, activation propagated from bottom to top before dying out, thus capturing the brief activation of cortical pathways during subliminal perception. A slightly longer stimulus led to global ignition: the top-down connections amplified the input and led to a second wave of long-lasting activation, thus capturing the activations observed during conscious perception.

connected several of them into functional long-distance brain networks. We simulated a hierarchy of four brain areas and assumed that each of them contained two columns coding for two target objects, a sound and a light. Our network could distinguish between only two perceptions—a massive oversimplification that was unfortunately needed in order to keep the simulation tractable. We simply assumed that the physiological properties would not be massively changed if a much broader set of states was included.[46]

At the periphery, perception operated in parallel: neurons coding for the sound and light could be simultaneously activated, without interfering with each other. At the higher levels of the cortical hierarchy, however, they actively inhibited each other, such that these regions could entertain only a single integrated state of neural firing—a single "thought."

As in the real brain, cortical areas projected serially onto one another in a feed-forward manner: the primary area received sensory inputs, then sent its spikes to the secondary area, which itself projected to a third and then a fourth region. Crucially, long-distance feedback projections folded the network back onto itself, by allowing the higher areas to send excitatory support to the very sensory areas that initially excited them. The result was a simplified global workspace: a tangle of feed-forward and feedback connections with multiple nested scales—neurons, columns, areas, and the long-distance connections between them.

After so much computer programming, it was fun to finally turn the simulation on and see how the virtual neurons lit up. To mimic perception, we injected a little current into the visual thalamic neurons—coarsely imitating what happens when, say, light receptors in the retina are activated and, after retinal preprocessing, excite the relay neurons in a subpart of the thalamus called the lateral geniculate body. We then let the simulation roll according to its equations. As we had hoped, although dramatically simplified, our mock-up exhibited many physiological properties that had been seen in real experiments and whose origins suddenly became open to investigation.

The first of these properties was global ignition. When we presented a pulse of stimulation, it slowly climbed its way up the cortical hierarchy in a fixed order, from the primary area to the second, then to the third and the fourth. This feed-forward wave mimicked the well-known transmission of neural activity across the hierarchy of visual areas. After a while, the entire set of columns coding for the perceived object began to ignite. As a result of massive feedback connections, neurons coding for the same perceptual input exchanged mutually reinforcing excitatory signals, thus leading to a sudden ignition of activity. Meanwhile, the alternate percept was actively inhibited. This sustained activation lasted for hundreds of milliseconds. Its duration was essentially unrelated to that of the initial stimulus; even a brief external pulse could lead to a sustained reverberating state. These experiments captured the essence of how the brain forms a long-lasting representation of a flashed picture and maintains it online.

The model's dynamics reproduced the properties that we had observed in our electroencephalographic and intracranial recordings. Most simulated neurons showed a late and sudden increase in the overall synaptic currents that they received. Excitation progressed forward but also returned to the original sensory areas that started it—mimicking the late amplification that we had seen in sensory areas during conscious access. In the simulation, the ignited state also led to a reverberation of neuronal activity across the model's many nested loops: within a cortical column, from cortex to thalamus and back, and across the long distances of the cortex. The net effect was an increase in oscillatory fluctuations in a broad band of frequencies, with a prominent peak in the gamma range (30 hertz and above). At the time of global ignition, spikes became strongly coupled and synchronized among the neurons coding for the conscious representation. In brief, the computer simulation mimicked our four empirical signatures of conscious access.

By simulating this process, we gained novel mathematical insights. Conscious access corresponded to what the theoretical physicist calls a "phase transition"—the sudden transformation of a physical system from one state to another. As I explained in Chapter 4, a phase transition occurs, for instance, when water turns into ice: the H_2O molecules suddenly assemble into a rigid structure with new emergent features. During a phase transition, the physical properties of the system often change suddenly and discontinuously. In our computer simulations, likewise, the spiking activity jumped from an ongoing state of low spontaneous activity to a temporary moment of elevated spiking and synchronized exchanges.

It is easy to see why this transition was nearly discontinuous. Since the neurons at the higher level sent excitation to the very units that activated them in the first place, the system possessed two stable states separated by an unstable ridge. The simulation either stayed at a low level of activity or, as soon as the input increased beyond a critical value, snowballed into an avalanche of self-amplification, suddenly plunging a subset of neurons into frantic firing. The fate of a stimulus of intermediate intensity was therefore unpredictable—activity either quickly died out or suddenly jumped to a high level.

This aspect of our simulations fits nicely with a 150-year-old concept in psychology: the idea that consciousness possesses a threshold that sharply delineates unconscious (subliminal) from conscious (supraliminal) thoughts. Unconscious processing corresponds to neuronal activation that propagates from one area to the next without triggering a global ignition. Conscious access, on the other hand, corresponds to the sudden transition toward a higher state of synchronized brain activity.

The brain, however, is vastly more complicated than a snowball. Reaching an adequate theory of the phase transitions that actually occur in the dynamics of actual neural networks will take many more years.[47] Actually, our simulations already contained two nested phase transitions. One of them, which I just explained, involved a global ignition. However, the threshold for this ignition was itself under the control of another phase transition, which corresponded to the "awakening" of the entire network. Each pyramidal neuron in our simulated cortex received a vigilance signal, a small amount of current that summarized, in highly simplified form, the well-known activating effects of acetylcholine, noradrenaline, and serotonin ascending from various nuclei in the brain stem, basal forebrain, and hypothalamus and turning the cortex "on." Our model thus captured changes in the *state* of consciousness—the switch from an unconscious to a conscious brain.

When the vigilance signal was low, spontaneous activity was drastically reduced, and the ignition property disappeared: even a strong sensory input, while activating thalamic and cortical neurons in primary and secondary areas, quickly fizzled without making it over the threshold for global ignition. In this state, our network therefore behaved as a sleepy or anesthetized brain.[48] It responded to stimuli but only in its peripheral sensory areas—activation typically failed to climb all the way up to workspace areas and ignite a full-blown cell assembly. As we increased the vigilance parameter, however, a structured electroencephalogram emerged in the model, and ignition by external stimuli suddenly recovered. The threshold for this ignition varied with the model's drowsiness, indicating how a heightened vigilance increases the probability that we detect even faint sensory inputs.

The Restless Brain

I tell you: one must still have chaos in one, to give birth to a dancing star. I tell you: ye have still chaos in you.

—Friedrich Nietzsche, *Thus Spoke Zarathustra* (1883–85)

Another fascinating phenomenon emerged in our simulation: spontaneous neuronal activity. We did not have to constantly stimulate our network. Even in the absence of any input, neurons would spontaneously fire, triggered by random events at their synapses—and this chaotic activity self-organized into recognizable patterns.

At high levels of the vigilance parameter, complex patterns of firing continuously waxed and waned on our computer screens. Within them, we occasionally recognized a global ignition—triggered in the absence of any stimulus. An entire set of cortical columns, all coding for the same stimulus, would activate for a short period, then fade away. A fraction of a second later, another global assembly would replace it. Without any cueing, the network self-organized into a series of random ignitions, closely resembling those evoked during the perception of external stimuli. The only difference was that spontaneous activity tended to start at the higher cortical levels, within workspace areas, and to propagate downward into the sensory regions—the converse of what happened during perception.

Do such bouts of endogenous activity exist in the real brain? Yes. In fact, organized spontaneous activity is omnipresent in the nervous system. Anyone who has ever seen an EEG knows this: the two hemispheres constantly generate massive high-frequency electrical waves, whether the person is awake or asleep. This spontaneous excitation is so intense that it dominates the landscape of brain activity. By comparison, the activation evoked by an external stimulus is barely detectable, and much averaging is needed before it can be observed. Stimulus-evoked activity accounts for only a very small amount of the total energy consumed by the brain, probably less than 5 percent. The nervous system primarily acts as an autonomous device that generates its own thought patterns. Even in the dark, while we rest and "think of nothing," our brain constantly produces complex and ceaselessly changing arrays of neuronal activity.

Organized patterns of spontaneous cortical activity were first observed in animals. Using voltage-sensitive dyes, which transform invisible voltages into visible changes in light reflectance, Amiram Grinvald and his colleagues at the Weizmann Institute recorded the electrical activity of a large patch of cortex for an extended period of time.[49] Fascinatingly, even though the animal was anesthetized, complex patterns emerged. In the dark, without any stimulation, a visual neuron would suddenly start to discharge at a higher rate. It was not alone: imaging showed that, at the very same moment, a whole assembly of neurons had spontaneously activated.

A similar phenomenon exists in the human brain.[50] Images of brain activation during quiet rest revealed that, far from remaining silent, the human brain exhibits constantly changing patterns of cortical activity. Global networks, often distributed across the two hemispheres, activate similarly in different people. Some of them correspond tightly to the patterns evoked by an external stimulation. For instance, a large subset of the language circuit activates when we listen to a story, but it also discharges spontaneously when we rest in darkness—giving support to the notion of "internal speech."

The meaning of this resting-state activity remains a matter of debate among neuroscientists. Some of it may simply indicate that the brain's random discharges follow the existing network of anatomical connections. Where else could they go? Indeed, part of the correlated activation remains present during sleep, under anesthesia, or in unconscious patients.[51] However, in awake and attentive participants, another part seems to directly betray the subject's ongoing thoughts. For instance, one of the resting-state networks, called the default-mode network, turns on whenever we reflect upon our personal situation, retrieve autobiographical memories, or compare our thoughts with those of others.[52] When people lie in the scanner, and we wait until their brain is in this default state before asking them what they were thinking of, they report that they had been mind-wandering into their own thoughts and memories—more so than when they were interrupted at other times.[53]

Thus, the particular network that is spontaneously activated predicts, at least in part, the mental state of the person.

In a nutshell, ceaseless neuronal discharges create our ruminating thoughts. Furthermore, this internal stream competes with the external world. During moments of high default-mode activity, the presentation of an unexpected stimulus such as a picture no longer evokes a large P3 brain wave, as it does in an attentive subject.[54] Endogenous states of consciousness interfere with our ability to become aware of external events. Spontaneous brain activity invades the global workspace and, if absorbing, can block access to other stimuli for extended periods of time. We met a variant of this phenomenon in Chapter 1 under the name of "inattentional blindness."

My colleagues and I were enormously pleased when our computer simulation exhibited the very same sort of endogenous activity.[55] Bouts of spontaneous ignition occurred in front of our eyes, and they were more likely to be globally coherent when the simulation's vigilance parameter was high. Crucially, during this period, if we stimulated the network with an external input, even way above the normal ignition threshold, its progression was blocked and did not lead to global ignition: internal activity competed with external drives. Our simulation could mimic inattentional blindness and the attentional blink—two phenomena that epitomize the brain's inability to consciously attend to two things at once.

Spontaneous activity also explains why the very same incoming stimulus sometimes leads to a full-blown ignition, and sometimes to only a trickle of activity. It all depends on whether the noisy pattern of activation *prior* to the stimulus is aligned with the incoming train of spikes or is incompatible with it. In our simulation, as in the living human brain, random fluctuations in activity bias the perception of a faint external stimulus.[56]

Darwin in the Brain

Spontaneous activity is one of the most frequently overlooked features of the global workspace model—yet I personally view it as one of its most original and important traits. Too many neuroscientists still adhere to the obsolete idea of the reflex arc as a fundamental model for the human brain.[57] This idea, which dates back to René Descartes, Charles

Sherrington, and Ivan Pavlov, depicts the brain as an input-output device that merely transfers data from the senses to our muscles, as in Descartes's famous schema of how the eye commands the arm (figure 2). We now know that this view is deeply wrong. Autonomy is the primary property of the nervous system. Intrinsic neuronal activity dominates over external excitation. As a result, our brain is never passively submitted to its environment but generates its own stochastic patterns of activity. During brain development, the relevant patterns are preserved while the inappropriate ones are weeded out.[58] This joyfully creative algorithm, particularly evident in young children, submits our thoughts to a Darwinian selection process.

This point lay at the heart of William James's view of the organism. "Why not say," he rhetorically asked, "that just as the spinal cord is a machine with few reflexes, so the hemispheres are a machine with many, and that that is all the difference?" Because, he answers, the evolved circuitry of the brain acts as "an organ whose natural state is one of unstable equilibrium," allowing its "possessor to adapt its conduct to the minutest alterations in the environing circumstances."

The crux of this faculty lies in the nerve cells' excitability: early on in evolution, neurons acquired the ability to self-activate and spontaneously discharge a spike. Filtered and amplified by brain circuits, this excitability turns into purposeful exploratory behavior. Any animal explores its environment in a partially random manner, thanks to hierarchically organized "central pattern generators"—neural networks whose spontaneous activity generates rhythmic walking or swimming movements.

I contend that, in the primate brain and probably in many other species, a similar exploration occurs inside the brain, at a purely cognitive level. By spontaneously generating fluctuating patterns of activity, even in the absence of external stimulation, the global workspace allows us to freely generate new plans, try them out, and change them at will if they fail to fulfill our expectations.

A Darwinian process of variation followed by selection occurs within our global workspace system.[59] Spontaneous activity acts as a "generator of diversity" whose patterns are constantly sculpted by the brain's evaluation of future rewards. Neuronal networks endowed with this idea can be very powerful. In computer simulations, Jean-Pierre Changeux and I showed that they resolve complex problems and mind teasers, such as the classical Tower of London problem.[60] The logic of learning by selection, when combined with classical synaptic learning rules, yields a robust architecture capable of learning from its mistakes and extracting the abstract rules behind a problem.[61]

Although "Generator of Diversity" can be abbreviated as GOD, there is nothing magical behind the notion of spontaneous activity—certainly not a dualistic action of mind on matter. Excitability is a natural, physical property of nerve cells. In every neuron, the membrane potential undergoes ceaseless fluctuations in voltage. Those fluctuations are due in large part to the random release of vesicles of neurotransmitters at some of the neuron's synapses. In the final analysis, this randomness arises from thermal noise, which constantly rocks and rolls our molecules around. One would think that evolution would minimize the impact of this noise, as engineers do in digital chips, when they set very distinct voltages for 0s and 1s, so that thermal noise cannot offset them. Not so in the brain: neurons not only tolerate noise but even amplify it—probably because some degree of randomness is helpful in many situations where we search for an optimal solution to a complex problem. (Many algorithms, such as the "MonteCarlo Markov chain" and "simulated annealing," require an efficient source of noise.)

Whenever a neuron's membrane fluctuations exceed a threshold level, a spike is emitted. Our simulations show that these random spikes can be shaped by the vast sets of connections that link neurons into columns, assemblies, and circuits, until a global activity pattern emerges. What starts out as local noise ends up as a structured avalanche of spontaneous activity that corresponds to our covert thoughts and goals. It is humbling to think that the "stream of consciousness," the words and images that constantly pop up in our mind and make up the texture of our mental life, finds its ultimate origin in random spikes sculpted by the trillions of synapses laid down during our lifelong maturation and education.

A Catalog of the Unconscious

In recent years, global workspace theory has become a major interpretive tool, a prism through which to revisit empirical observations. One of its successes has been to clarify the various types of unconscious processes in the human brain. Much as the eighteenth-century Swedish scholar Carl Linnaeus conceived a "taxonomy" of all living species (an organized classification of plants and animals into types and subtypes), we can now begin to propose a taxonomy of the unconscious.

Remember the main message from Chapter 2: most of the brain's operations are unconscious. We are unaware of most of what we do and know, from respiration to posture control, from low-level vision to fine hand movements, from letter statistics to grammatical rules—and during inattentional blindness, we may even miss a gorilla-clad youngster banging his chest. A wild profusion of unconscious processors weaves the texture of who we are and how we act.

Global workspace theory helps bring some order to this jungle.[62] It leads us to pigeonhole our unconscious feats in distinct bins whose brain mechanisms differ radically (figure 28). Consider first what happens during inattentional blindness. Here a visual stimulus is presented way above the normal threshold for conscious perception—yet we fail to notice it because our mind is entirely set on a different task. I write these words in my wife's birth home, a seventeenth-century farmhouse whose charming living room features a huge long-case grandfather clock. The pendulum swings right in front of me, and I can easily hear it ticking. But whenever I concentrate on writing, the rhythmic noise vanishes from my mental world: inattention prevents awareness.

In our catalog of the unconscious, my colleagues and I have proposed to label this kind of unconscious information with the adjective *preconscious*.[63] It is consciousness-in-waiting: information that is already encoded by an active assembly of firing neurons and that thus could become conscious at any time, if only it were attended—yet it isn't. Actually, we borrowed the word from Sigmund Freud. In his *Outline of Psychoanalysis*, he observed that "some processes . . . may cease to be conscious, but can become conscious once more without any trouble. . . . Everything unconscious that behaves in this way, that can easily exchange the unconscious condition for the conscious one, is therefore better described as 'capable of entering consciousness' or as *preconscious*."

Simulations of the global workspace point to a putative neuronal mechanism for the preconscious state.[64] When a stimulus enters our

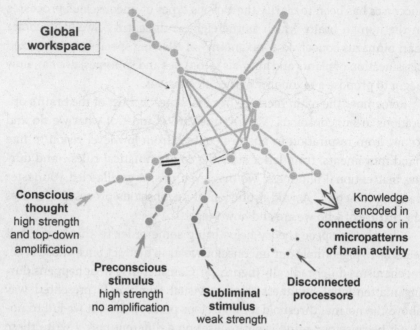

Global workspace

Conscious thought
high strength
and top-down
amplification

Preconscious stimulus
high strength
no amplification

Subliminal stimulus
weak strength

Knowledge encoded in connections or in micropatterns of brain activity

Disconnected processors

FIGURE 28. Knowledge may remain unconscious for several different reasons. At any given moment, only a single thought ignites the workspace. Other objects fail to gain access to consciousness, either because they are unattended and are therefore denied entry in the workspace (preconscious) or because they are too weak to cause a full-blown avalanche of activation, all the way up to the workspace level (subliminal). We also remain unaware of information that is encoded in processors disconnected from the workspace. Finally, vast amounts of unconscious information rest in our brain connections and micropatterns of brain activity.

simulation, its activation propagates and ultimately ignites the global workspace. In turn, this conscious representation creates a fringe of surrounding inhibition that prevents a second stimulus from entering at the same time. This central competition is unavoidable. I noted earlier that a conscious representation is defined as much by what it is *not* as by what it is. According to our hypothesis, some workspace neurons must be actively silenced in order to delimit the current conscious content and signal what it is *not*. This diffuse inhibition creates a bottleneck within the higher centers of the cortex. The neuronal silencing that forms an inescapable part of any conscious state prevents us from seeing two things at once and from performing two effortful tasks at the same time. It does not, however, preclude the activation of early sensory areas—they clearly light up, virtually at the same level as usual, even when the workspace is already occupied by a first stimulus. Preconscious information is temporarily buffered in such transient memory stores, outside the global workspace. There it will slowly decay to oblivion—unless we decide to orient our attention to it. For a brief period, the decaying preconscious information can still be recovered and brought to consciousness, in which case we experience it in retrospect, long after the fact.[65]

The preconscious state contrasts sharply with a second type of unconsciousness, which we dubbed the *subliminal* state. Consider an image that is flashed so briefly or so weakly that we cannot see it. Here the situation is very different. However hard we attend, we are unable to perceive the hidden stimulus. Sandwiched between geometrical shapes, the masked word forever escapes us. Such a subliminal stimulus does induce detectable activity in visual, semantic, and motor areas of the brain, but this activation is too short-lived to cause a global ignition. My lab's simulations again capture this state of affairs. In the computer, a brief pulse of activity can fail to trigger a global ignition, because by the time the top-down signals from higher areas return to early sensory areas and have a chance to amplify the incoming activity, the original activation is already gone and replaced by the mask.[66] Playing tricks on the brain, the astute psychologist easily designs stimuli so weak, so short, or so cluttered that they systematically prevent global ignition. The term *subliminal* applies to this category of situations where the incoming sensory wave dies out before creating a tsunami on the shores of the global neuronal workspace. However hard we try to perceive it, a subliminal stimulus will never become conscious, whereas a preconscious stimulus will, if only we find time to attend it. This is a key difference, with many consequences at the brain level.

The preconscious/subliminal distinction does not exhaust the stock of unconscious knowledge in our brains. Consider breathing. Every minute of your life, harmonious patterns of neural firing, generated deep in your brain stem and sent to your chest muscles, shape the ventilation rhythms that keep you alive. Ingenious feedback loops adapt them to the levels of oxygen and carbon dioxide in your blood. This sophisticated neuronal machinery remains totally unconscious. Why? Its neural firing is strong and extended in time, so it is not subliminal; yet no amount of attention can bring it to mind, so it is not preconscious either. Within our taxonomy, this case corresponds to a third category of unconscious representation: *disconnected patterns*. Encapsulated in your brain stem, the firing patterns that control your breathing are disconnected from the global workspace system in prefrontal and parietal cortex.

To become conscious, information inside a neural assembly has to be communicated to workspace neurons in the prefrontal cortex and associated sites. Respiration data, however, are forever locked in your brain stem neurons. The firing patterns that signal your blood CO_2 level cannot be transmitted to the rest of your cortex. As a result, you remain unaware of them. Many of our specialized neuronal circuits are so deeply entrenched that they simply lack the connections needed to reach our awareness. The only way to bring them to mind, interestingly, consists in recoding them via another sensory modality—we become aware of how we breathe only indirectly, when we attend to our chest movements.

Although we all feel that we are in control of our bodies, hundreds of neuronal signals constantly traffic through our brain modules without reaching our awareness, disconnected as they are from the appropriate higher-level cortical regions. In some stroke patients, the situation gets even worse. A lesion to the brain's white matter pathways can disconnect specific sensory or cognitive systems, suddenly rendering them inaccessible to consciousness. A spectacular case is the disconnection syndrome that occurs when a stroke affects the corpus callosum, the vast bundle of connections that links the two hemispheres. A patient with such a lesion may lose any awareness of his own motor plan. He will even disown the movements of his left hand, commenting that it behaves randomly and out of control. What happens is that the motor command of the left hand arises from the right hemisphere, while verbal comments are made by the left hemisphere. Once these two systems are disconnected, the patient hosts two impaired workspaces, each partially unconscious of what the other is brooding.

Beyond disconnection, a fourth way in which neural information can remain unconscious, according to workspace theory, is to be *diluted* into a complex pattern of firing. To take a concrete example, consider a visual grating that is so finely spaced, or that flickers so fast (50 hertz and above), that you cannot see it. Although you perceive only a uniform gray, experiments show that the grating is actually encoded inside your brain: distinct groups of visual neurons fire for different orientations of the grating.[67] Why can't this pattern of neuronal activity be brought to consciousness? Probably because it makes use of an extremely tangled spatiotemporal pattern of firing in the primary visual area, a neural cipher too complex to be explicitly recognized by global workspace neurons higher up in the cortex. Although we do not yet fully understand the neural code, we believe that, in order to become conscious, a piece of information first has to be re-encoded in an explicit form by a compact assembly of neurons. The anterior regions of the visual cortex must dedicate specific neurons to meaningful visual inputs, before their own activity can be amplified and cause a global workspace ignition that brings the information into awareness. If the information remains diluted in the firing of myriad unrelated neurons, then it cannot be made conscious.

Any face that we see, any word that we hear, begins in this unconscious manner, as an absurdly contorted spatiotemporal train of spikes in millions of neurons, each sensing only a minuscule part of the overall scene. Each of these input patterns contains virtually infinite amounts of information about the speaker, message, emotion, room size . . . if only we could decode it—but we can't. We become aware of this latent information only once our higher-level brain areas categorize it into meaningful bins. Making the message explicit is an essential role of the hierarchical pyramid of sensory neurons that successively extract increasingly abstract features of our sensations. Sensory training makes us aware of faint sights or sounds because, at all levels, neurons reorient their properties to amplify these sensory messages.[68] Prior to learning, a neuronal message was already present in our sensory areas, but only implicitly, in the form of a diluted firing pattern inaccessible to our awareness.

This fact has a fascinating consequence: the brain contains signals that even its owner ignores—for instance, about flashed visual gratings and faint intentions.[69] Brain imaging is beginning to decode these cryptic forms. A program by the U.S. military involves flashing satellite photos at the amazing rate of ten per second to a trained observer, and monitoring his brain potentials for any unconscious hunch that an enemy plane is present. Within our unconscious lies an unimaginable richness waiting to be tapped. In the future, by amplifying those faint micropatterns that our senses detect but that our consciousness overlooks, computer-assisted brain decoding may grant us a rigorous form of extrasensory perception—a heightened sense of our surroundings.

Finally, a fifth category of unconscious knowledge lies dormant in our nervous system, in the form of latent connections. According to workspace theory, we become aware of neuronal firing patterns only if they form active brain-scale assemblies. Inordinately larger amounts of information, however, are stored in our quiescent synaptic connections. Even prior to birth, our neurons sample the statistics of the world and adapt their connections accordingly. Cortical synapses, numbering in the hundred thousand billions in the human brain, contain dormant memories of our entire life. Millions of synapses are formed or destroyed every day, particularly during the first few years of our lives, when our brain adapts the most to its environment. Each synapse stores a minuscule bit of statistical wisdom: How likely is my presynaptic neuron to fire just before my postsynaptic one?

Everywhere in the brain, such connection strengths lie at the foundation of our learned unconscious intuitions. In early vision, cortical connections compile statistics of how adjacent lines connect to form the contours of objects.[70] In auditory and motor areas, they store our covert knowledge of sound patterns. There, years of piano practice induce a detectable change in gray matter density, presumably due to changes in synaptic densities, dendritic sizes, white matter structure, and the supporting glial cells.[71] And in the hippocampus (a curly structure underneath the temporal lobes), synapses gather our episodic memories: where, when, and with whom an event happened.

Our memories may lie dormant for years, their content compressed into a distribution of synaptic spines. We cannot tap this synaptic wisdom directly, because its format is quite different from the pattern of neuronal firing that supports conscious thoughts. To retrieve our memories, we need to convert them from dormant to active. During memory retrieval, our synapses promote the reenactment of a precise pattern of neuronal firing—and only then do we consciously recall. A conscious memory is just an old conscious moment, the approximate reconstruction of a precise pattern of activation that once existed. Brain imaging shows that memories have to be transformed into explicit neuronal

activity patterns that invade the prefrontal cortex and the interconnected cingulate regions before we regain consciousness of a specific episode of our lives.[72] Such reactivation of distant cortical areas during
conscious recall fits perfectly with our workspace theory.

The distinction between latent connections and active firing explains
why we remain utterly unaware of the grammatical rules by which we
process speech. In the sentence "John believes that he is clever," can the
pronoun *he* refer to John himself? Yes. What about "He believes that John
is clever"? No. And "The speed with which he solved the problem pleased
John"? Yes. We know the answers, but we have no idea of the rules by
which we get them. Our language networks are wired to process words
and phrases, but this wiring diagram is permanently inaccessible to our
awareness. Global workspace theory can explain why: the knowledge is
in the wrong format for conscious access.

Grammar contrasts dramatically with arithmetic. When we multiply
24 by 31, we are supremely conscious. Each intermediate operation, its
nature and order, and even the occasional errors that we make are accessible to our introspection. When we process speech, by contrast, we remain paradoxically speechless about our internal processes. The problems
cracked by our syntax processor are just as hard as arithmetic, but we are
clueless as to how we solve them. Why this difference? Complex arith-
metical computations are performed step by step, under the direct control of key nodes of the workspace network (prefrontal, cingulate, and
parietal areas). Such complex sequences are explicitly coded in the firing
of prefrontal neurons. Individual cells encode our intentions, our plans,
individual steps, their number, and even our errors and their correc-
tions.[73] Thus, for arithmetic, both the plan and how it unfolds are explicitly coded in neural firing, within the neuronal network that supports
consciousness. Grammar, by contrast, is implemented by bundles of
connections linking the left superior temporal lobe and inferior frontal
gyrus, and it spares the networks for conscious effortful processing in
dorsolateral prefrontal cortex.[74] During anesthesia, a large part of the
temporal language cortex continues to process speech in an autonomous
manner, without awareness.[75] We do not know how neurons encode
grammatical rules—but once we do, I predict that their coding scheme
will differ radically from that of mental arithmetic.

Subjective States of Matter

In summary, the theory of a global neuronal workspace makes sense of a large number of observations about consciousness and its brain mechanisms. It explains why we become aware of only a scrawny portion of the knowledge stored in our brains. To be consciously accessible, information must be encoded as an organized pattern of neuronal activity in higher cortical regions, and this pattern must, in turn, ignite an inner circle of tightly interconnected areas forming a global workspace. The characteristics of this long-distance ignition account for the signatures of consciousness identified in brain-imaging experiments.

Although my lab's computer simulations reproduce some features of conscious access, they are a long way from mimicking the actual brain—the simulation is far from being conscious. In principle, however, I do not doubt that a computer program could capture the details of a conscious state. A more appropriate simulation would have billions of differentiated neuronal states. Instead of merely propagating activation around, it would perform useful statistical inferences on its inputs—for instance, by computing the likelihood that a specific face is present or the probability that a motor gesture will successfully reach its target.

We begin to envision how networks of neurons can be wired to perform such statistical computations.[76] Elementary perceptual decisions arise through the accumulation of the noisy evidence provided by specialized neurons.[77] During conscious ignition, a subset of them collapses into a unified interpretation, leading to an internal decision about what to do next. Picture a large internal arena where multiple brain regions, like the daemons in Selfridge's pandemonium, struggle for coherence. Their operating rules make them constantly search for a single coherent interpretation of the diverse messages that they receive. Through long-distance connections, they confront their piecemeal knowledge and accumulate evidence, this time at a global level, until a coherent answer is reached that satisfies the organism's current goals.

The entire machine is only partially affected by external inputs. Autonomy is its motto. It generates its own goals, thanks to spontaneous activity, and these patterns in turn shape the rest of the brain's activity in a top-down manner. They induce other areas to retrieve long-term memories, generate a mental image, and transform it according to linguistic or logical rules. A constant flux of neuronal activation circulates within the internal workspace, carefully sifting through millions of parallel processors. Each coherent result moves us one step forward in a mental algorithm that never stops—the flux of conscious thought.

Simulating such a massively parallel statistical machine, based on realistic neuronal principles, would be fascinating. In Europe research forces are gathering for the Human Brain Project, an epic attempt at understanding and simulating human-size cortical networks. Simulations of networks comprising millions of neurons and billions of synapses are already within reach, based on dedicated "neuromorphic" silicon chips.[78] In the next decade, these computational tools will paint a much more detailed picture of how brain states cause our conscious experience.

6

THE ULTIMATE TEST

Any theory of consciousness must face the ultimate test: the clinic. Every year thousands of patients fall into a coma. Many will remain permanently unresponsive, in a dreaded condition called the "vegetative state." Can our budding science of consciousness help them? The answer is a tentative yes. The dream of a "consciousness-o-meter" is within reach. Sophisticated mathematical analysis of brain signals is beginning to reliably sort out which patients retain a conscious life and which do not. Clinical interventions are also in sight. Stimulation of the brain's deep nuclei may speed up the recovery of consciousness. Brain-computer interfaces may even restore a form of communication to locked-in patients who are conscious but fully paralyzed. Future neurotechnologies will forever change the clinical handling of diseases of consciousness.

> How frozen and how faint I then became,
> Ask me not, reader! for I write it not,
> Since words would fail to tell thee of my state.
> I was not dead nor living.
>
> —Dante Alighieri, *The Divine Comedy* (ca. 1307–21)

Every year a tremendous number of car crashes, strokes, failed suicides, carbon monoxide poisonings, and drowning accidents leave adults and children terribly crippled. Comatose and quadriplegic, unable to move and speak, they seem to have lost the very spark of mental life. And yet deep inside, consciousness may still linger. In *The Count of Monte Cristo* (1844), Alexandre Dumas painted a dramatic picture of how an intact consciousness may be buried alive inside the tomb of a paralyzed body:

> Monsieur Noirtier, although almost as immovable as a corpse, looked at the newcomers with a quick and intelligent expression. . . . Sight and hearing were the only senses remaining, and they, like two solitary sparks, remained to animate the miserable body which seemed fit for nothing but the grave; it was only, however, by means of one of these senses that he could reveal the thoughts and feelings that still occupied his mind, and the look by which he gave expression to his inner life was like the distant gleam of a candle which a traveler sees by night across some desert place, and knows that a living being dwells beyond the silence and obscurity.

Monsieur Noirtier is a fictional character—probably the first literary description of a locked-in syndrome. His medical condition, however, is all too real. Jean-Dominique Bauby, the editor of the French fashion magazine *Elle*, was only forty-three when his life took a sudden turn. "Until then," he writes, "I had never even heard of the brain stem. I've learned since that it is an essential component of our internal computer, the inseparable link between the brain and the spinal cord. I was brutally introduced to this vital piece of anatomy when a cerebrovascular accident took my brain stem out of action."

On December 8, 1995, a stroke plunged Bauby into a twenty-day coma. He awakened to find himself in a hospital ward, fully paralyzed except for one eye and part of his head. He survived for fifteen months, enough time to conceive, memorize, dictate, and publish an entire book. A moving testimony of the inner life of a patient with locked-in syndrome, *The Diving Bell and the Butterfly* (1997) instantly became a best seller. Imprisoned in a body that would not move, like a modern Noirtier, Jean-Dominique Bauby dictated his book one character at a time by blinking his left eyelid while an assistant recited the letters E, S, A, R, I, N, T, U, L, O, M. . . . Two hundred thousand blinks tell the story of a beautiful mind shattered by a cerebral stroke. Pneumonia took his life a mere three days after the book was published.

In a sober, though at times humorous, manner, the ex-editor of *Elle* magazine describes his daily ordeal, infused with frustration, isolation, incommunicability, and occasional despair. Although he was imprisoned in a motionless body, which he aptly likens to a diving bell, his concise and elegant prose springs as lightly as a butterfly—his metaphor for the fully intact meanderings of his mind. There is no better proof of the autonomy of consciousness than Jean-Dominique Bauby's vivid imagination and alert writing. Clearly, a full repertoire of mental states, from vision to touch, from savory smell to drowning emotion, can flow as freely as ever, even from the jail of a forever-locked body.

In many patients similar to Bauby, however, the presence of a rich mental life goes undetected.[1] According to a recent survey by the French Association of Locked-In Syndrome (founded by Bauby and run by patients themselves, using state-of-the-art computer interfaces), the person who first detects the patient's consciousness is usually not the physician. More than half the time, it is a family member.[2] Worse, following brain injury, an average duration of 2.5 months elapses before the correct diagnosis is established. Some patients are not diagnosed until four years later. Because their paralyzed body occasionally spurts out involuntary twitches and stereotyped reflexes, their voluntary eye movements and blinks, if noticed at all, are often dismissed as reflexive. Even in the best hospitals, about 40 percent of the patients who are initially classified as utterly unresponsive and "vegetative" turn out, upon closer examination, to present signs of minimal consciousness.[3]

Patients who are unable to express their consciousness present an urgent challenge to neuroscience. A good theory of consciousness should explain why some patients lose that ability while others do not. Above all, it should provide concrete help. If the signatures of consciousness are detectable, they should be applied to those who need it the most: crippled patients for whom the detection of a sign of consciousness is, literally, a matter of life or death. In intensive care units all over the world, half of the deaths result from a clinical decision to withdraw life support.[4] One is left wondering how many Noirtiers and Baubys died just because medicine lacked the means to detect their residual consciousness or to foresee that they would ultimately emerge from coma and regain a valuable mental life.

Today, however, the future looks resolutely brighter. Neurologists and brain-imaging scientists are making significant progress in identifying conscious states. The field is now moving toward simpler and cheaper methods to detect consciousness and restore communication with aware patients. In this chapter, we will look at this exciting new frontier of science, medicine, and technology.

How to Lose Your Mind

Let us start by sorting out the different kinds of neurological disorders of consciousness or communication with the external world (figure 29).[5] We may begin with the familiar term *coma* (from the Ancient Greek κωμα, "deep sleep"), since most patients start in that state. Coma typically occurs within minutes to hours following damage to the brain. Its causes are diverse and include head trauma (typically from a car

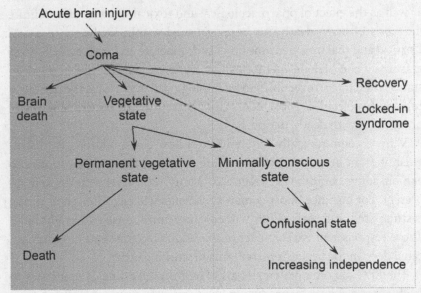

FIGURE 29. Brain injury may cause a variety of disorders of consciousness and communication. In this illustration, the main categories of patients are ordered from left to right in rough correspondence with the presence of consciousness and its stability during the day. Arrows indicate how a patient's condition may evolve over time. A minimal contrast separates vegetative-state patients, who show no clinical signs of consciousness, from minimally conscious patients, who may still perform some voluntary acts.

accident), stroke (the rupture or plugging of a brain artery), anoxia (the loss of oxygen supply to the brain, typically due to cardiac arrest, carbon monoxide poisoning, or drowning), and poisoning (sometimes caused by binge drinking). Coma is defined clinically as a prolonged loss of the capacity to be roused. The patient lies unresponsive, with his eyes closed. No amount of stimulation can awaken him, and he shows no signs of awareness of himself or his environment. For the term *coma* to apply, clinicians further require that this state last for an hour or more (thus distinguishing it from a transient syncope, concussion, or stupor).

Coma patients are not brain-dead, however. *Brain death* is a distinct state, characterized by a total absence of brain stem reflexes, together with a flat EEG and an inability to initiate breathing. In brain-dead patients, positron emission tomography (PET) and other measures such as Doppler ultrasonography show that cortical metabolism and the perfusion of blood to the brain are annihilated. Once hypothermia is excluded, as well as the effect of pharmacological and toxical substances, a definite diagnosis of brain death can be established within six hours to a day. Cortical and thalamic neurons quickly degenerate and melt away, forever erasing all the lifelong memories that define a person. The brain-dead state is therefore irreversible: no technology will ever revive the dissolved cells and molecules. Most countries, as well as the Vatican,[6] identify brain death with death, period.

Why is coma radically different? And how can a neurologist distinguish it from brain death? First of all, in coma, the body continues to exhibit some coordinated reactions. Many high-level reflexes remain present. For instance, most comatose patients will gag when their throat is stimulated, and their pupils will contract in response to a bright light. Those responses prove that part of the brain's unconscious circuitry, located deep in the brain stem, remains in working order.

The EEG of coma patients is also far from a flat line. It continues to fluctuate at a slow rate, producing low-frequency waves somewhat similar to those seen during sleep or anesthesia. Many cortical and thalamic cells are still alive and active but in an inappropriate network state. Some rare cases even show high-frequency theta and alpha rhythms ("alpha coma") but with an unusual regularity, as if large chunks of the brain, instead of showing the desynchronized rhythms that characterize a well-functioning thalamic-cortical network, were invaded by exceedingly synchronous waves.[7] My colleague the neurologist Andreas Klein-schmidt likens the alpha rhythm to the "brain's windshield wiper"—and even in the normal conscious brain, alpha waves are used to shut off specific regions, such as the visual areas when we concentrate on a sound.[8] During some comas, much as in anesthesia with propofol (the sedative that killed Michael Jackson),[9] a giant alpha rhythm seems to invade the cortex and wipe out the very possibility of a conscious state. Yet because the cells are still active, their normal coding rhythms may one day return.

Comatose patients thus possess a demonstrably active brain. Their cortex generates a fluctuating EEG but lacks the ability to emerge from "deep sleep" and elicit a conscious state. Fortunately, coma rarely lasts long. Within days or weeks, if medical complications such as infection are prevented, the vast majority of patients gradually recover. The first sign is usually the return of the sleep-wake cycle. Most coma patients then regain consciousness, communication, and intentional behavior.

In unfortunate cases, however, recovery stops in a very strange state of arousal without awareness.[10] Every day the patient awakens—but during these waking moments, he remains unresponsive and seemingly unaware of his surroundings, somehow lost in Dante's infernal limbo, "not dead nor living." A preserved sleep-wake cycle with no signs of consciousness is the hallmark of the vegetative state, also known as "unresponsive wakefulness," a condition that may persist for many years. The patient breathes spontaneously and, when fed artificially, does not die. American readers may remember Terri Schiavo, who spent fifteen years in a vegetative state while her family, the state of Florida, and even President George W. Bush fought legal battles; she was finally left to die in March 2005, when her feeding tube was ordered disconnected.

Exactly what does *vegetative* mean? The term is somewhat unfortunate, as it brings to mind an impotent "vegetable"—and sadly, in poorly tended wards, this nickname sticks. The neurologists Jennett and Plum coined the adjective from the verb *vegetate,* which according to the Oxford English Dictionary means to "live a merely physical life devoid of intellectual activity or social intercourse."[11] Functions that depend on the autonomous nervous system, such as the regulation of cardiac frequency, vascular tone, and body temperature, are generally intact. The patient is not immobile and will occasionally make slow and spectacular movements with the body or the eyes. A smile, cry, or frown may suddenly illuminate the patient's face with no obvious cause. Such behavior can create considerable confusion in the family. (In Terri Schiavo's case, it persuaded her parents that she could still be helped.) But neurologists know that such bodily responses may arise reflexively. The spinal cord and the brain stem often generate purely involuntary movements, undirected toward a specific goal. Crucially, the patient never responds to verbal orders, nor does she say a word, though she may emit random grunts.

Once a month has elapsed since the initial damage, doctors speak of a "persistent vegetative state," and after three to twelve months, depending on whether the brain damage is due to anoxia or to a cranial trauma, the diagnosis of "permanent vegetative state" is posed. Those terms are debated, however, because they imply a lack of recovery, suggest a fixed condition of unconsciousness, and may therefore lead to a premature decision to discontinue life support. Several clinicians and researchers favor the neutral expression "unresponsive wakefulness," a factual phrase that leaves open the exact nature of the patient's present and future state. The truth of the matter, as we shall shortly see, is that the vegetative state is a mixed bag of poorly understood conditions that even include rare cases of conscious but noncommunicating patients.

In some patients with severe brain damage, consciousness may fluctuate broadly, even within the space of a few hours. During some periods, they regain a degree of voluntary control over their actions, which justifies placing them in a distinct category: the "minimally conscious state." In 2005 a working group of neurologists introduced this term to refer to patients with rare, inconsistent, and limited responses that suggest residual comprehension and volition.[12] Minimally conscious patients may respond to a verbal order by blinking or may follow a mirror with their eyes. Some form of communication can usually be established: many patients can answer yes or no by speaking the words aloud or just by nodding. Unlike a vegetative patient, who smiles or cries at random times, a minimally conscious patient may also express emotions that are appropriately linked to the current context.

A single cue does not suffice to make a sure diagnosis; signs of consciousness have to be observed with a certain consistency. And yet paradoxically, minimally conscious patients are in a state that may prevent them from expressing their thoughts in a consistent manner. Their behavior can be highly variable. On some days, no consistent signs of consciousness are observed, or the signs may be seen in the morning but not in the afternoon. Furthermore, the observer's assessment of whether a patient laughed or cried at the right moment may be highly subjective. In order to improve the reliability of the diagnosis, the neuropsychologist Joseph Giacino created the Coma Recovery Scale, a series of objective bedside tests that are applied in a precisely controlled manner.[13] The probes assess simple functions, such as the capacity to recognize and manipulate objects, to orient gaze spontaneously or in response to verbal commands, and to react to an unexpected noise. The medical team is trained to query the patient in a persistent manner and to carefully watch for any behavioral response, even if it is extremely slow or barely appropriate. The tests are generally administered repeatedly, at different times of the day.

Using this scale, the medical team can distinguish between a vegetative patient and a minimally conscious one with much greater accuracy.[14] This information is crucial, of course, not only for making any end-of-life decision but also for anticipating the possibility of recovery. Statistically speaking, patients who are diagnosed as minimally conscious have a better chance of regaining stable consciousness than do those who stay in a vegetative state for years (although the fate of any single person remains very difficult to predict). Recovery is often painfully slow: week after week the patient's responses become increasingly consistent and reliable. In a few dramatic cases, a sudden awakening occurs over the course of just a few days. Once they regain a stable capacity to communicate with others, patients are no longer considered minimally conscious.

What is it like to be in a minimally conscious state? Do these patients live a fairly normal inner life, ripe with past memories, future hopes, and perhaps most important, a rich consciousness of the present, possibly full of suffering and despair? Or are they mostly in a haze and unable to muster enough energy to blurt out a detectable response? We do not know, but huge fluctuations in responsiveness suggest that the latter may be closer to the truth. Perhaps an appropriate analogy is the confused, sluggish state of mind that we all experience after being knocked out, anesthetized, or severely inebriated.

In this respect, minimal consciousness is probably very different from the last medical condition on our list: the "locked-in syndrome" that Jean-Dominique Bauby experienced. The locked-in state typically results from a well-delimited lesion, usually on the protuberance of the brain stem. With excruciating accuracy, such a lesion disconnects the cortex from its output pathways in the spinal cord. By sparing the cortex and the thalamus, it often leaves consciousness entirely intact. The patient awakens from coma only to find himself imprisoned in a paralyzed body, unable to move or talk. His eyes are still. Only small vertical eye movements and blinks, generated by distinct neuronal pathways, are generally spared and open up a channel of communication with the outside world.

In *Thérèse Raquin* (1867), the French naturalistic novelist Émile Zola vividly captured the mental life of Madame Raquin, a locked-in and quadriplegic old lady. Zola carefully noted that the eyes remained the only window into the poor soul's mind:

> This face looked like that of a dead person in the centre of which two living eyes had been fixed. These eyes alone moved, rolling rapidly in their orbits. The cheeks and mouth maintained such appalling immobility that they seemed as though petrified. . . . Each day the sweetness and brightness of her eyes became more penetrating. She had reached the point of making them perform the duties of a hand or mouth, in asking for what she required and in expressing her thanks. In this way she replaced the organs that were wanting, in a most peculiar and charming manner. Her eyes, in the centre of her flabby and grimacing face, were of celestial beauty.

In spite of their communication impairment, locked-in patients may keep a crystal-clear mind, vividly aware not only of their deficit but also of their own mental abilities and the care they receive. Once their condition is detected and their pain alleviated, they may live a fulfilling life. Proof that an intact cortex and thalamus suffice to generate autonomous mental states, locked-in brains continue to experience the full gamut of life's experiences. In Zola's novel, Madame Raquin savors sweet revenge as her niece and her lover, whom she hates for killing her son, commit a double suicide before her ever-watchful eyes. In Dumas's *Count of Monte Cristo*, a paralyzed Noirtier manages to warn his granddaughter that she is about to wed the son of a man he killed many years earlier.

The lives of actual locked-in patients are less eventful, perhaps, but no less extraordinary. With the help of computerized eye-tracking devices, some locked-in patients manage to answer their e-mails, head a non-profit organization, or, like the French executive Philippe Vigand, write two books and father a child. Unlike comatose, vegetative, and minimally conscious patients, they cannot be considered as suffering from a disorder of consciousness. Even their spirits can be high: a recent survey of their subjective quality of life revealed that the vast majority of them, once they moved past the first few horrific months, gave happiness ratings that matched the average of the normal, unimpaired population.[15]

Cortico Ergo Sum

In 2006 the subdivision of noncommunicative patients into coma, vegetative, minimally conscious, and locked-in states seemed well established when a shocking report, published in the prestigious journal *Science*, suddenly shattered the clinical consensus. The British neuroscientist Adrian Owen described a patient who showed all the clinical signs of a vegetative state but whose brain activity suggested a considerable degree of consciousness.[16] Horrifyingly, the report implied the existence of patients in a state worse than the usual locked-in syndrome: conscious but without *any* means of expressing it to the outside world, not even through the batting of an eyelid. While demolishing established clinical rules, this research also carried a message of hope: brain imaging was now sensitive enough to detect the presence of a conscious mind and even, as we shall see, to reconnect it with the outside world.

The patient that Adrian Owen and his colleagues studied in their *Science* paper was a twenty-three-year-old woman who had been involved in a traffic accident and suffered from bilateral damage to the frontal lobes. Five months later, in spite of a preserved sleep-wake cycle, she remained fully unresponsive—the very definition of the vegetative state: Even an experienced team of clinicians could detect no signs of residual awareness, communication, or voluntary control.

The surprise came from visualizing her brain activity. As part of a research protocol for monitoring the state of the cortex in vegetative patients, she underwent a series of fMRI exams. When she listened to sentences, the researchers were astounded to observe that her cortical language network was fully active. Both superior and middle temporal gyri, which house the circuits for hearing and speech comprehension, fired quite strongly. There was even a strong activation in the left inferior frontal cortex (Broca's area) when the sentences were made more difficult by including ambiguous words (e.g., "the *creak* came from a *beam* in the ceiling").

Such elevated cortical activity suggested that her speech processing included stages of word analysis and sentence integration. But did she really understand what was being said to her? By itself, the activation of the language network did not provide conclusive evidence of awareness; several prior studies had shown that that network could be largely preserved during sleep or anesthesia.[17] To figure out whether the patient understood anything, Owen therefore ran a second series of scans, in which the spoken sentences that were played to her conveyed complex instructions. She was told to "imagine playing tennis," "imagine visiting the rooms in your home," and "just relax." The instructions asked her to start and stop these activities at precise times. Thirty-second bouts of vivid imagination, cued by the spoken word "tennis" or "navigation," alternated with thirty seconds of rest, cued by the spoken word "relax."

Outside the scanner, Owen had no way of knowing whether the mute and motionless patient understood these commands, let alone whether she followed them. However, fMRI readily provided the answer: her brain activity closely tracked the spoken instructions. When she was asked to imagine playing tennis, the supplementary motor area went on and off every thirty seconds, exactly as requested. And when she mentally visited her apartment, a distinct brain network lit up, involving areas engaged in the representation of space: the parahippocampal gyrus, the posterior parietal lobe, and the premotor cortex. Amazingly, she activated the very same brain regions as healthy control subjects performing the same mental imagery tasks.

So was she conscious? A few scientists played devil's advocate.[18] Perhaps, they argued, it was possible to activate these areas in a totally unconscious manner, without the patient consciously understanding the instructions. Simply hearing the noun *tennis* might suffice to activate motor areas, just because action is an integral part of this word's meaning. Likewise, perhaps hearing the word *navigation* sufficed to trigger a sense of space. Conceivably, then, brain activation might occur automatically, without the presence of a conscious mind. More philosophically, could any brain image ever prove or disprove the existence of a mind? Commenting negatively upon this issue, the American neurologist Allan Ropper expressed his pessimistic conclusion with a clever witticism: "Physicians and society are not ready for 'I have brain activation, therefore I am.' That would seriously put Descartes before the horse."[19]

Pun aside, this conclusion is wrong. Brain imaging has truly come of age, and even a problem as complex as the identification of residual consciousness from purely objective images of the brain is now on the brink of being solved. Critiques, even logically sound ones, were shattered when Owen performed an elegant control experiment. He scanned normal volunteers while they merely listened to the words *tennis* and *navigation;* they had not received any instructions as to what they should do when they heard them.[20] Unsurprisingly perhaps, the activations evoked by those two words were not detectably different from each other. In these passive listeners, the landscape of brain activity differed from the network that activated when Owen's patient or the controls received the imagination instructions. This finding clearly refuted the devil's advocates. When activating her premotor, parietal, and hippocampal areas in a task-relevant manner, Owen's patient did much more than react unconsciously to a single word—she appeared to be *thinking* about the task.

As Owen and his colleagues pointed out, hearing a single word seemed unlikely to trigger brain activity for a full thirty seconds, unless the patient were somehow using the word as a cue to perform the requested mental task. From the theoretical perspective of my global neuronal workspace model, if the word had triggered only an unconscious activation, we would have expected it to quickly dissipate and return to baseline after a few seconds at most. On the contrary, the observation of a sustained activation of specific prefrontal and parietal regions for thirty seconds almost surely reflected the presence of conscious thoughts in working memory. Although Owen and his colleagues could be criticized for selecting a rather arbitrary task, their choice was intelligent and pragmatic: the imagination task was easy for the patient to perform, yet it was hard to see how the brain activity it evoked could occur without consciousness.

Freeing the Inner Butterfly

If any doubts lingered that vegetative patients could be conscious, a second paper, published in the high-profile *New England Journal of Medicine*, fully dispelled them.[21] It provided proof that brain imaging could open up a communication channel with a vegetative patient. The experiment was surprisingly simple. First, the researchers replicated Owen's imagination study. Out of fifty-four patients with disorders of consciousness, five showed distinctive brain activity when asked to imagine a game of tennis or a visit to their home. Four of them were vegetative. One of them was then invited to a second MRI session. Prior to each scan, he was asked a personal question such as "Do you have any brothers?" He could not move or speak—but Martin Monti and his collaborators asked him for a purely mental answer. "If you want to respond 'yes,'" they said, "please imagine playing tennis in your head. If you want to respond 'no,' please imagine visiting your apartment instead. Start when you hear the word 'answer,' and stop when you hear the word 'relax.'"

This clever strategy worked remarkably well (figure 30). For five out of six questions, one of the two brain networks previously identified showed a significant activation. (For the sixth question, neither was activated, so no response was scored.) The researchers were unaware of the correct answers—but when they compared the brain activity that they had detected with the ground truth provided by the patient's family, they were pleased to see that all five were correct.

Let us pause for a moment to digest the implications of these amazing findings. In the patient's brain, a long chain of mental processes must have been intact. First, the patient understood the question, retrieved the correct answer, and held it in mind for several minutes prior to the scan. This implied intact language comprehension, long-term memory, and working memory. Second, he willfully followed the experimenter's instructions, which arbitrarily mapped the yes answer to tennis playing and the no answer to mental navigation. Thus the patient could still flexibly route information through an arbitrary set of brain modules—a finding that, in and of itself, suggests that his global neuronal workspace was intact. Finally, the patient applied the instructions at the appropriate time and readily changed his response across the five successive scans.

Patient in an apparent vegetative state

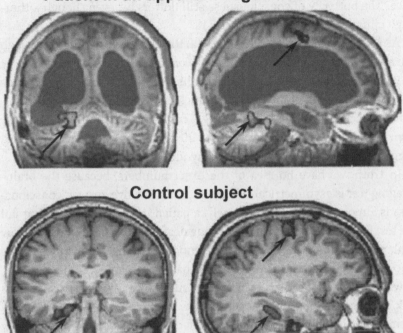

Control subject

FIGURE 30. Some patients in an apparent vegetative state show virtually normal brain activity during complex mental tasks, suggesting that they are in fact conscious. The patient in the top image could no longer move or speak, but he correctly responded to verbal questions by activating his brain. To respond no, he was asked to imagine visiting his apartment, and to respond yes, to imagine playing tennis. When asked whether his father's name was Thomas, his brain regions for spatial navigation lit up exactly as in a normal subject, thus giving the correct response: no. Because the patient showed absolutely no signs of overt communication or consciousness, he was considered in a vegetative state. The patients' massive lesions are clearly visible.

This capacity for executive attention and task switching hints at a preserved central executive system. Although the evidence remains scarce, and a demanding statistician would wish that this patient had answered twenty questions instead of five, it is hard to escape the conclusion that he still possessed a conscious, willful mind.

This conclusion shatters established clinical categories and forces us to confront a hard reality: some patients are vegetative only in appearance. The butterfly of consciousness still flutters by, even though a thorough clinical examination may miss it.

As soon as Owen's research came out, the news quickly spread through the media. Unfortunately, the findings were often misinterpreted. One of the silliest conclusions that some journalists drew was that "coma patients are conscious." Not at all! The study included only vegetative and minimally conscious state cases, not a single comatose patient. Even then only a small fraction, on the order of 10 or 20 percent, responded to the test, suggesting that this "super-locked-in" syndrome is relatively rare.

In truth, we have no idea of the exact numbers, because the brain-imaging test is asymmetrical. When it gives a positive answer, consciousness is almost a certainty; conversely, a patient may be conscious but fail the test for all sorts of reasons, including deafness, language disorders, low vigilance, or an inability to sustain attention. Strikingly, the only patients who responded were survivors of a traumatic brain injury. Other patients, in whom the loss of consciousness was caused by a massive stroke or by a lack of oxygen, showed no capacity to perform the task, perhaps because their brain, like Terri Schiavo's, had suffered from diffuse and essentially irreversible damage to cortical neurons. The "miracle" of finding intact consciousness inside a vegetative patient concerned only a small subset of cases, and to use it as a pro-life argument for providing unlimited medical support to all coma patients would be utterly irrational.

Even more surprising perhaps is that thirty out of thirty-one *minimally conscious* patients failed the test. On bedside testing, all these patients occasionally manifested signs of preserved volition and awareness. But through a terrible irony, all but one missed their chance to definitely prove it during the brain-imaging test. Who knows why? Perhaps the test came at a time when their vigilance was low. Perhaps they were unable to concentrate in the strange and noisy environment of the MRI machine. Or perhaps their cognitive functions were too weak to perform this complex task. At the very least, two conclusions ensue: first, the clinical diagnosis of "minimal consciousness" certainly does not imply that these patients possess a fully normal conscious mind; and second, Owen's imagination test probably underestimates consciousness by a wide margin.

Because of such issues, no single test will ever prove, once and for all, whether consciousness is present. The ethical approach would be to develop a whole battery of them and see which, if any, managed to establish communication with a patient's inner butterfly. In an ideal world, these tests should be much simpler than having to imagine a tennis game. Furthermore, they should be repeated on multiple days, so as not to miss a locked-in patient whose consciousness fluctuates over time. Unfortunately, fMRI is a terrible tool for this purpose, because the equipment is so complex and expensive that patients typically undergo only one or two scans. As Adrian Owen himself noted, "It's hard to open up a channel of communication with a patient and then not be able to follow up immediately with a tool for them and their families to be able to do this routinely."[22] Even Owen's second patient, who gave such clear signs of willful responding, could be tested only once before being sent back to the jail of his locked-in state.

Realizing how urgent it is to go beyond this frustrating state of affairs, several research teams are now developing brain-computer interfaces that are based on the much simpler technology of electroencephalography—a cheap technique, routinely available in clinics, that requires only the amplification of electrical signals from the head surface.[23]

Unfortunately, playing tennis and navigating one's apartment are rather difficult to track with EEG. In one study, the researchers therefore relied on a much simpler instruction to patients: "Every time you hear a beep, try to imagine that you are squeezing your right-hand into a fist and then relaxing it. Concentrate on the way your muscles would feel if you were really performing this movement."[24] On another trial, the patients had to imagine wiggling their toes. While the patients mentally performed these actions, the researchers looked for distinct patterns of oscillatory EEG activity over the motor cortex. For each patient, a computerized machine-learning algorithm attempted to sort the signals into fist versus toes trials. And in three out of sixteen vegetative patients, it seemed to work—but the technique remains too unreliable to fully exclude the possibility of a chance finding.[25] (Even in healthy, conscious participants, it worked on only nine out of twelve occasions.) Another team, led by Nicholas Schiff in New York, conducted a test in which five healthy volunteers and three patients had to imagine either swimming or visiting their apartment.[26] Again, although the test seemed to give reliable results, the numbers were too small to be conclusive.

In spite of its current shortcomings, such EEG-based communication represents the most practical path for future research.[27] Many engineers are strongly attracted to the challenge of wiring a computer to the brain, and they are developing increasingly sophisticated systems. While most are still based on gaze and visual attention, which is awkward for many patients, progress is also being made in decoding auditory attention and motor imagery. The game industry is joining in with lighter, wireless recording devices. Electrodes may even be surgically implanted directly on the cortex of paralyzed patients. Using such a device, a quadriplegic patient managed to mentally control a robotic arm.[28] Perhaps if the device was placed over language areas, a speech synthesizer might one day be able to turn the patient's intended speech into actual words.[29]

Broad research avenues have opened. Not only will they lead to better communication devices for locked-in patients, but they will also provide new means for detecting residual consciousness. In advanced clinical research centers such as the Coma Science Group, led by Steven Laureys in Liège, Belgium, brain-computer interfaces are already included in the battery of tests that are systematically deployed whenever a vegetative patient is admitted. I surmise that, twenty years from now, it will be perfectly banal to see quadriplegic and locked-in patients drive their wheelchair by a pure act of will.

Conscious Novelty Detection

Although I admire Adrian Owen's pioneering research, the theorist in me remains frustrated. Passing his test undoubtedly requires a conscious mind—but the assay does not easily relate to any specific theory of consciousness. Since it involves language, memory, and imagination, there are many ways that a patient could fail it and still be conscious. Can we design a much simpler litmus test for consciousness? Thanks to advances in brain imaging, we have now identified many signatures of consciousness. Couldn't we monitor them in order to decide whether or not a patient is conscious? Such a minimal, theory-driven test would also have the advantage of helping with the difficult issue of determining whether young children, premature babies, and even rats and monkeys possess a form of consciousness.

In 2008, over a memorable lunch in Orsay, south of Paris, my colleagues Tristan Bekinschtein, Lionel Naccache, Mariano Sigman, and I asked ourselves this naïve question: If we were to design the simplest possible detector of consciousness, how would we proceed? We quickly decided that it should be based on EEG—the simplest and cheapest brain-imaging technique. We also decided that it should be based on auditory stimuli, because hearing is preserved in most patients, whereas their vision is often impeded. Our decision to use audio raised some issues, because the signatures of consciousness that we had discovered were based primarily on visual experiments. Still, we were confident that the broad principles of conscious access that we had uncovered would generalize to the auditory modality.

We decided to capitalize on the clearest signature that we had recorded in experiment after experiment: the massive P3 wave, which indexes the synchronous ignition of a brain web of cortical regions. Eliciting an auditory P3 wave is remarkably easy. Imagine listening to a quiet symphony concert when suddenly somebody's cell phone rings. This unexpected sound triggers a massive P3 wave, as you reorient your attention and become aware of this odd event.[30]

In our design, we would present a series of regularly repeating sounds: *beep beep beep beep* . . . At an unpredictable moment, an oddball sound would arise: *boop*. When a subject is awake and attentive, this deviant event systematically generates a P3-like event, our proxy for consciousness. To ensure that this brain response was not just due to sound intensity or some other low-level feature, in a separate set of trials we would reverse the items: *boop* would become the standard, and *beep* the deviant. Using this trick, we could prove that the P3 occurred solely because of the improbability of the sound in the current context.

The scenario, however, had a lingering complication. Deviant sounds trigger not only a P3 wave but also a series of earlier brain responses that are known to reflect unconscious processing. As early as 100 milliseconds after the onset of the sound, the auditory cortex is generating a large response to the deviant. This response has been called the "mismatch response" or "mismatch negativity" (MMN for short) because it shows up as a negative voltage at the top of the head.[31] The problem is that this MMN is not a signature of consciousness; it is an automatic response to auditory novelty that occurs whether the person is attending, mind-wandering, reading a book, watching a movie, or even falling asleep or lying in a coma. Effectively, our nervous system contains an unconscious novelty detector. To quickly detect deviant sounds, it unconsciously compares the current stimulus to a prediction based on past sounds. This sort of prediction is ubiquitous: any patch of cortex probably houses a simple network of neurons that predicts and compares.[32] These operations are automatic, and only their outcome grabs our attention and awareness.

What this means is that as a signature of consciousness, the oddball paradigm fails: even a comatose brain may jump to a novel sound. The MMN response shows merely that the auditory cortex is fit enough to detect novelty, not that the patient is conscious.[33] It belongs to the catalog of early sensory operations that are sophisticated yet operate outside awareness. What my team and I needed was to evaluate the subsequent brain events: Would a patient's brain generate the late avalanche of neuronal activity that indexes consciousness?

To create a version of the oddball test that specifically elicits a late and conscious response to novelty, we invented a new trick—pitting local and global novelty against each other. Imagine that you hear a sequence of five tones ending with a different sound: *beep beep beep beep boop*. In response to the final deviant, your brain initially generates both an early MMN and a late P3. Now repeat this sequence several times. Your brain quickly gets used to hearing four *beeps* followed by a *boop*—at a conscious level, the surprise is gone. Remarkably, the final deviant continues to generate an early MMN response. The auditory cortex clearly houses a rather stupid novelty detection device. Instead of noticing the global pattern, it sticks to the shortsighted prediction that beeps are followed by *beeps*—which is of course violated by the final *boop*.

Interestingly, the P3 wave is a much cleverer beast. Once again it closely tracks awareness: as soon as the subject notices the global pattern of five sounds and is no longer surprised by the final change, the P3 vanishes. Once this conscious expectation is set, we can violate it by presenting, on rare occasions, five identical sounds: *beep beep beep beep beep*. Such a rare deviant does indeed evoke a late P3 wave. Note how curious this is—the brain classifies a perfectly monotonic stream of tones as novel. It does so only because it detects that this sequence deviates from the one previously registered in working memory.

Our goal is achieved: we can elicit a pure P3 wave in the absence of earlier unconscious responses. We can even amplify it by asking our subjects to count the deviant sequences. Explicit counting greatly enhances the observed P3 wave, turning it into an easily detectable marker (figure 31). When we see it, we can be pretty sure that the patient is aware and able to follow our instructions.

Empirically, the local-global test works fine. My team and I easily detected the global P3 response in every normal person, even after a very short recording session. Furthermore, it was present only when subjects were attentive and aware of the overall rule.[34] When we distracted them with a difficult visual task, the auditory P3 vanished. When we let them mind-wander, the P3 was present only in those who, at the end of the experiment, were capable of reporting the auditory regularity and its violations. Participants who were oblivious to the rule had no P3.

The network of areas that are activated by global deviants also suggests a conscious ignition. Using EEG, fMRI, and intracranial recordings in epileptic patients, we confirmed that the global workspace network lights up whenever the globally deviant sequence appears. Upon hearing such a deviant sequence, brain activity does not remain confined to the auditory cortex but invades a broad workspace circuit that comprises the

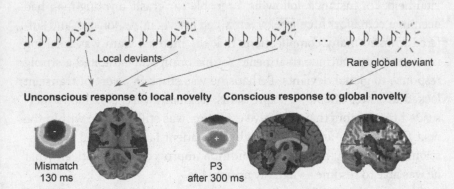

FIGURE 31. The local-global test can detect residual consciousness in injured patients. The test consists of repeating, many times, an identical sequence of five sounds. When the last sound differs from the first four, auditory areas react with a "mismatch response"—an automatic reaction to local novelty that is fully unconscious and that persists even in deep sleep or in coma. Consciously, however, the brain quickly adapts to the repeating melody. After adaptation, it is the absence of the final novelty that now triggers a response to novelty. Crucially, this higher-order response seems to exist only in conscious patients. It presents all the signatures of consciousness, including a P3 wave and a synchronous activation of distributed parietal and prefrontal areas.

bilateral prefrontal cortex, the anterior cingulate, the parietal, and even some occipital areas. This implies that the information about sound novelty is being broadcast globally—a sign that this information is conscious.

Would the test also work in a clinical setting? Would conscious patients react to global auditory novelty? Our initial trial with eight patients was quite successful.[35] In all four of the vegetative patients, the response to global deviants was absent, but in three of the four minimally conscious ones, it was present (and those three patients later regained consciousness).

My colleague Lionel Naccache then began to apply this test routinely in the Salpêtrière Hospital in Paris, with very positive results.[36] Whenever the global response was present, the patient seemed to be conscious. Out of twenty-two vegetative-state patients, only two exceptional subjects ever showed a global P3 wave, and they recovered some degree of minimal consciousness in the next few days, thus suggesting that they might have already been conscious during the test, much like Owen's responsive patients.

In the intensive care unit, our local-global test occasionally provides vital help. For instance, following a terrible car crash, a young man had been in a coma for three weeks, remained utterly unresponsive, and suffered from so many complications that the medical team was debating whether to discontinue treatment. Yet his brain still exhibited a strong response to global deviants. Perhaps he was stuck in a sort of transient locked-in state, unable to express his residual awareness? Lionel persuaded the doctors that a positive evolution was still possible within the next few days . . . and sure enough, the patient later regained full consciousness. In fact, his medical condition improved so dramatically that he was able to resume a virtually normal life.

Global workspace theory helps explain why the test works. In order to detect the repeated sequence, participants must store a sequence of five tones in their memory. Then they must compare it with the next sequence, which arrives more than a second later. As we discussed in Chapter 3, the ability to hold information in mind for a few seconds is a hallmark of the conscious mind. In our test, this function plays out in two different ways: the mind must integrate the individual notes into an overall pattern, and it must compare several such patterns.

Our test also taps a second level of information processing. Think of the operations needed to decide that a perfectly monotonous sequence of beeps is, in fact, novel. Upon hearing the standard sequence *beep beep beep beep boop,* our brain gets used to the final deviant sound. Although that sound still generates a first-order novelty signal in auditory areas, a second-order system manages to predict it.[37] On the rare occasions when the monotonous sequence of five *beeps* is heard instead, this second-order system is surprised. The novelty, indeed, is that there is no final novelty. Our test works because it sidesteps the first-order novelty detector and selectively taps a second-order stage, closely related to global ignition of the prefrontal cortex, and therefore to consciousness.

Pinging the Cortex

My research group and I now have enough success stories to believe that our local-global test indexes consciousness. Nevertheless, the test is still far from perfect. We have had too many false negatives—patients who have recovered from coma and are now clearly conscious, but in whom our test fails. We do get some additional mileage by applying a sophisticated machine-learning algorithm to our data.[38] This Google-like tool allows us to search the brain for *any* response to global novelty, even if it is unusual and unique to a single patient. Still, in about half the patients who are minimally conscious or whose communication abilities have returned, we remain unable to detect any reaction to the rare sequences.

Statisticians describe this as a case of high specificity but poor sensitivity. Simply put, our test, like Owen's, is asymmetrical: if it gives a positive answer, we are almost sure that the patient *is* conscious, but if it gives a negative answer, we cannot use it to conclude that a patient is *not* conscious. There are many possible reasons for this reduced sensitivity. Our EEG recordings could be too noisy; it is notoriously difficult to get a clean signal from a hospital bed, surrounded by piles of electronic equipment, and with a patient who is often unable to stay still or keep his gaze fixed. More likely, some of the patients are conscious but are unable to understand the test. Their lesions are so extensive that they cannot count the deviants or perhaps detect them—or even simply focus their attention on the sounds for more than a few seconds.

Still, these patients have an ongoing mental life. If our theory is correct, this means that their brain remains able to propagate global information over long cortical distances. So how can researchers detect it? In the late 2000s, Marcello Massimini, from the University of Milan, had a clever idea.[39] While my lab's tests of consciousness all involved monitoring the progression of a sensory signal into the brain, Massimini proposed to use an internal stimulus. Let's trigger electrical activity directly into the cortex, he thought. Like the *ping* of a sonar pulse, this intense stimulus would propagate into the cortex and thalamus, and the strength and duration of its echo would indicate the integrity of the areas it traversed. If the activity got broadcast to distant regions, and if it reverberated for a long time, then the patient would probably be conscious. Remarkably, the patient would not even have to attend to the stimulus or understand it. A pulse could probe the state of long-distance cortical highways even if the patient remained unaware of it.

To implement his idea, Massimini relied on a sophisticated combination of two technologies: TMS and EEG. Transcranial magnetic stimulation, as I explained in Chapter 4, uses magnetic induction to stimulate the cortex by discharging current into a coil placed near the head; EEG, as the reader knows by now, is just good old recording of brain waves. Massimini's trick would be to "ping the cortex" using TMS, then use EEG to record the propagation of brain activity elicited by this magnetic pulse. This required special amplifiers that would quickly recover from the intense current delivered by TMS, and paint an accurate picture of the ensuing activity only a few milliseconds later.

Massimini's results to date are exciting. He first applied the technique in normal subjects during wakefulness, sleep, and anesthesia. During the loss of consciousness, the TMS pulse caused only a short and focal activation, which remained confined to the first 200 milliseconds or so. By contrast, whenever the participant was conscious—or even dreaming—the very same pulse caused a complex and long-lasting sequence of brain activity. The precise location of the stimulation site did not seem to matter: wherever the trigger pulse initially entered the cortex, the complexity and duration of the subsequent response provided an excellent index of consciousness.[40] This observation seemed highly compatible with what my team and I had found with sensory stimuli: the diffusion of signals into a brain-scale network, beyond 300 milliseconds, indexes the conscious state.

Crucially, Massimini went on to test his stimulator on five vegetative, five minimally conscious, and two locked-in syndrome patients.[41] Although these numbers are small, the test was 100 percent correct: all the conscious patients showed complex and lasting responses to a cortical impulse. Five additional vegetative patients were then followed up for several months. During this period, three of them moved to the "minimally conscious" category, as they gradually recovered some degree of communication. Those were precisely the three patients in whom the brain signals regained in complexity. And in agreement with the global workspace model, the progression of the signals into the prefrontal and parietal regions was a particularly good index of the patients' level of consciousness.

Detecting Spontaneous Thought

Only the future can tell whether Massimini's pulse test is as good as it seems and will become a standard clinical tool to detect consciousness in individual patients. What is most exciting is that it seems to work in every single case. The technology remains complex, however. Not every hospital has a high-density EEG system capable of absorbing the large shocks generated by a transcranial magnetic stimulator. In theory, there should be a much easier solution. If the global workspace hypothesis is correct, then even in the dark, in the absence of any external stimulation, a conscious person should exhibit detectable signatures of long-distance cerebral communication. A constant stream of brain activity should travel between the prefrontal and parietal lobes, generating fluctuating periods of synchrony with distant brain regions. This activity should be associated with a heightened state of electrical activity, especially in the medium (beta) and high (gamma) frequencies. Such long-distance broadcasting should consume a lot of energy. Can't we simply detect it?

We have known for many years that global brain metabolism, as measured by positron emission tomography (PET), drops during loss of consciousness. A PET scanner is a sophisticated detector of high-energy gamma rays that can be used to measure how much glucose (a chemical source of energy) is consumed anywhere in the body. The trick is to inject the patient with a precursor of glucose, labeled with traces of a radioactive compound, and use the scanner to detect the peaks of radioactive disintegration. The locations of the radioactivity peaks indicate where in the brain the glucose is being consumed. The striking result is that, in normal people, anesthesia and deep sleep cause a 50 percent reduction in glucose consumption throughout the cortex. A similar state of depressed energy consumption occurs during coma and a vegetative state. As early as the 1990s, Steven Laureys's team, in Liège, produced striking images of anomalies in brain metabolism in the vegetative state (figure 32).[42]

Importantly, the reduction in glucose consumption, as well as oxygen metabolism, differs across brain areas. The loss of consciousness seems specifically associated with a depressed activity of the bilateral prefrontal and parietal regions, as well as midline areas of the cingulate and precuneus. These regions overlap almost exactly with our global workspace network, the regions richest in long-distance cortical projections— another confirmation that this workspace system is crucial to conscious experience. Other isolated regions of the sensory and motor cortex may remain anatomically intact and metabolically active even in the absence of any conscious response.[43] For instance, vegetative patients who make

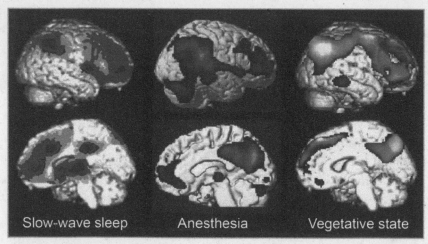

Slow-wave sleep   Anesthesia   Vegetative state

FIGURE 32. Reductions in frontal and parietal metabolism underlie the loss of consciousness in slow-wave sleep, anesthesia, and vegetative-state patients. Although other regions may also show reduced activity, the areas that form the global neuronal workspace exhibit a reproducible drop in energy consumption when consciousness is lost.

occasional facial movements show preserved activity in focal motor areas. For the past twenty years, one patient had been spurting out an occasional word, apparently unconsciously and without any relevance to his surroundings. His neuronal activity and metabolism were confined to a few islands of preserved cortex in the language areas of the left hemisphere. Clearly, such splattered activation did not suffice to sustain a conscious state: broader communication was needed.

Unfortunately, brain metabolism per se does not suffice to infer the presence or absence of residual consciousness. Some vegetative patients have virtually normal cortical metabolism; presumably their lesion affected only the ascending structures of the diencephalon rather than the cortex itself. Conversely, and more important, many vegetative patients who partially recover and move into the "minimally conscious" category do not exhibit normal metabolism. A comparison of pre- versus postrecovery images does show increased energy consumption in workspace regions, but the gain is modest. Metabolism generally fails to return to normal, presumably because the cortex remains damaged beyond repair. Even fine-grained images of the lesions, using the best magnetic resonance imagers, are only indicative:[44] they fail to provide a set of foolproof predictors of consciousness. By using only metabolic or anatomical images, it has not yet been possible to accurately gauge the circulation of neuronal information that underlies the conscious state.

In order to build a better detector of residual consciousness, my colleagues Jean-Rémi King, Jacobo Sitt, and Lionel Naccache and I returned to the idea of using the raw EEG as a marker of cortical communication.[45] Naccache's team had obtained close to 200 high-density recordings, with 256 electrodes monitoring the electrical activity of vegetative, minimally conscious, and conscious patients. Could we use these measures to quantify the amount of information exchange in the cortex? By digging in the literature, Sitt, who is at once a genial physicist, a computer scientist, and a psychiatrist, came up with a brilliant idea. He concocted a fast program to compute a mathematical quantity called "weighted symbolic mutual information," designed to evaluate how much information was being shared between two brain sites.[46]

When applied to our patients' data, this measure tightly separated the vegetative patients from everybody else (figure 33). Compared to conscious subjects, the vegetative group showed a much-reduced amount of

| Vegetative-state patients | Minimally conscious patients | Conscious patients | Conscious controls |

FIGURE 33. Information exchanged over long cortical distances is an excellent index of consciousness in patients with brain lesions. To create this image, electroencephalographic brain signals were recorded from 256 electrodes in nearly 200 patients with or without a loss of consciousness. For each pair of electrodes, symbolized by an arc, we computed a mathematical index of the amount of information shared by the underlying brain areas. Vegetative-state patients showed a much smaller amount of shared information than conscious patients and control subjects. This finding fits with a central tenet of global workspace theory, that information exchange is an essential function of consciousness. A follow-up study showed that the few vegetative patients who showed high information sharing had a better chance of regaining consciousness within the next days or months.

information sharing. This was particularly true when we restricted the analysis to pairs of electrodes that were separated by at least 7 or 8 centimeters—once again, long-distance broadcasting was the privilege of conscious brains. Using another directional measure, we saw that the brain's conversation was bidirectional: specialized areas of the back of the brain were talking to the generalist areas of the parietal and prefrontal lobes, which returned backward signals.

The patients' consciousness was also reflected in many other features of the EEG.[47] Mathematical measures of the amount of energy in different frequency bands showed, unsurprisingly, that loss of consciousness led to the disappearance of the high frequencies that characterize neural coding and processing, to the benefit of very low frequencies typical of sleep or anesthesia.[48] Measures of the synchrony among these brain oscillations confirmed that, during the conscious state, cortical regions tended to harmonize their exchanges.

Each of these mathematical quantities shed a slightly different light on consciousness, thus providing complementary vistas into the same conscious state. In order to combine them, Jean-Rémi King designed a program that learned, quite automatically, which blend of measures provided an optimal prediction of the patients' clinical state. Twenty minutes of EEG recording provided an excellent diagnosis. We almost never mistook a vegetative-state patient for a conscious person. Most of our program's errors consisted in labeling a minimally conscious patient as vegetative. We cannot guarantee that they were not, in fact, accurate: during those twenty minutes, a minimally conscious patient could have slipped out—so repeating the measure on another day would probably have improved the diagnosis.

The converse error also occurred: our program occasionally labeled a patient as minimally conscious, while clinical examination put him in the vegetative category. Was it a genuine mistake? Or might these patients be those paradoxical ones who seem vegetative but are in fact conscious and completely locked in? When we looked at the clinical outcome of our vegetative patients in the months following EEG recording, we were in for a very exciting result. For two-thirds of them, our computer program agreed with the clinical diagnosis of a vegetative state—and only 20 percent of them recovered and moved to the minimally conscious category. In the remaining third, however, our system detected a hint of consciousness where the clinician saw none—and among those cases, a full 50 percent recovered a clinically obvious state of consciousness within the next few months.

This difference in prognosis has huge implications. It means that, using automated brain measures, we can now detect traces of consciousness long before they manifest in overt behavior. Our theory-driven signatures of consciousness have become more sensitive than the experienced clinician. The new science of consciousness is yielding its first fruits.

Toward Clinical Interventions

Canst thou not minister to a mind diseased,
Pluck from the memory a rooted sorrow,
Raze out the written troubles of the brain . . . ?

—Shakespeare, *Macbeth* (1606)

Detecting a tinge of consciousness is only a start. What patients and their families long for is an answer to the Shakespearean query: "Canst thou not minister to a mind diseased?" Can we help coma and vegetative-state patients regain consciousness? Their mental faculties sometimes return suddenly, years after the original accident. Can we accelerate this recovery process?

When devastated families ask these questions, the medical community generally gives a pessimistic answer. Once a whole year has elapsed and the patient still remains unconscious, he or she is said to be in a "permanent vegetative state." This clinical label comes with a plain subtext: very little change will occur, no matter how much stimulation is provided. And in many patients, this is the sad truth.

In 2007, however, Nicholas Schiff and Joseph Giacino published a spectacular paper in the high-profile journal *Nature*, suggesting that this issue should be revisited.[49] For the first time, they presented a treatment that slowly brought a minimally conscious patient back to a more stable conscious state. Their intervention consisted in inserting long electrodes into the brain and stimulating a location of central importance: the aptly named central thalamus and the surrounding intralaminar nuclei.

Thanks to the pioneering research of Giuseppe Moruzzi and Horace Magoun in the 1940s, these regions were already known as essential nodes of the ascending system that regulates the overall vigilance level of the cortex.[50] Central thalamic nuclei contain a high density of projection neurons, marked by a particular protein (calcium-binding protein), that are known to project broadly toward the cortex, particularly to the frontal lobes. Interestingly, their axons specifically target the pyramidal neurons in the upper layers of the cortex—precisely those with long-distance projections that underlie the global neuronal workspace. In animals, activating the central thalamus can modulate the overall activity of the cortex, enhance motor activity, and boost learning.[51]

In a normal brain, the activity of the central thalamus is, in turn, modulated by the prefrontal and cingulate areas of the cortex. This feedback loop probably allows us to dynamically adjust cortical excitation as a function of task demands: an attention-grabbing task turns it on, enhancing the brain's processing capacity.[52] In the severely injured brain, however, a global reduction in the overall level of circulating neuronal activity may disrupt this essential loop that constantly regulates our level of arousal. Thus, Schiff and Giacino predicted that stimulating the central thalamus may "reawaken" the cortex. It would restore, from the outside, the sustained level of arousal that the patient's brain had become unable to control from inside.

As we have already discussed, vigilance is not the same as conscious access. Vegetative-state patients often have a partially preserved vigilance system: they awaken in the morning and open their eyes, but this does not suffice to bring the cortex back to the conscious mode. Indeed, most patients in a persistent vegetative state show little benefit from a thalamic stimulator. Terri Schiavo had one, yet she showed no long-term improvement, probably because her cortex and especially the underlying white matter were dramatically injured. In the few cases where it seemed to work, spontaneous recovery could not be excluded.

Well aware of this grim baseline, Schiff and Giacino nevertheless mapped out a plan to increase their chances of success. First they specifically targeted the central lateral nucleus of the thalamus, which enters into those direct loops with the prefrontal cortex. Second, they selected a patient in whom they thought the intervention was likely to succeed because he was already on the brink of consciousness. Remember that Joseph Giacino himself had played an instrumental role in defining the minimally conscious state: a category of patients who show fleeting signs of conscious processing and intentional communication yet are unable to manifest it in a systematic and reproducible manner. Schiff's team identified one such patient in whom brain imaging showed that the cortex was remarkably spared. Although he had been in a stable state of minimal consciousness for many years, both of his hemispheres were still activating in response to speech. His global cortical metabolism, however, was dramatically reduced, suggesting that arousal was poorly regulated. Could thalamic stimulation provide the missing kick that would bring him back to a stable state of consciousness?

Schiff and Giacino proceeded in several careful steps. Before implanting the patient with electrodes, they carefully monitored him for months. They repeatedly tested him with the same battery (the coma recovery scale) until they had a stable estimate of his abilities and their fluctuations. Importantly, several tests gave intermediate results: the patient exhibited a few signs of acting intentionally, and he even blurted out an occasional word, but this behavior was erratic. This meant that he was minimally conscious and that there was much room for improvement.

With these observations in mind, Schiff and Giacino proceeded to implant the electrodes. During surgery, they carefully guided two long wires all the way through the left and right cortex and into the central thalamus. Forty-eight hours later the electrodes were turned on. Immediately, the results were dramatic: the patient, who had been minimally conscious for six years, opened his eyes, his heart rate increased, and he spontaneously turned in response to voices. His responses remained limited, though; when asked to name objects, his speech remained "unintelligible and was limited to episodes of incomprehensible word-mouthing."[53] As soon as the stimulator was turned off, these behaviors vanished.

To establish a post-intervention baseline, the researchers let two months pass without applying any further stimulation. During that time, there was no improvement. Then, every other month, in a double-blind study, they turned the stimulator either on or off, in an alternating pattern. The patient improved spectacularly. On all measures of arousal, communication, motor control, and object naming, the test scores shot up during the period when the stimulator was turned on. Furthermore, and crucially, those measures dropped only slightly when it was turned off—the patient did not return all the way back to baseline. The effect was slow but cumulative, and six months later he could feed himself by bringing a cup to his mouth. His family noted a marked improvement in his social interactions. He remained severely handicapped, but he could now take an active part in his life and even discuss his medical treatment.

This success story raises great hopes. Deep brain stimulation, by increasing the level of cortical arousal and therefore bringing neuronal activity closer to its normal operating level, may help the brain recover its autonomy.

Even in patients with a long history of a vegetative state or minimal consciousness, the brain remains plastic, and spontaneous recovery can never be excluded. Indeed, strange reports of sudden remissions abound in medical records. One man remained in a minimally conscious state for nineteen years, then suddenly recovered language and memory. Images of his brain, created using the technique of diffusion tensor imaging, suggested that several of his long-distance brain connections had slowly regrown.[54] In another patient, communication between the frontal cortex and the thalamus had been depressed when he was vegetative, but it returned to normal after he spontaneously recovered.[55]

We do not expect such a recovery to be possible in every patient—but can we understand why some patients recover while others don't? Clearly, if too many of the prefrontal cortex neurons are dead, no amount of stimulation will revive them. In some cases, however, the neurons are intact but have lost many of their connections. In yet others, the self-sustained dynamics of brain circuits seem to be the culprit: although connections are still present, the circulating information no longer suffices to maintain a sustained state of activity, and the brain switches itself off. If the circuit is sufficiently spared to be switched back on, such patients may exhibit a surprisingly fast recovery.

But how can we flip the cortical switch back to the "on" position? Pharmacological agents that act on the dopamine circuits of the brain are prime candidates. Dopamine is a neurotransmitter that is primarily involved in the brain's reward circuits. Neurons that use dopamine send massive modulatory projections to the prefrontal cortex and the deep gray nuclei that control our voluntary actions. Stimulating the dopamine circuits may therefore help them restore a normal level of arousal. Indeed, three patients in a persistent vegetative state suddenly regained consciousness after administration of a drug called levodopa, a chemical precursor of dopamine that is typically given to Parkinsonian patients.[56] Amantadine is another stimulant of the dopamine system that, in controlled clinical tests, has been found to slightly speed up the recovery of vegetative and minimally conscious patients.[57]

Other cases on record are much stranger. Most paradoxical is the effect of Ambien, a sleeping pill that, bizarrely, may revive consciousness. One patient had been totally mute and immobile for months, in a neurological syndrome called "akinetic mutism." To ease his sleep, he was given a pill of Ambien, a well-known hypnotic—and all of a sudden he awoke, moved, and began to speak.[58] In another case, a woman who had suffered a left-hemispheric stroke and was dramatically aphasic, unable to say more than the occasional random syllable, was also prescribed Ambien because she had trouble falling asleep. The first time she took it, she immediately resumed speaking for a few hours. She could answer questions, count, and even name objects. She then fell asleep, and sure enough, the next morning her aphasia had returned. The phenomenon repeated every evening, whenever her family gave her the sleeping pill.[59] Not only did it fail to put her to sleep, it had the paradoxical effect of re-awakening her dormant cortical circuitry for language.

These phenomena are only beginning to receive an explanation. They seem to arise from the multiple loops that link the cortical workspace network, the thalamus, and two of the basal ganglia (the striatum and the pallidum). Via these loops, the cortex can indirectly excite itself, as activation propagates in a circular path from frontal cortex to striatum, pallidum, thalamus, and back to cortex. However, two of these connections rely on inhibition rather than excitation: the striatum inhibits the pallidum, and the pallidum in turn inhibits the thalamus. When the brain loses its oxygen supply, the inhibitory cells of the striatum seem to be among the first to suffer. As a result, the pallidum is insufficiently inhibited. Its activity is free to shoot up, thus shutting down the thalamus and the cortex and preventing them from sustaining any conscious activity.

These pathways are still largely intact, however; they are only massively inhibited. They can be switched back on by inserting a circuit breaker into this vicious circle. Many solutions seem to be available. An electrode that is planted deep in the thalamus may counteract the excessive inhibition of thalamic neurons and thus switch them on again. Alternatively, dopamine or amantadine may be used to excite the cortex, either directly or through the remaining neurons in the striatum. Finally, a drug such as Ambien may inhibit the inhibition: by binding to the many inhibitory receptors in the pallidum, it forces its overexcited inhibitory cells to switch off, thus releasing the cortex and thalamus from their unwanted quiescence. All these mechanisms, although still hypothetical, may explain why these drugs have similar effects in the end: they all bring cortical activity closer to its normal level.[60]

The above tricks will work only if the cortex itself is not exceedingly damaged. A favorable sign is when the prefrontal cortex seems intact on an anatomical image yet shows a dramatically reduced metabolism; the cortex may have been simply switched off and may be reawakened. Once it is switched on, it will slowly return to a self-regulating state. In their normal range of operation, many of the brain's synapses are plastic and can increase their weight to help stabilize the active neuronal assemblies. Thanks to such brain plasticity, a patient's workspace connections may progressively gain strength and become increasingly able to sustain a durable state of conscious activity.

Even for patients whose cortical circuits have been damaged, we may envisage futuristic solutions. If the workspace hypothesis is correct, consciousness is nothing but the flexible circulation of information within a dense switchboard of cortical neurons. Is it too far-fetched to imagine that some of its nodes and connections might be replaced by external loops? Brain-computer interfaces, particularly using implanted devices, have the potential to restore long-distance communication in the brain. We will soon be able to collect spontaneous brain discharges in the prefrontal or premotor cortex and play them back to other, distant regions—either directly in the form of electrical discharges or perhaps more simply by recoding them into visual or auditory signals. Such sensory substitution is already used to make the blind "see" by training them to recognize audio signals that encrypt the image from a video camera.[61] Following the same principle, sensory substitution could help reconnect the brain with itself, restoring a denser form of inner communication. Denser loops may provide the brain with the critical amount of self-excitation needed to maintain an active state and remain conscious.

Time will tell whether this idea is far-fetched. What is certain is that, in the next decades, the renewed interest in coma and vegetative states, based on an increasingly solid theory of how neuronal circuits engender conscious states, will lead to massive improvements in medical care. We are in for a revolution in the treatment of disorders of consciousness.

7

THE FUTURE OF CONSCIOUSNESS

The emerging science of consciousness still faces many challenges. Can we determine the precise moment when consciousness first emerges in babies? Can we figure out whether a monkey, or a dog, or a dolphin is conscious of its surroundings? Can we solve the riddle of self-consciousness, our surprising ability to think about our own thinking? Is the human brain unique in this respect? Does it host distinctive circuits, and if so, can their dysfunction explain the origins of uniquely human diseases such as schizophrenia? And if we manage to analyze those circuits, could we ever duplicate them in a computer, thus giving rise to artificial consciousness?

I sort of resent the idea of science poking its nose into this business, my business. Hasn't science already appropriated enough of reality? Must it lay claim to the intangible invisible essential self as well?
—David Lodge, *Thinks... (2001)*

In point of fact, the greater one's science,
the deeper the sense of mystery.
—Vladimir Nabokov, *Strong Opinions* (1973)

The black box of consciousness is now cracked open. Thanks to a variety of experimental paradigms, we have learned to make pictures visible or invisible, then track the patterns of neuronal activity that occur only when conscious access happens. Understanding how the brain handles seen and unseen images has turned out not to be as subtle as we initially feared. Many electrophysiological signatures have manifested the presence of a conscious ignition. These signatures of consciousness have proved solid enough that they are now being used in clinics to probe residual consciousness in patients with massive brain lesions.

No doubt this is only a beginning. The answers to many questions still elude us. In this closing chapter, I would like to outline what I see as the future of consciousness research—the outstanding questions that will keep neuroscientists at work for many more years.

Some of these questions are thoroughly empirical and have already received an inkling of an answer. For instance, when does consciousness emerge—in development as well as in evolution? Are newborns conscious? What about premature infants or fetuses? Do monkeys, mice, and birds share a workspace similar to ours?

Other problems border on the philosophical—and yet I firmly believe that they will ultimately receive an empirical answer, once we find an experimental line of attack. For instance, what is self-consciousness? Surely something particular about the human mind allows it to turn the flashlight of consciousness onto itself and think about its own thinking. Are we unique in this respect? What makes human thought so powerful but also uniquely vulnerable to psychiatric diseases such as schizophrenia? Will this knowledge allow us to build an artificial consciousness—a sentient robot? Would it have feelings, experiences, and even a sense of free will?

No one can claim to know the answers to these conundrums, and I will not pretend that I can resolve them. But I would like to show how we might begin to address them.

Conscious Babies?

Consider the onset of consciousness in childhood. Are babies conscious? What about newborns? Premature infants? Fetuses inside the womb? Surely some degree of brain organization is needed before a conscious mind is born—but exactly how much?

For decades, this controversial question has pitted defenders of the sanctity of human life against rationalists. Provocative statements abound on both sides. For instance, the University of Colorado philosopher Michael Tooley bluntly writes that "new-born humans are neither persons nor quasi-persons, and their destruction is in no way intrinsically wrong."[1]

According to Tooley, up to the age of three months at least, infanticide is morally justified because a newborn infant "does not possess the concept of a continuing self, any more than a newborn kitten" does, and therefore it has "no right to life."[2] Continuing this grim message, the Princeton bioethics professor Peter Singer argues that "life only begins in the morally significant sense when there is awareness of one's existence over time":

> The fact that a being is a human being, in the sense of a member of the species Homo sapiens, is not relevant to the wrongness of killing it; it is, rather, characteristics like rationality, autonomy, and self-consciousness that make a difference. Infants lack these characteristics. Killing them, therefore, cannot be equated with killing normal human beings, or any other self-conscious beings.[3]

Such assertions are preposterous for many reasons. They clash with the moral intuition that all human beings, from Nobel Prize winners to handicapped children, have equal rights to a good life. They also conflict head-on with our intuitions of consciousness—just ask any mother who has exchanged eye contact and goo-goo-ga-gas with her newborn baby. Most shocking, Tooley and Singer pronounce their confident ukases without the slightest supporting evidence. How do they know that babies have no experiences? Are their views founded upon a firm scientific basis? Not at all—they are purely a priori, detached from experimentation—and, in fact, are often demonstrably wrong. For instance, Singer writes that "in most respects, [coma and vegetative patients] do not differ importantly from disabled infants. They are not self-conscious, rational, or autonomous . . . their lives have no intrinsic value. Their life's journey has come to an end." In Chapter 6, we saw that this view is dead wrong: brain imaging reveals residual consciousness in a fraction of adult vegetative patients. Such an arrogant view, denying the complexity of life and consciousness, is appalling. The brain deserves a better philosophy.

The alternative path that I propose is simple: we must learn to do the right experiments. Although the infant mind remains a vast terra incognita, behavior, anatomy, and brain imaging can provide much information about conscious states. The signatures of consciousness, once validated in human adults, can and should be searched for in human babies of various ages.

To be sure, this strategy is imperfect, because it is built upon an analogy. We hope to find, at some point in early childhood development, the same objective markers that we know index subjective experience in adults. If we find them, we will conclude that at this age, children possess a subjective viewpoint on the outside world. Of course, nature could be more complex; the markers of consciousness could change with age. Also, we may not always get an unambiguous answer. Different markers may disagree, and the workspace that operates as an integrated system in adulthood may consist of fragments or pieces that develop at their own pace during infancy. Still, the experimental method has a unique capacity to inform the objective side of the debate. Any scientific knowledge will be better than the a priori proclamations of philosophical and religious leaders.

So do infants possess a conscious workspace? What does brain anatomy say? In the past century, babies' immature cortex, replete with scrawny neurons, puny dendrites, and skinny axons that lack their insulating sheet of myelin, led many pediatricians to believe that the mind was not operative at birth. Only a few islands of visual, auditory, and motor cortex, they thought, were sufficiently mature to provide infants with primitive sensations and reflexes. Sensory inputs fused to create "one great blooming, buzzing confusion," in the famous words of William James. It was widely believed that the higher-level reasoning centers in the babies' prefrontal cortex remained silent at least until the end of the first year of life, when they finally began to mature. This virtual frontal lobotomy explained infants' systematic failure on behavioral tests of motor planning and executive control, such as Piaget's famous A-not-B test.[4] To many a pediatrician, it was perfectly obvious, then, that newborns did not experience pain—so why anesthetize them? Injections and even surgeries were routinely performed without any regard for the possibility of infant consciousness.

Recent advances in behavioral testing and brain imaging, however, refute this pessimistic view. The great mistake, indeed, was to confuse immaturity with dysfunction. Even in the womb, starting at around six and a half months of gestation, a baby's cortex starts to form and to fold. In the newborn, distant cortical regions are already strongly interconnected by long-distance fibers.[5] Although they are not covered with myelin, these connections process information, albeit at a much slower pace than in adults. Right from birth, they already promote a self-organization of spontaneous neuronal activity into functional networks.[6]

Consider speech processing. Babies are immensely attracted to language. They probably begin to learn it inside the womb, because even newborns can distinguish sentences in their mother tongue from those in a foreign language.[7] Language acquisition happens so fast that a long line of prestigious scientists, from Darwin to Chomsky and Pinker, has postulated a special organ, a "language acquisition device" specialized for language learning and unique to the human brain. My wife, Ghislaine Dehaene-Lambertz, and I tested this idea directly, by using fMRI to look inside babies' brains while they listened to their maternal language.[8] Swaddled onto a comfortable mattress, their ears protected from the machine's noise by a massive headset, two-month-old infants quietly listened to infant-directed speech while we took snapshots of their brain activity every three seconds.

To our amazement, the activation was huge and definitely not restricted to the primary auditory area. On the contrary, an entire network of cortical regions lit up (figure 34). The activity nicely traced the contours of the classical language areas, at exactly the same place as in the adult's brain. Speech inputs were already routed to the left hemisphere's temporal and frontal language areas, while equally complex stimuli such as Mozart music were channeled to other regions of the right hemisphere.[9] Even Broca's area, in the left inferior prefrontal cortex, was already stirred up by language. This region was mature enough to activate in two-month-old babies. It was later found to be one of the earliest-maturing and best-connected regions of the baby's prefrontal cortex.[10]

By measuring the speed of activation with MRI, we confirmed that a baby's language network is working—but at a speed much slower than in an adult, especially in the prefrontal cortex.[11] Does this slowness prevent the emergence of consciousness? Do infants process speech in a "zombie mode," much as a comatose brain unconsciously responds to novel tones? The mere fact that an attentive two-month-old, during language processing, activates the same cortical network as an adult is unfortunately inconclusive, because we know that much of this network (though perhaps not Broca's area) can activate unconsciously—for instance, during anesthesia.[12] Crucially, however, our experiment also showed that babies possess a rudimentary form of verbal working memory. When we repeated

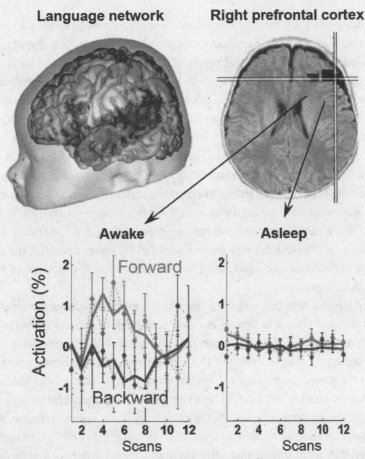

FIGURE 34. The prefrontal cortex is already active in awake infants. Two-month-old infants listened to sentences in their maternal language while their brain was scanned with fMRI. Speech activated a broad language network, including the left inferior frontal region known as Broca's area. Playing the same tape backward, thus destroying most speech cues, caused a much-reduced activation. Awake infants also activated their right prefrontal cortex. This activity was related to consciousness, because it vanished when the infants fell asleep.

the same sentence after a fourteen-second interval, our two-month-olds gave evidence of remembering:[13] their Broca's area lit up much more strongly on the second occasion than on the first. Already at two months, their brain bore one of the hallmarks of consciousness, the capacity to hold information in working memory for a few seconds.

Equally crucially, infants' responses to speech differed when they were awake and asleep. Their auditory cortex always lit up, but the activity cascaded into the dorsolateral prefrontal cortex only in awake babies; in sleeping babies we saw a flat curve in this area (figure 34). The prefrontal cortex, this crucial node of the adult workspace, therefore seems to already contribute primarily to conscious processing in awake infants.

A much tighter proof that few-month-olds are conscious comes from the application of the local-global test that I described in Chapter 6 and that probes residual consciousness in vegetative-state adult patients. In that simple test, patients listen to repeated series of sounds such as *beep beep beep beep boop* while we record their brain waves, using EEG. Occasionally, a rare sequence violates the rule, ending for instance with a fifth *beep*. When this novelty evokes a global P3 wave, invading the prefrontal cortex and the associated workspace areas, the patient is very likely to be conscious.

Undergoing this test requires no education, no language, and no instruction, and thus it is simple enough to be run in infants (or virtually any animal species). Any child can listen to a sequence of tones and, if its brain is smart enough, work out the regularities. Event-related potentials can be recorded from the first few months of life. The only problem is that babies quickly get fussy when the test is too repetitive. In order to probe this signature of consciousness in babies, my wife Ghislaine, who is a neuropediatrician and a specialist of infant cognition, therefore adapted our local-global test. She turned it into a multimedia show in which attractive faces articulated a sequence of vowels: *aa aa aa ee*. The constantly changing faces, with their moving mouths, fascinated the babies—and once we managed to capture their attention, we were pleased to see that, at two months of age, their brain already emitted a global conscious response to novelty—a signature of consciousness.[14]

Most parents will not be surprised to learn that their two-month-old baby already scores high on a test of consciousness—yet our tests also showed that their consciousness differs in one important respect from that in adults: in infants, the latency of the brain responses is dramatically slower than in adults. Every processing step seems to take a disproportionately longer time. Our babies' brain needed one-third of a second to register the vowel change and to generate an unconscious mismatch response. And a full second was needed before their prefrontal cortex reacted to global novelty—about three to four times longer than in adults. Thus, the architecture of the baby's brain, in the first weeks of life, includes a functional global workspace, albeit a very slow one.

My colleague Sid Kouider replicated and extended this finding, this time using vision. He focused on face processing, another domain for which even newborn babies have an innate competence.[15] Babies love faces and magnetically orient toward them from birth. Kouider capitalized on this natural tropism to study whether babies are sensitive to visual masking and exhibit the same sort of threshold for conscious access as adults. He adapted, to five-month-olds, the masking paradigm that we had used to study conscious vision in adults.[16] An attractive face was flashed for a brief and variable duration, immediately followed by an ugly scrambled picture that served as a mask. The question was, did infants see the face? Were they conscious of it?

You might remember from Chapter 1 that during masking, adult viewers report seeing nothing unless the target picture lasts more than about one-twentieth of a second. Although speechless babies cannot report what they see, their eyes, like those of a locked-in patient, tell a similar story. When the face is flashed below a minimal duration, Kouider found, they do not gaze at it, suggesting that they fail to see it. Once the face is exposed for some threshold duration, however, they orient toward it. Just like adults, they suffer from masking and perceive the face only when it is "supraliminal," presented above the perception threshold. Critically, the threshold duration turns out to be two to three times longer in infants than in adults. Five-month-olds detect the face only when it is shown for more than 100 milliseconds, whereas in adults the masking threshold typically falls between 40 and 50 milliseconds. Very interestingly, the threshold drops to its adult value when babies reach ten to twelve months of age, precisely the time when behaviors that depend on the prefrontal cortex begin to emerge.[17]

Having shown the existence of a threshold for conscious access in babies, Sid Kouider, Ghislaine Dehaene-Lambertz, and I went on to record the infants' brain responses to flashed faces. We saw exactly the same series of cortical processing stages that we had found in adults: a subliminal linear phase followed by a sudden nonlinear ignition (figure 35). During the first phase, activity in the back of the brain increases steadily with face duration, regardless of whether images are below or above

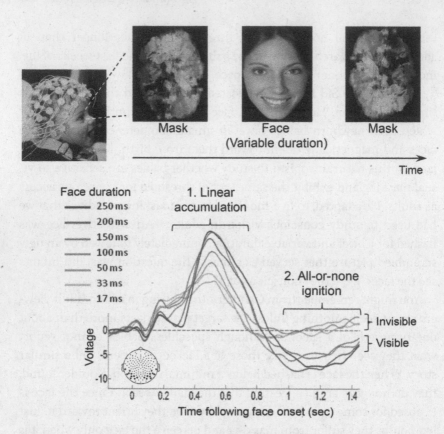

FIGURE 35. Infants exhibit the same signatures of conscious perception as adults, but they process information at a much slower speed. In this experiment, twelve- to fifteen-month-old infants were flashed attractive faces that were masked to render them visible or invisible. The infant brain exhibited two stages of processing: first a linear accumulation of sensory evidence, then a nonlinear ignition. The late ignition may reflect conscious perception, because it occurred only when the face was presented for 100 milliseconds or more, precisely the duration needed for infants to orient their gaze. Note that conscious ignition started 1 second after the face appeared, which is about three times longer than in adults.

threshold: the infant's brain clearly accumulates the available evidence about the flashed face. During the second phase, only above-threshold faces trigger a slow negative wave over the prefrontal cortex. Functionally and topographically, this late activation shares a lot of similarity with the adult P3 wave. Clearly, if enough sensory evidence is available, even the infant brain can propagate it all the way into the prefrontal cortex, although at a much-reduced speed. Because this two-stage architecture is essentially the same as in conscious adults, who can report what they see, we can assume that babies already enjoy conscious vision, although they cannot yet report it by speaking aloud.

In fact, a very slow frontal negativity shows up in all sorts of infant experiments that involve the orienting of attention toward a novel stimulation, be it auditory or visual.[18] Other researchers have noticed its similarity to the adult P3 wave,[19] which shows up whenever conscious access occurs, regardless of the sensory modality. For instance, the frontal negativity occurs when infants attend to deviant sounds,[20] but only when they are awake, not when they are asleep.[21] In experiment after experiment, this slow frontal response behaves as a marker of conscious processing.

We can now safely conclude that conscious access exists in babies as in adults, but in a dramatically slower form, perhaps up to four times slower. Why this sluggishness? Remember that the infant brain is immature. The major long-distance fiber tracts that form the adult global workspace are already present at birth,[22] but they are not yet electrically insulated. The sheaths of myelin, the fatty membrane that surrounds the axons, continue to mature well into childhood and even adolescence. Their main role is to provide electrical insulation and, as a result, increase the speed and fidelity with which neuronal discharges propagate to distant sites. The baby's brain web is wired but not yet insulated; information integration therefore operates at a much slower pace. An infant's sluggishness is perhaps comparable to that of a patient returning from coma. In both cases, adaptive responses can be evoked, but it takes one or two seconds before a smile, a frown, or a stammering syllable emerges from their lips. Think it of it as a foggy, dawdling, but definitely conscious mind.

Because the youngest subjects we tested were two-month-olds, we still do not know the exact moment at which consciousness emerges. Is a newborn already conscious, or does it take a few weeks before his or her cortical architecture starts to function properly? I will hedge my bets until all the evidence is in, but I would not be surprised if we discovered that consciousness exists at birth. Long-distance anatomical connections already crisscross the newborn baby's brain, and their processing depth should not be underestimated. A few hours after birth, infants already exhibit sophisticated behavior, such as the capacity to distinguish sets of objects based on their approximate number.[23]

The Swedish pediatrician Hugo Lagercrantz and the French neurobiologist Jean-Pierre Changeux have proposed a very interesting hypothesis: birth would coincide with the first access to consciousness.[24] In the womb, they argue, the fetus is essentially sedated, bathed in a drug stream that includes "the neurosteroid anesthetics pregnanolone and the sleep-inducing prostaglandin D2 provided by the placenta." Birth coincides with a massive surge of stress hormones and stimulating neurotransmitters such as catecholamines; in the following hours, the newborn baby is usually awake and energized, his eyes wide open. Is he having his first conscious experience? If these pharmacological inferences turn out to be valid, delivery is an even more significant event than we thought: the genuine birth of a conscious mind.

Conscious Animals?

He who understands baboon would do more towards metaphysics than Locke.

—Charles Darwin, *Notebooks* (1838)

The same questions that we ask concerning infants should also be asked about our speechless cousins—animals. Animals cannot describe their conscious thoughts, but does that mean they have none? An extraordinary diversity of species has evolved on earth, from patient predators (cheetahs, eagles, moray eels) to careful route planners (elephants, geese), playful characters (cats, otters), clever problem solvers (magpies, octopuses), vocal geniuses (parakeets), and social grandmasters (bats, wolves). I would be very surprised if none of them shared at least part of our conscious experiences. My theory is that the architecture of the conscious workspace plays an essential role in facilitating the exchange of information among brain areas. Thus, consciousness is a useful device that is likely to have emerged a long time ago in evolution and perhaps more than once.

Why should we naïvely suppose that the workspace system is unique to humans? It isn't. The dense network of long-distance connections that links the prefrontal cortex with other associative cortexes is evident in macaque monkeys, and this workspace system may well be present in all mammals. Even the mouse has tiny prefrontal and cingulate cortexes that get activated when it keeps visual information in mind for a second.[25] An exciting question is whether some birds, especially those with vocal communication and imitation, may possess analogous circuitry with a similar function.[26]

Attributing consciousness to animals should not be based solely on their anatomy. Although they lack language, monkeys can be trained to report what they see by pressing keys on a computer. This approach is providing mounting evidence that they have subjective experiences that are very similar to ours. For instance, they can be rewarded to press one key if they see a light and another if they do not. This motor act can then be used as a proxy for a minimal "report": a nonverbal gesture equivalent to the animal's saying "I think I saw a light" or "I didn't see anything." A monkey can also be trained to classify the images it perceives, pressing one key for faces and the other for nonfaces. Once trained, the animal can then be tested with the same variety of visual paradigms that probe conscious and unconscious processing in humans.

The results of these behavioral studies prove that monkeys, like us, experience visual illusions. If we show them two different images, one to each eye, they report binocular rivalry: they press the keys in alternation, indicating that they too see only one of the two images at a given time. The images ceaselessly wax and wane in and out of their consciousness at the same rhythm as in any of us.[27] Masking also works in monkeys. When we flash them a picture and follow it by a random mask, macaques report that they did not see the hidden image, although their visual cortex still shows a transient and selective neuronal discharge.[28] Thus like us, they possess a form of subliminal perception, as well as a precise threshold beyond which the image becomes visible.

Finally, when their primary visual cortex is damaged, monkeys too develop a form of blindsight. In spite of the lesion, they can still accurately point to a light source in their impaired visual field. However, when trained to report the presence or absence of light, they label a stimulus presented in their impaired visual field by using the "no light" key, suggesting that, like human blindsight patients, their perceptual awareness is gone.[29]

There is little doubt that macaque monkeys can use their rudimentary workspace to think about the past. They easily pass the delayed response task, which requires holding information in mind long after the stimulus is gone. Like us, they do so by maintaining a sustained discharge in their prefrontal and parietal neurons.[30] If anything, when passively watching a movie, they tend to activate their prefrontal cortex more than humans.[31] We may be superior to monkeys in our ability to inhibit distraction, and when we are watching a movie, our prefrontal cortex can therefore decouple from the incoming stream, letting our mind wander freely.[32] But macaque monkeys too possess a spontaneous "default mode" network of regions that activate during rest[33]—regions similar to those activated when we introspect, remember, or mind-wander.[34]

What about our litmus test of conscious auditory perception: the local-global test that we used to reveal a residual consciousness in patients recovering from coma? My colleagues Bechir Jarraya and Lynn Uhrig tested whether monkeys notice that *beep beep beep beep* is an anomalous sequence when it occurs amid a flurry of frequent *beep beep beep boop* sounds. They clearly do. Functional MRI shows that the monkeys' prefrontal cortex lights up only to the globally deviant sequences.[35] As in humans, this prefrontal response goes away when the monkeys are anesthetized. Once again, a signature of consciousness seems to exist in monkeys.

In pilot research conducted by Karim Benchenane, even mice seem to pass this elementary test. In future years, as we systematically test a variety of species, I would not be surprised if we discovered that all mammals, and probably many species of birds and fish, show evidence of a convergent evolution to the same sort of conscious workspace.

Self-Conscious Monkeys?

Macaque monkeys undoubtedly possess a global workspace largely similar to ours. But is it identical? In this book, I have focused on the most basic aspect of consciousness: conscious access, or the ability to become aware of selected sensory stimuli. This competence is so basic that we share it with monkeys and probably a great many other species. When it comes to higher-order cognitive functions, however, humans are clearly very different. We have to ask whether the human conscious workspace possesses additional properties that radically set us apart from all other animals.

Self-awareness seems a prime candidate for human uniqueness. Aren't we *sapiens sapiens*—the only species who know that they know? Isn't the capacity to reflect upon our own existence a uniquely human feat? In *Strong Opinions* (1973), Vladimir Nabokov, a superb novelist but also a passionate entomologist, made precisely this point:

> Being aware of being aware of being . . . if I not only know that I *am* but also know that I know it, then I belong to the human species. All the rest follows—the glory of thought, poetry, a vision of the universe. In that respect, the gap between ape and man is immeasurably greater than the one between amoeba and ape.

Nabokov was wrong, however. "Know thyself," the famous motto inscribed in the *pronaos* of the Temple of Apollo at Delphi, is not the privilege of mankind. In recent years, research has revealed the amazing sophistication of animal self-reflection. Even in tasks that require second-order judgments, as when we detect our errors or ponder our success or failure, animals are not as incompetent as we might think.

This domain of competence is called "metacognition"—the capacity to entertain thoughts about our thoughts. Donald Rumsfeld, George W. Bush's secretary of defense, neatly outlined it when, in a briefing to the Department of Defense, he famously distinguished among the known knowns ("things we know we know"), the known unknowns ("we know there are some things we do not know"), and the unknown unknowns ("the ones we don't know we don't know"). Metacognition is about knowing the limits of one's own knowledge—assigning degrees of belief or confidence to our own thoughts. And evidence suggests that monkeys, dolphins, and even rats and pigeons possess the rudiments of it.

How do we know that animals know what they know? Consider Natua, a dolphin swimming freely in his home coral pool at the Dolphin Research Center in Marathon, Florida.[36] The animal has been trained to classify underwater sounds according to their pitch. This he does extremely well, pressing a paddle on the left wall for low pitches, and one on the right wall for high pitches.

The experimenter set the boundary between low and high pitches at a frequency of 2,100 hertz. When the sound is far enough from this reference, the animal quickly swims to the correct side. When the sound frequency is very close to 2,100 hertz, however, Natua's responses become very slow. He shakes his head before hesitantly swimming to one side, often the wrong one.

Does this hesitant behavior suffice to indicate that the animal "knows" that he is having a hard time deciding? No. In itself, the increase in difficulty at short distances is quite banal. In humans as in many other animals, decision time and error rate typically increase whenever the difference that must be discriminated is reduced. But crucially, in humans a smaller perceptual distance also elicits a second-order feeling of lack of confidence. When the sound is too close to the boundary, we realize that we face a difficulty. We feel unsure, and we know that our decision may well turn out to be wrong. If we can, we bail out, openly reporting that we have no idea of the correct answer. This is typical metacognitive knowledge: *I know that I don't know.*

Does Natua have any such knowledge of his own uncertainty? Can he tell whether he knows the correct response or whether he is unsure? Does he have a sense of confidence in his own decisions? To answer these questions, J. David Smith, from the State University of New York, designed a clever trick: the "escape" response. After the initial perceptual training, he introduced the dolphin to a third response paddle. By trial and error, Natua learned that, whenever he presses it, the stimulus sound is immediately replaced by an easy low-pitch sound (at 1,200 hertz), which earns him a small reward. Whenever the third paddle is present, Natua has the option to escape from the main task. However, he is not allowed to opt out on every trial: the escape paddle must be used sparingly; otherwise, the reward is dramatically delayed.

Here is the beautiful experimental finding: during the pitch task, Natua spontaneously decides to use the opt-out response only on difficult trials. He presses the third paddle only when the stimulation frequency is close to the reference of 2,100 hertz—precisely those trials where he is likely to make an error. It looks as if he uses the third key as a second-order "commentary" on his first-order performance. By pressing it, he "reports" that he finds it too hard to respond to the primary task and that he prefers an easier trial. A dolphin is smart enough to discern his own lack of confidence. Like Rumsfeld, he knows what he doesn't know.

Some researchers dispute this mentalist interpretation. They point out that the task can be described in much simpler behaviorist terms: the dolphin merely exhibits a trained motor behavior that maximizes reward. Its only unusual feature is to allow for three responses instead of two. As usual in a reinforcement learning task, the animal has discovered exactly which stimuli make it more advantageous to press the third key—nothing more than rote behavior.

While many past experiments fall prey to this low-level interpretation, new research in monkeys, rats, and pigeons addresses this criticism and strongly tips the scales toward genuine metacognitive competence. Animals often use the opt-out response more intelligently than reward alone would predict.[37] For instance, when given the option to escape *after* making a choice, but *before* being told whether they were right or wrong, they finely monitor which trials are subjectively difficult for them. We know this because they indeed perform worse on trials where they opt out than on trials where they stick to their initial response, even when the very same stimulus is presented on both occasions. They seem to internally monitor their mental state and sift out precisely those trials where, for one reason or the other, they were distracted and the signal that they processed was not as crisp as usual. It looks as if they can truly evaluate their self-confidence on every trial and opt out only when they feel unconfident.[38]

How abstract is animal self-knowledge? In monkeys at least, a recent experiment shows that it is not tied to a single overtrained context; macaques spontaneously generalize the use of the opt-out key beyond the bounds of their initial training. Once they figure out what this key means in a sensory task, they immediately use it appropriately in the novel context of a memory task. Having learned to report *I didn't perceive well*, they generalize to *I don't remember well*.[39]

These animals clearly possess some degree of self-knowledge, but might it all be unconscious? We have to be careful here, because as you may remember from Chapter 2, much of our behavior stems from unconscious mechanisms. Even self-monitoring mechanisms may unfold unconsciously. When I mistype a letter on the keyboard or when my eyes are attracted to the wrong target, my brain automatically registers these errors and corrects them, and I may never become aware of them.[40] Several arguments, however, suggest that the monkeys' self-knowledge is not based only on such subliminal automatisms. Their opt-out judgments are flexible and generalize to an untrained task. They involve pondering a past decision for several seconds, a long-term reflection whose duration is unlikely to be within the reach of unconscious processes. They require the use of an arbitrary response signal, the opt-out key. At the neurophysiological level, they involve a slow accumulation of evidence and recruit high-level areas of the parietal and prefrontal lobes.[41] If we extrapolate from what we know of the human brain, it seems unlikely that such slow and complicated second-order judgments could unfold in the absence of awareness.

If this inference is correct (and it certainly needs to be validated by more research), then animal behavior bears the hallmark of a conscious and reflexive mind. We are probably not alone in knowing that we know, and the adjective *sapiens sapiens* should no longer be uniquely attached to the genus *Homo*. Several other animal species can genuinely reflect upon their state of mind.

Uniquely Human Consciousness?

Although monkeys clearly possess a conscious neuronal workspace and may use it to ponder themselves and the external world, humans undoubtedly exhibit superior introspection. But what exactly sets the human brain apart? Is it sheer brain size? Language? Social cooperation? Long-lasting plasticity? Education?

Answering these questions is one of the most exciting tasks for future research in cognitive neuroscience. Here I will venture only a tentative answer: although we share most if not all of our core brain systems with other animal species, the human brain may be unique in its ability to combine them using a sophisticated "language of thought." René Descartes was certainly right about one thing: only *Homo sapiens* "use[s] words or other signs by composing them, as we do to declare our thoughts to others." This capacity to *compose* our thoughts may be the crucial ingredient that boosts our inner thoughts. Human uniqueness resides in the peculiar way we explicitly formulate our ideas using nested or recursive structures of symbols.

According to this argument, and in agreement with Noam Chomsky, language evolved as a representational device rather than a communication system—the main advantage that it confers is the capacity to *think* new ideas, over and above the ability to share them with others. Our brain seems to have a special knack for assigning symbols to any mental representation and for entering these symbols into entire novel combinations. The human global neuronal workspace may be unique in its capacity to formulate conscious thoughts such as "taller than Tom," "left of the red door," or "not given to John." Each of these examples combines several elementary concepts that lie in utterly different domains of competence: size (tall), person (Tom, John), space (left), color (red), object (door), logic (not), or action (give). Although each is initially encoded by a distinct brain circuit, the human mind assembles them at will—not only by associating them, as animals undoubtedly do, but by composing them using a sophisticated syntax that carefully distinguishes, for instance, "my wife's brother" from "my brother's wife," or "dog bites man" from "man bites dog."

I speculate that this compositional language of thought underlies many uniquely human abilities, from the design of complex tools to the creation of higher mathematics. And when it comes to consciousness, this capacity may explain the origins of our sophisticated capacity for self-consciousness. Humans possess an incredibly refined sense of the mind—what psychologists call a "theory of mind," an extensive set of intuitive rules that allow us to represent and reason about what others think. Indeed, all human languages have an elaborate vocabulary for mental states. Among the ten most frequent verbs in English, six refer to knowledge, feelings, or goals (*find, tell, ask, seem, feel, try*). Crucially, we apply them to ourselves as well as to others, using identical constructions with pronouns (*I* is the tenth most frequent word in English, and *you* is the eighteenth). Thus, we can represent what we know in the same exact format as what others know ("I believe X, but you believe Y"). This mentalist perspective is present right from the start: even seven-month-old infants already generalize from what they know to what others know.[42] And it may well be unique to humans: two-and-a-half-year-old children already surpass adult chimpanzees and other primates in their understanding of social events.[43]

The recursive function of human language may serve as a vehicle for complex nested thoughts that remain inaccessible to other species. Without the syntax of language, it is unclear that we could even entertain nested conscious thoughts such as *He thinks that I do not know that he lies.* Such thoughts seem to be vastly beyond the competence of our

252 Consciousness and the Brain

primate cousins.[44] Their metacognition seems to include only two steps (a thought and a degree of belief in it) rather than the potential infinity of concepts that a recursive language affords.

Alone in the primate lineage, the human neuronal workspace system may possess unique adaptations to the internal manipulation of compositional thoughts and beliefs. Neurobiological evidence, although scarce, fits with this assumption. As we discussed in Chapter 5, the prefrontal cortex, a pivotal hub of the conscious workspace, occupies a sizable portion of any primate's brain—but in the human species, it is vastly expanded.[45] Among all primates, human prefrontal neurons are the ones with the largest dendritic trees.[46] As a result, our prefrontal cortex is probably much more agile in collecting and integrating information from processors elsewhere in the brain, which may explain our uncanny ability for introspection and self-oriented thinking, detached from the external world.

Regions of the midline and anterior frontal lobe systematically activate whenever we deploy our talents for social or self-oriented reasoning.[47] One of these regions, called the frontopolar cortex, or Brodmann's area 10, is larger in *Homo sapiens* than in any other ape. (Experts debate whether it exists at all in macaque monkeys.) The underlying white matter, which supports the brain's long-distance connections, is disproportionately larger in humans compared with any other primate, even after correcting for the massive change in overall brain size.[48] All these findings make the anterior prefrontal cortex a major candidate for the locus for our special introspective skills.

Another special region is Broca's area, the left inferior frontal region that plays a critical role in human language. Its layer-3 neurons, which send long-distance projections, are more broadly spaced in humans than in other apes, again permitting a greater interconnection.[49] In this area, as well as in the midline anterior cingulate, another crucial region for self-control, Constantin von Economo discovered giant neurons that may well be unique to the brains of humans and great apes such as chimps and bonobos, as they seem to be absent in other primates, such as macaques.[50] With their giant cell bodies and long axons, these cells probably make a very significant contribution to the broadcasting of conscious messages in the human brain.

All these adaptations point to the same evolutionary trend. During hominization, the networks of our prefrontal cortex grew denser and denser, to a larger extent than would be predicted by brain size alone. Our workspace circuits expanded way beyond proportion, but this increase is probably just the tip of the iceberg. We are more than just primates with larger brains. I would not be surprised if, in the coming years, cognitive neuroscientists find that the human brain possesses unique microcircuits that give it access to a new level of recursive, language-like operations. Our primate cousins certainly possess an internal mental life and a capacity to consciously apprehend their surroundings, but our inner world is immensely richer, perhaps because of a unique faculty for thinking nested thoughts.

In summary, human consciousness is the unique outcome of two nested evolutions. In all primates, consciousness initially evolved as a communication device, with the prefrontal cortex and its associated long-distance circuits breaking the modularity of local neuronal circuits and broadcasting information across the entire brain. In humans alone, the power of this communication device was later boosted by a second evolution: the emergence of a "language of thought" that allows us to formulate sophisticated beliefs and to share them with others.

Diseases of Consciousness?

The two successive evolutions of the human workspace must rely on specific biological mechanisms laid down by particular genes. A natural question therefore is: Do diseases target the human conscious machinery? Can genetic mutations or brain impairments invert the evolutionary trend and induce a failure of the global neuronal workspace?

The long-distance cortical connections that support consciousness are likely to be fragile. Compared to any other cell type in the body, neurons are monster cells, since their axon can easily span tens of centimeters. Supporting such a long appendix, more than a thousand times larger than the cell's main body, poses unique problems of gene expression and molecular trafficking. DNA transcription always takes places in the cell's nucleus, yet somehow its end products must be routed to synapses located centimeters away. Complex biological machinery is needed to solve this logistics problem. We might therefore expect the evolved system of long-distance workspace connections to be the target of specific impairments.

Jean-Pierre Changeux and I speculate that the mysterious cluster of psychiatric symptoms called schizophrenia may begin to find an explanation at this level.[51] Schizophrenia is a common ailment affecting about 0.7 percent of adults. It is a devastating mental illness in which adolescents and young adults lose touch with reality, develop delusions and hallucinations (so-called positive symptoms), and simultaneously experience a general reduction in intellectual and emotional capacity, including disorganized speech and repetitive behaviors (the "negative" symptoms).

It long proved difficult to identify a single principle underlying this variety of manifestations. It is striking, however, that these deficits always seem to affect functions hypothetically associated with the conscious global workspace in humans: social beliefs, self-monitoring, metacognitive judgments, and even elementary access to perceptual information.[52]

Clinically, schizophrenic patients exhibit a dramatic overconfidence in their bizarre beliefs. Metacognition and theory of mind can be so seriously impaired that patients fail to distinguish their own thoughts, knowledge, actions, and memories from those of others. Schizophrenia drastically alters the conscious integration of knowledge into a coherent belief network, leading to delusions and confusions. As an example, patients' conscious memories can be flagrantly wrong—minutes after seeing a list of pictures or words, they often do not remember seeing some items, and their metacognitive knowledge of whether, when, and where they saw or learned something is often terrible. Yet, remarkably, their implicit unconscious memories may remain completely intact.[53]

Given this background, my colleagues and I wondered whether there might be a basic deficit of conscious perception in schizophrenia. We investigated schizophrenics' experience of masking—the subjective disappearance of a word or picture when it is followed, at a short interval, by another image. Our findings were very clear: the minimal duration of presentation needed to see a masked word was strongly altered in schizophrenics.[54] The threshold for conscious access was elevated: schizophrenics stayed in the subliminal zone for much longer, and they needed much more sensory evidence before they reported the experience of conscious seeing. Remarkably, their unconscious processing was intact. A subliminal digit flashed for only 29 milliseconds led to a detectable unconscious priming effect, exactly as in normal subjects. The preservation of such a subtle measure indicates that the feed-forward chain of unconscious processing, from visual recognition to the attribution of meaning, remains largely unaffected by the disease. Schizophrenics' main problem seems to lie in the global integration of incoming information into a coherent whole.

My colleagues and I have observed a similar dissociation between intact subliminal processing and impaired conscious access in patients with multiple sclerosis, a disease that affects the white matter connections of the brain.[55] At the very onset of the disease, before any other major symptoms arise, patients fail to consciously see flashed words and digits, but they still process them unconsciously. The severity of this deficit in conscious perception can be predicted from the amount of damage to the long-distance fibers that link the prefrontal cortex to the posterior regions of the visual cortex.[56] These findings are important, first, because they confirm that white matter impairments can selectively affect conscious access; and second, because a small fraction of patients with multiple sclerosis develops psychiatric disorders akin to schizophrenia, again suggesting that the loss of long-distance connections may play a crucial role in the onset of mental illness.

Brain imaging of schizophrenic patients proves that their capacity for conscious ignition is dramatically reduced. Their early visual and attentional processes can be largely intact, but they lack the massive synchronous activation that creates a P3 wave at the surface of the head and signals a conscious percept.[57] Another signature of conscious access, the sudden emergence of a coherent brain web with massive correlations between distant cortical regions in the range of beta frequencies (13–30 hertz), is also characteristically deficient.[58]

Is there even more direct evidence for an anatomical alteration of global workspace networks in schizophrenia? Yes. Diffusion tensor imaging reveals massive anomalies of the long-distance bundles of axons that link cortical regions. The fibers of the corpus callosum, which interconnect the two hemispheres, are particularly impaired, as are the connections that link the prefrontal cortex with distant regions of the cortex, hippocampus, and thalamus.[59] The outcome is a severe disruption of resting-state connectivity: during quiet rest, in schizophrenic patients the prefrontal cortex loses its status as a major interconnected hub, and activations are much less integrated into a functional whole than in normal controls.[60]

At a more microscopic level, the huge pyramidal cells in the dorsolateral prefrontal cortex (layers 2 and 3), with their extensive dendrites capable of receiving thousands of synaptic connections, are much smaller in schizophrenic patients. They exhibit fewer spines, the terminal sites of excitatory synapses whose enormous density is characteristic of the human brain. This loss of connectivity may well play a major causal role in schizophrenia. Indeed, many of the genes that are disrupted in schizophrenia affect either or both of two major molecular neurotransmission systems, the dopamine D2 and glutamate NMDA receptors, which play a key role in prefrontal synaptic transmission and plasticity.[61]

Most interesting, perhaps, is that normal adults experience a transient schizophrenia-like psychosis when taking drugs such as phencyclidine (better known as PCP, or angel dust) and ketamine. These agents act by blocking neuronal transmission, quite specifically, at excitatory synapses of the NMDA type, which are now known to be essential for the transmission of top-down messages across the long distances of the cortex.[62] In my computer simulations of the global workspace network, NMDA synapses were essential for conscious ignition: they formed the long-distance loops that linked high-level cortical areas, in a top-down manner, to the lower-level processors that originally activated them. Removing NMDA receptors from our simulation resulted in a dramatic loss of global connectivity, and ignition disappeared.[63] Other simulations show that NMDA receptors are equally important for the slow accumulation of evidence underlying thoughtful decision making.[64]

A global loss of top-down connectivity may go a long way toward explaining the negative symptoms of schizophrenia. It would not affect the feed-forward transmission of sensory information, but it would selectively prevent its global integration via long-distance top-down loops. Thus schizophrenic patients would present entirely normal feed-forward processing, including the subtle operations that induce subliminal priming. They would experience a deficit only in the subsequent ignition and information broadcasting, disrupting their capacities for conscious monitoring, top-down attention, working memory, and decision making.

What about patients' positive symptoms, their bizarre hallucinations and delusions? The cognitive neuroscientists Paul Fletcher and Chris Frith have proposed a precise explanatory mechanism, also based on an impaired propagation of information.[65] As we discussed in Chapter 2, the brain acts like a Sherlock Holmes, a sleuth that draws maximal inferences from its various inputs, whether perceptual or social. Such statistical learning requires a bidirectional exchange of information:[66] sensory regions send their messages upward in the hierarchy, and higher regions respond with top-down predictions, as part of a learning algorithm that constantly strives to account for the information arising from the senses. Learning stops when the higher-level representations are so accurate that their predictions fully match the bottom-up inputs. At this point, the brain perceives a negligible error signal (the difference between predicted and observed signals), and as a consequence, surprise is minimal: the incoming signal is no longer interesting and thus no longer triggers any learning.

Now imagine that, in schizophrenia, the top-down messages are reduced, because of impaired long-distance connections or dysfunctional NMDA receptors. This, Fletcher and Frith argue, would result in a strong mistuning of the statistical learning mechanism. Sensory inputs would never be satisfactorily explained. Error signals would forever remain, triggering an endless avalanche of interpretations. Schizophrenics would continually feel that something remains to be explained, that the world contains many hidden layers of meaning, deep levels of explanation that only they can perceive and compute. As a result, they would continually concoct far-fetched interpretations of their surroundings.

Consider, for instance, how the schizophrenic brain would monitor its own actions. Normally, whenever we move, a predictive mechanism cancels out the sensory consequences of our actions. Thanks to it, we are not surprised when we grab a coffee cup: the warm touch and light weight that our hand senses are highly predictable, and even before we act, our motor areas send a top-down prediction to our sensory areas to inform them that they are about to experience a grabbing action. This forecast works so well that when we act, we are generally unaware of touch—we become keenly aware only when our prediction goes wrong, as when we grab an unexpectedly hot mug.

Next, imagine living in a world in which top-down prediction systematically fails. Even your coffee cup feels wrong: when you grab it, its touch subtly deviates from your expectations, causing you to wonder who or what is altering your senses. Above all, speaking feels strange. You can hear your own voice while speaking, and it sounds funny. Oddities in the incoming sound constantly draw your attention. You begin to think that someone is tampering with your speech. From there it is a short step to becoming convinced that you hear voices in your head, and that evil agents, perhaps your neighbor or the CIA, control your body and perturb your life. You find yourself constantly searching for hidden causes of mysterious events that others do not even notice—a pretty accurate picture of schizophrenic symptoms.

In brief, schizophrenia seems to be a strong candidate for a disease of the long-distance connections that broadcast signals throughout the brain and form the conscious workspace system. I am not suggesting, of course, that patients with schizophrenia are unconscious zombies. My view is simply that, in schizophrenia, conscious broadcasting is much more egregiously impaired than other automatic processes. Diseases tend to respect the boundaries of the nervous system, and schizophrenia may specifically affect the biological mechanisms that sustain long-distance top-down neuronal connections.

In schizophrenics, this breakdown is not complete; otherwise, the patient would simply fall unconscious. Can such a dramatic medical condition exist? In 2007 neurologists at the University of Pennsylvania discovered an amazing new disease.[67] Young people were entering the hospital with a variety of symptoms. Many were women with ovarian cancer, but others just complained of headache, fever, or flu-like symptoms. Quickly, their disease took an unexpected turn. They developed "prominent psychiatric symptoms, including anxiety, agitation, bizarre behavior, delusional or paranoid thoughts, and visual or auditory hallucinations"—an acute, acquired, and quickly evolving form of schizophrenia. Within three weeks, the patients' consciousness started to decline. Their EEG began to exhibit slow brain waves, as when people fall asleep or into a coma. They became motionless and ceased to respond to stimulation or even to breathe by themselves. Several of them died within a few months. Others later recovered and had a normal life and mental health but confirmed that they had no memories of the unconscious episode.

What was happening? A careful inquiry revealed that all these patients suffered from a massive autoimmune disease. Their immune system, instead of watching for external intruders such as viruses or bacteria, had turned onto itself. It was selectively destroying a molecule inside the patients' body: the NMDA receptor for the neurotransmitter glutamate. As we saw earlier, this essential element of the brain plays a key role in the top-down transmission of information at cortical synapses. When a culture of neurons was exposed to serum from the patients, its NMDA synapses literally vanished within hours—but the receptor returned as soon as the lethal serum was removed.

It is fascinating that a single molecule, when wiped out, suffices to cause a selective loss of mental health and, eventually, consciousness itself. We may be witnessing the first medical condition in which a disease selectively disrupts the long-distance connections that, according to my global neuronal workspace model, underlie any conscious experience. This focused attack quickly disrupts consciousness, first inducing an artificial form of schizophrenia, then destroying the very possibility of maintaining a vigilant state. In future years, this medical condition may serve as a model disease whose molecular mechanisms shed light on psychiatric diseases, their onset, and their link to conscious experience.

Conscious Machines?

Now that we are beginning to understand the function of consciousness, its cortical architecture, its molecular basis, and even its diseases, can we envisage simulating it in the computer? Not only do I fail to see any logical problem with this possibility, I consider it to be an exciting avenue of scientific research—a grand challenge that computer science may resolve over the next decades. We are nowhere near having the capacity to build such a machine yet, but the very fact that we can make a concrete proposal about some of its key features indicates that the science of consciousness is moving forward.

In Chapter 5, I outlined a general scheme for a computer simulation of conscious access. Those ideas could serve as a basis for a new kind of software architecture. Much as a modern computer runs many special-purpose programs in parallel, our software would contain a great many specialized programs, each dedicated to a certain function, such as face recognition, movement detection, spatial navigation, speech production, or motor guidance. Some of these programs would take their inputs from inside rather than from outside the system, thus providing it with a form of introspection and self-knowledge. For instance, a specialized device for error detection might learn to predict whether the organism is likely to deviate from its current goal. Current computers possess the rudiments of this idea, since they increasingly come equipped with self-monitoring devices that probe remaining battery life, disk space, memory integrity, or internal conflicts.

I see at least three critical functions that current computers miss: flexible communication, plasticity, and autonomy. First, the programs should flexibly communicate with one another. At any given time, the output of one of the programs would be selected as the focus of interest for the entire organism. The selected information would enter the workspace, a limited-capacity system that operates in a slow and serial manner but has the huge advantage of being able to broadcast the information back to any other program. In current computers, such exchanges are usually forbidden: each application executes in a separate memory space, and its outputs cannot be shared. Programs have no general means of exchanging their expert knowledge—aside from the clipboard, which is rudimentary and under user control. The architecture I have in mind would dramatically enhance the flexibility of information exchanges by providing a sort of universal and autonomous clipboard—the global workspace.

How would the receiving programs use the information broadcast by the clipboard? My second key ingredient is a powerful learning algorithm. The individual programs would not be static but would be endowed with a capacity to discover the best use for the information they receive. Each program would adjust itself according to a brainlike learning rule that would capture the many predictive relationships that exist among its inputs. Thus, the system would adapt to its environments and even to the quirks of its own architecture, rendering it robust, for instance, to the failure of a subprogram. It would discover which of its inputs are worthy of attention and how to combine them to compute useful functions.

And that leads me to my third desired feature: autonomy. Even in the absence of any user interaction, the computer would use its own value system to decide which data are worthy of a slow conscious examination in the global workspace. Spontaneous activity would constantly let random "thoughts" enter the workspace, where they would be retained or rejected depending on their adequacy to the organism's basic goals. Even in the absence of inputs, a serial stream of fluctuating internal states would arise.

The behavior of such a simulated organism would be reminiscent of our own variety of consciousness. Without any human intervention, it would set its own goals, explore the world, and learn about its own internal states. And at any time, it would focus its resources on a single internal representation—what we may call its conscious content.

Admittedly, these ideas remain vague. Much work will be needed to turn them into a detailed blueprint. But at least in principle, I see no reason why they would not lead to an artificial consciousness.

Many thinkers disagree. Let us briefly consider their arguments. Some believe that consciousness cannot be reduced to information processing, because no amount of information processing will ever cause a subjective experience. The NYU philosopher Ned Block, for instance, concedes that the workspace machinery may explain conscious access, but he argues that it is inherently incapable of explaining our qualia—the subjective states or raw feelings of "what it is like" to experience a feeling, a pain, or a beautiful sunset.[68]

David Chalmers, a philosopher at the University of Arizona, similarly maintains that even if workspace theory explains which operations may or may not be performed consciously, it will never explain the riddle of first-person subjectivity.[69] Chalmers is famous for introducing a distinction between the easy and the hard problems of consciousness. The easy problem of consciousness, he argues, consists in explaining the many functions of the brain: How do we recognize a face, a word, or a landscape? How do we extract information from the senses and use it to guide our behavior? How do we generate sentences to describe what we feel? "Although all these questions are associated with consciousness," Chalmers argues, "they all concern the objective mechanisms of the cognitive system, and consequently, we have every reason to expect that continued work in cognitive psychology and neuroscience will answer them."[70] By contrast, the hard problem is

> the question of how physical processes in the brain give rise to subjective experience ... : the way things feel for the subject. When we see, for example, we experience visual sensations, such as that of vivid blue. Or think of the ineffable sound of a distant oboe, the agony of an intense pain, the sparkle of happiness or the meditative quality of a moment lost in thought. . . . It is these phenomena that pose the real mystery of the mind.

My opinion is that Chalmers swapped the labels: it is the "easy" problem that is hard, while the hard problem just seems hard because it engages ill-defined intuitions. Once our intuition is educated by cognitive neuroscience and computer simulations, Chalmers's hard problem will evaporate. The hypothetical concept of qualia, pure mental experience detached from any information-processing role, will be viewed as a peculiar idea of the prescientific era, much like vitalism—the misguided nineteenth-century thought that, however much detail we gather about the chemical mechanisms of living organisms, we will never account for the unique qualities of life. Modern molecular biology shattered this belief, by showing how the molecular machinery inside our cells forms a self-reproducing automaton. Likewise, the science of consciousness will keep eating away at the hard problem until it vanishes. For instance, current models of visual perception already explain not only why the human brain suffers from a variety of visual illusions but also why such illusions would appear in any rational machine confronted with the same computational problem.[71] The science of consciousness already explains significant chunks of our subjective experience, and I see no obvious limits to this approach.

A related philosophical argument proposes that, however hard we try to simulate the brain, our software will always lack a key feature of human consciousness: free will. To some people, a machine with free will is a contradiction in terms, because machines are deterministic; their behavior is determined by their internal organization and their initial state. Their actions may not be predictable, due to measurement imprecision and chaos, but they cannot deviate from the causal chain that is dictated by their physical organization. This determinism seems to leave no room for personal freedom. As the poet and philosopher Lucretius wrote in the first century BC:

> If all movement is always interconnected, the new arising from the old in a determinate order—if the atoms never swerve so as to originate some new movement that will snap the bonds of fate, the everlasting sequence of cause and effect—what is the source of the free will possessed by living things throughout the earth?[72]

Even top-notch contemporary scientists find this problem so insuperable that they search for new laws of physics. Only quantum mechanics, they argue, introduces the right element of freedom. John Eccles (1903–1997), who received the Nobel Prize in 1963 for his major discoveries on the chemical basis of signal transmission at synapses, was one of these neuroskeptics. For him, the main problem of neuroscience was to figure out "how the self controls its brain," as the title of one of his numerous books put it[73]—a questionable expression that smacks of dualism. He ended up gratuitously supposing that the mind's immaterial thoughts act on the material brain by tweaking the probabilities of quantum events at synapses.

Another brilliant contemporary scientist, the accomplished physicist Sir Roger Penrose, agrees that consciousness and free will require quantum mechanics.[74] Penrose, together with the anesthesiologist Stuart Hameroff, developed the fanciful view of the brain as a quantum computer. The ability of a quantum physical system to exist in multiple superimposed states would be exploited by the human brain to explore nearly infinite options in finite time, somehow explaining the mathematician's ability to see through Gödel's theorem.

Unfortunately, these baroque proposals rest on no solid neurobiology or cognitive science. Although the intuition that our mind chooses its actions "at will" begs for an explanation, quantum physics, the modern version of Lucretius's "swerving atoms," is no solution. Most physicists agree that the warm-blooded bath in which the brain soaks is incompatible with quantum computing, which requires cold temperatures to avoid a quick loss of quantum coherence. And the time scale at which we become aware of aspects of the external world is largely unrelated to the femtosecond (10^{-15}) scale at which this quantum decoherence typically occurs.

Most crucially, even if quantum phenomena influenced some of the brain's operations, their intrinsic unpredictability would not satisfy our notion of free will. As convincingly argued by the contemporary philosopher Daniel Dennett, a pure form of randomness in the brain does not provide us with any "kind of freedom worth having."[75] Do we really want our bodies to be haphazardly shaken around by uncontrollable swerves generated at the subatomic level—like the random twitches and tics of a patient with Tourette syndrome? Nothing could be further from our concept of freedom.

When we discuss "free will," we mean a much more interesting form of freedom. Our belief in free will expresses the idea that, under the right circumstances, we have the ability to guide our decisions by our higher-level thoughts, beliefs, values, and past experiences, and to exert control over our undesired lower-level impulses. Whenever we make an autonomous decision, we exercise our free will by considering all the available options, pondering them, and choosing the one that we favor. Some degree of chance may enter in a voluntary choice, but this is not an essential feature. Most of the time our willful acts are anything but random: they consist in a careful review of our options, followed by the deliberate selection of the one we favor.

This conception of free will requires no appeal to quantum physics and can be implemented in a standard computer. Our global neuronal workspace allows us to collect all the necessary information, both from our current senses and from our memories, synthesize it, evaluate its consequences, ponder them for as long we want, and eventually use this internal reflection to guide our actions. This is what we call a willed decision.

In thinking about free will, we therefore need to sharply distinguish two intuitions about our decisions: their fundamental indeterminacy (a dubious idea) and their autonomy (a respectable notion). Our brain states are clearly not uncaused and do not escape the laws of physics—nothing does. But our decisions are genuinely free whenever they are based on a conscious deliberation that proceeds autonomously, without any impediment, carefully weighting the pros and cons before committing to a course of action. When this occurs, we are correct in speaking of a voluntary decision—even if it is, of course, ultimately caused by our genes, our life history, and the value functions they have inscribed in our neuronal circuits. Because of fluctuations in spontaneous brain activity, our decisions may remain unpredictable, even to us. Yet this unpredictability is not a defining feature of free will; nor should it be confused with absolute indeterminacy. What counts is the autonomous decision making.

In my opinion, a machine with free will is therefore not a contradiction in terms, just a shorthand description of what we are. I have no problem imagining an artificial device capable of willfully deciding on its course of action. Even if our brain architecture were fully deterministic, as a computer simulation might be, it would still be legitimate to say that it exercises a form of free will. Whenever a neuronal architecture exhibits autonomy and deliberation, we are right in calling it "a free mind"— and once we reverse-engineer it, we will learn to mimic it in artificial machines.

In brief, neither qualia nor free will seems to pose a serious philosophical problem for the concept of a conscious machine. Reaching the end of our journey into consciousness and the brain, we realize how carefully we should treat our intuitions of what a complex neuronal machinery can achieve. The richness of information processing that an evolved network of sixteen billion cortical neurons provides lies beyond our current imagination. Our neuronal states ceaselessly fluctuate in a partially autonomous manner, creating an inner world of personal thoughts. Even when confronted with identical sensory inputs, they react differently depending on our mood, goals, and memories. Our conscious neuronal codes also vary from brain to brain. Although we all share the same overall inventory of neurons coding for color, shape, or movement, their detailed organization results from a long developmental process that sculpts each of our brains differently, ceaselessly selecting and eliminating synapses to create our unique personalities.

The neuronal code that results from this crossing of genetic rules, past experiences, and chance encounters is unique to each moment and to each person. Its immense number of states creates a rich world of inner representations, linked to the environment but not imposed by it. Subjective feelings of pain, beauty, lust, or regret correspond to stable neuronal attractors in this dynamic landscape. They are inherently subjective, because the dynamics of the brain embed its present inputs into a tapestry of past memories and future goals, thus adding a layer of personal experience to raw sensory inputs.

What emerges is a "remembered present,"[76] a personalized cipher of the here and now, thickened with lingering memories and anticipated forecasts, and constantly projecting a first-person perspective on its environment: a conscious inner world.

This exquisite biological machinery is clicking right now inside your brain. As you close this book to ponder your own existence, ignited assemblies of neurons literally make up your mind.

ACKNOWLEDGMENTS

My views on consciousness did not develop in a void. Over the past thirty years, I have been immersed in a cloud of ideas and surrounded by a dream team of colleagues who often became close friends. I am especially indebted to three of them. In the early 1990s, my mentor Jean-Pierre Changeux first suggested to me that the problem of consciousness may not be out of reach, and that we could attack it jointly from the empirical and theoretical sides. My friend Laurent Cohen then pointed me to a variety of highly pertinent neuropsychological cases. He also introduced me to Lionel Naccache, then a young medical student, now a brilliant neurologist and cognitive neuroscientist, with whom we explored the depth of subliminal processing. Our collaborations and discussions have never ceased. Jean-Pierre, Laurent, Lionel, thank you for your constant encouragement and friendship.

Paris has become an important hub for consciousness research. My laboratory greatly benefited from this stimulating environment, and I am especially grateful for the enlightening discussions I shared with Patrick Cavanagh, Sid Kouider, Jérôme Sackur, Etienne Koechlin, Kevin O'Regan, and Mathias Pessiglione. Many bright students and postdocs, often supported by the Fyssen Foundation or the excellent master's program in cognitive science at the École Normale Supérieure, enriched my lab with their energy and creativity. My gratitude goes to my Ph.D. students Lucie Charles, Antoine Del Cul, Raphael Gaillard, Jean-Rémi King, Claire Sergent, Mélanie Strauss, Lynn Uhrig, Catherine Wacongne, and Valentin Wyart, and to my postdoctoral colleagues Tristan Bekinschtein, Floris de Lange, Sébastien Marti, Kimihiro Nakamura, Moti Salti, Aaron Schurger, Jacobo Sitt, Simon van Gaal, and Filip Van Opstal for their ceaseless questions and ideas. Special thanks to Mariano Sigman for ten years of fruitful collaboration, generous sharing, and simple friendship.

Insights on consciousness come from a great variety of disciplines, labs, and researchers throughout the world. I am particularly glad to acknowledge conversations with Bernard Baars (the mind behind the first version of global workspace theory), Moshe Bar, Edoardo Bisiach, Olaf

268 *Acknowledgments*

Blanke, Ned Block, Antonio Damasio, Dan Dennett, Derek Denton, Gerry Edelman, Pascal Fries, Karl Friston, Chris Frith, Uta Frith, Mel Goodale, Tony Greenwald, John-Dylan Haynes, Biyu Jade He, Nancy Kanwisher, Markus Kiefer, Christof Koch, Victor Lamme, Dominique Lamy, Hakwan Lau, Steve Laureys, Nikos Logothetis, Lucia Melloni, Earl Miller, Adrian Owen, Josef Parvizi, Dan Pollen, Michael Posner, Alex Pouget, Marcus Raichle, Geraint Rees, Pieter Roelfsema, Niko Schiff, Mike Shadlen, Tim Shallice, Kimron Shapiro, Wolf Singer, Elizabeth Spelke, Giulio Tononi, Wim Vanduffel, Larry Weiskrantz, Mark Williams, and many others.

My research received long-term support from the Institut National de la Santé et de la Recherche Médicale (INSERM), the Commissariat à l'Energie Atomique et aux Energies Alternatives (CEA), the Collège de France, Université Paris Sud, and the European Research Council. The NeuroSpin center, located south of Paris and headed by Denis Le Bihan, provided a stimulating environment in which to pursue this highly speculative topic, and I am thankful for the support and advice of my local colleagues, including Gilles Bloch, Jean-Robert Deverre, Lucie Hertz-Pannier, Bechir Jarraya, Andreas Kleinschmidt, Jean-François Mangin, Bertrand Thirion, Gaël Varoquaux, and Virginie van Wassenhove.

While writing this book, I benefited from the hospitality of several other institutions, notably the Peter Wall Institute of Advanced Studies in Vancouver, Macquarie University in Sydney, the Institute for Advanced Studies IUSS in Pavia, the Fondation des Treilles in southern France, the Pontifical Academy of Sciences in the Vatican, . . . and La Chouannière and La Trinitaine, my family's hideaways, where many of these lines were written.

My agent John Brockman, together with his son Max Brockman, initially played an instrumental role in pushing me to write this book. Melanie Tortoroli, at Viking, patiently corrected its many successive versions. I also benefited from two readings by the sharp yet benevolent eyes of Sid Kouider and Lionel Naccache.

Last but not least, my wife Ghislaine Dehaene-Lambertz shared with me not only her amazing knowledge of all that concerns the baby's brain and mind but also the love and tenderness that make life worth living and consciousness worth having.

NOTES

INTRODUCTION: THE STUFF OF THOUGHT

1. Jouvet 1999, 169–71.
2. Damasio 1994.
3. James 1890, chap. 5.
4. Descartes's quotes are from the *Treatise on Man*, written ca. 1632–33 and first published in 1662. English translation: Descartes 1985.
5. Undoubtedly, another factor was Descartes's fear of conflict with the Church. He was only four in 1600, when Giordano Bruno was burned at the stake, and he was thirty-seven in 1633, when Galileo narrowly escaped the same fate. Descartes made sure that his masterpiece, *Le monde* (The World), which contained the highly reductionist section *L'homme* (Man), remained unpublished during his lifetime; it was not published until 1664, long after his death in 1650. Only partial allusions to it appeared in the *Discourse on Method* (1637) and *Passions of the Soul* (1649). And he was right to be wary: in 1663, the Holy See officially placed his works on the Index of Prohibited Books. So Descartes's insistence on the immateriality of the soul was perhaps, in part, a facade, a protective measure to save his life.
6. Michel de Montaigne, *The Complete Essays*, trans. Michael Andrew Screech (New York: Penguin, 1987), 2:12.
7. E.g., Posner and Snyder 1975/2004; Shallice 1979; Shallice 1972; Marcel 1983; Libet, Alberts, Wright, and Feinstein 1967; Bisiach, Luzzatti, and Perani 1979; Weiskrantz 1986; Frith 1979; Weiskrantz 1997.
8. Baars 1989.
9. Watson 1913.
10. Nisbett and Wilson 1977; Johansson, Hall, Sikstrom, and Olsson 2005.
11. The philosopher Daniel Dennett calls this approach "heterophenomenology" (Dennett 1991).

1 CONSCIOUSNESS ENTERS THE LAB

1. Crick and Koch 1990a; Crick and Koch 1990b. To be sure, many other psychologists and neuroscientists had previously emphasized a reductionist research agenda for consciousness (see Churchland 1986; Changeux 1983; Baars 1989; Weiskrantz 1986; Posner and Snyder 1975/2004; Shallice 1972). But in my opinion, the Crick and Koch papers, with their down-to-earth approach focused on vision, played an essential role in attracting experimental scientists to the field.
2. Kim and Blake 2005.
3. Posner 1994.
4. Wyart, Dehaene, and Tallon-Baudry 2012; Wyart and Tallon-Baudry 2008.
5. Gallup 1970.
6. Plotnik, de Waal, and Reiss 2006; Prior, Schwarz, and Gunturkun 2008; Reiss and Marino 2001.
7. Epstein, Lanza, and Skinner 1981.
8. For in-depth discussion of the mirror test, see Suddendorf and Butler 2013.
9. Hofstadter 2007.
10. Comte 1830–42.

11. Some scientists use the term *awareness* to refer specifically to the simple form of consciousness in which we gain access to a sensory state—what I term "conscious access to sensory information." Most dictionary definitions, however, do not agree with this restricted use of the term, and even contemporary writers tend to treat *awareness* and *consciousness* as synonyms. In this book I have been using both words synonymously, while proposing a more precise subdivision in terms of *conscious access, wakefulness, vigilance, self-consciousness,* and *metacognition*.
12. Baars 1989.
13. Schneider and Shiffrin 1977; Shiffrin and Schneider 1977; Posner and Snyder 1975/2004; Raichle, Fiesz, Videen, and MacLeod 1994; Chein and Schneider 2005.
14. New and Scholl 2008; Ramachandran and Gregory 1991.
15. Leopold and Logothetis 1996; Logothetis, Leopold, and Sheinberg 1996; Leopold and Logothetis 1999. These pioneering studies have since been replicated and extended with the more sophisticated technique of "flash suppression," which provides a much tighter control over when an image is suppressed (see, e.g., Maier, Wilke, Aura, Zhu, Ye, and Leopold 2008; Wilke, Logothetis, and Leopold 2006; Fries, Schroder, Roelfsema, Singer, and Engel 2002). Several experimenters also used brain-imaging techniques to explore the neural fate of seen and extinguished images in humans (e.g., Srinivasan, Russell, Edelman, and Tononi 1999; Lumer, Friston, and Rees 1998; Haynes, Deichmann, and Rees 2005; Haynes, Driver, and Rees 2005).
16. Wilke, Logothetis, and Leopold 2003; Tsuchiya and Koch 2005.
17. Chong, Tadin, and Blake 2005; Chong and Blake 2006.
18. Zhang, Jamison, Engel, He, and He 2011; Brascamp and Blake 2012.
19. Zhang, Jamison, Engel, He, and He 2011.
20. Brascamp and Blake 2012.
21. Raymond, Shapiro, and Arnell 1992.
22. Marti, Sigman, and Dehaene 2012.
23. Chun and Potter 1995.
24. Telford 1931; Pashler 1984; Pashler 1994; Sigman and Dehaene 2005.
25. Marti, Sackur, Sigman, and Dehaene 2010; Dehaene, Pegado, Braga, Ventura, Nunes Filho, Jobert, Dehaene-Lambertz, et al. 2010; Corallo, Sackur, Dehaene, and Sigman 2008.
26. Marti, Sigman, and Dehaene 2012; Wong 2002; Jolicoeur 1999.
27. Mack and Rock 1998.
28. Simons and Chabris 1999. See the movie at http://www.youtube.com/watch?v=vJG698U2Mvo.
29. Rensink, O'Regan, and Clark 1997. For more recent work exploiting this technique to study the behavioral and brain correlates of change detection, see Beck, Rees, Frith, and Lavie 2001; Landman, Spekreijse, and Lamme 2003; Simons and Ambinder 2005; Beck, Muggleton, Walsh, and Lavie 2006; Reddy, Quiroga, Wilken, Koch, and Fried 2006.
30. Johansson, Hall, Sikstrom, and Olsson 2005.
31. See the movie at http://www.youtube.com/watch?v=ubNF9QNEQLA.
32. For discussion, see Simons and Ambinder 2005; Landman, Spekreijse, and Lamme 2003; Block 2007.
33. Woodman and Luck 2003; Giesbrecht and Di Lollo 1998; Di Lollo, Enns, and Rensink 2000.
34. Del Cul, Dehaene, and Leboyer 2006; Gaillard, Del Cul, Naccache, Vinckier, Cohen, and Dehaene 2006; Del Cul, Baillet, and Dehaene 2007; Del Cul, Dehaene, Reyes, Bravo, and Slachevsky 2009; Sergent and Dehaene, 2004.
35. Dehaene, Naccache, Cohen, Le Bihan, Mangin, Poline, and Rivière 2001.
36. Del Cul, Dehaene, Reyes, Bravo, and Slachevsky 2009; Charles, Van Opstal, Marti, and Dehaene 2013.

37. Dehaene and Naccache 2001.
38. Ffytche, Howard, Brammer, David, Woodruff, and Williams 1998.
39. Kruger and Dunning 1999; Johansson, Hall, Sikstrom, and Olsson 2005; Nisbett and Wilson 1977.
40. Dehaene 2009; Dehaene, Naccache, Cohen, Le Bihan, Mangin, Poline, and Rivière 2001.
41. Blanke, Landis, Spinelli, and Seeck 2004; Blanke, Ortigue, Landis, and Seeck 2002.
42. Lenggenhager, Mouthon, and Blanke 2009; Lenggenhager, Tadi, Metzinger, and Blanke 2007. See also Ehrsson 2007. A precursor of this experiment is the famous "rubber hand" illusion. See Botvinick and Cohen 1998; Ehrsson, Spence, and Passingham 2004.
43. An important recent finding is that different paradigms may not block conscious access at the same processing stage. For instance, interocular competition interferes with visual processing at an earlier stage than masking (Almeida, Mahon, Nakayama, and Caramazza 2008; Breitmeyer, Koc, Ogmen, and Ziegler 2008). Comparing multiple paradigms is thus essential if the goal is to understand the necessary and sufficient conditions for conscious access.

2 FATHOMING UNCONSCIOUS DEPTHS

1. For a detailed history of ideas on the unconscious, see Ellenberger 1970.
2. Gauchet 1992.
3. For a lucid, detailed, and accessible account of the history of neuroscience, see Finger 2001.
4. Howard 1996.
5. Ibid.
6. Maudsley 1868.
7. James 1890, 211 and 208. See Ellenberger 1970 and Weinberger 2000.
8. Vladimir Nabokov, *Strong Opinions* (1973, 1990), 66.
9. Ledoux 1996.
10. Weiskrantz 1997.
11. Sahraie, Weiskrantz, Barbur, Simmons, Williams, and Brammer 1997. See also Morris, DeGelder, Weiskrantz, and Dolan 2001.
12. Morland, Le, Carroll, Hoffmann, and Pambakian 2004; Schmid, Mrowka, Turchi, Saunders, Wilke, Peters, Ye, and Leopold 2010; Schmid, Panagiotaropoulos, Augath, Logothetis, and Smirnakis 2009; Goebel, Muckli, Zanella, Singer, and Stoerig 2001.
13. Goodale, Milner, Jakobson, and Carey 1991; Milner and Goodale 1995.
14. Marshall and Halligan 1988.
15. Driver and Vuilleumier 2001; Vuilleumier, Sagiv, Hazeltine, Poldrack, Swick, Rafal, and Gabrieli 2001.
16. Sackur, Naccache, Pradat-Diehl, Azouvi, Mazevet, Katz, Cohen, and Dehaene 2008; McGlinchey-Berroth, Milberg, Verfaellie, Alexander, and Kilduff 1993.
17. Marcel 1983; Forster 1998; Forster and Davis 1984. Many subliminal priming experiments are reviewed by Kouider and Dehaene 2007.
18. Bowers, Vigliocco, and Haan 1998; Forster and Davis 1984.
19. Dehaene, Naccache, Le Clec'H, Koechlin, Mueller, Dehaene-Lambertz, van de Moortele, and Le Bihan 1998; Dehaene, Naccache, Cohen, Le Bihan, Mangin, Poline, and Rivière 2001.
20. Dehaene 2009.
21. Dehaene and Naccache 2001 or Dehaene, Naccache, Cohen, Le Bihan, Mangin, Poline, and Rivière 2001; Dehaene, Jobert, Naccache, Ciuciu, Poline, Le Bihan, and Cohen 2004.
22. Goodale, Milner, Jakobson, and Carey 1991; Milner and Goodale 1995.
23. Kanwisher 2001.
24. Treisman and Gelade 1980; Kahneman and Treisman 1984; Treisman and Souther 1986.

25. Crick 2003; Singer 1998.
26. Finkel and Edelman 1989; Edelman 1989.
27. Dehaene, Jobert, Naccache, Ciuciu, Poline, Le Bihan, and Cohen 2004.
28. Henson, Mouchlianitis, Matthews, and Kouider 2008; Kouider, Eger, Dolan, and Henson 2009; Dell'Acqua and Grainger 1999.
29. de Groot and Gobet 1996; Gobet and Simon 1998.
30. Kiesel, Kunde, Pohl, Berner, and Hoffmann 2009.
31. McGurk and MacDonald 1976.
32. A demonstration of the McGurk illusion can be found at http://www.youtube.com/watch?v=jtsfidRq2tw.
33. Hasson, Skipper, Nusbaum, and Small 2007.
34. Singer 1998.
35. Tsunoda, Yamane, Nishizaki, and Tanifuji 2001; Baker, Behrmann, and Olson 2002; Brincat and Connor 2004.
36. Dehaene 2009; Dehaene, Pegado, Braga, Ventura, Nunes Filho, Jobert, Dehaene-Lambertz, et al. 2010.
37. Davis, Coleman, Absalom, Rodd, Johnsrude, Matta, Owen, and Menon 2007.
38. A much earlier precursor is Sidis's demonstration that a letter or digit can still be named with above-chance accuracy when it is placed so far away that the viewer denies seeing anything. Sidis 1898.
39. Broadbent 1962.
40. Moray 1959.
41. Lewis 1970.
42. Marcel 1983.
43. Marcel 1980.
44. Schvaneveldt and Meyer 1976.
45. Holender 1986; Holender and Duscherer 2004.
46. Dell'Acqua and Grainger 1999; Dehaene, Naccache, Le Clec'H, Koechlin, Mueller, Dehaene-Lambertz, van de Moortele, and Le Bihan 1998; Naccache and Dehaene 2001b; Merikle 1992; Merikle and Joordens 1997.
47. Abrams and Greenwald 2000.
48. In principle, the association might even run from the letters *h-a-p-p-y* to the motor response itself. Anthony Greenwald and his colleagues refuted this interpretation, however. When the hands assigned to the "positive" and "negative" response categories were switched, the word *happy* still primed the "positive" category, even though it was now associated with a different hand. See Abrams, Klinger, and Greenwald 2002.
49. Dehaene, Naccache, LeClec'H, Koechlin, Mueller, Dehaene-Lambertz, van de Moortele, and Le Bihan 1998; Naccache and Dehaene 2001a, Naccache and Dehaene 2001b; Greenwald, Abrams, Naccache, and Dehaene 2003; Kouider and Dehaene 2009.
50. Kouider and Dehaene 2009.
51. Naccache and Dehaene 2001b; Greenwald, Abrams, Naccache, and Dehaene 2003.
52. Naccache and Dehaene 2001a.
53. Dehaene 2011.
54. Nieder and Miller 2004; Piazza, Izard, Pinel, Le Bihan, and Dehaene 2004; Piazza, Pinel, Le Bihan, and Dehaene 2007; Nieder and Dehaene 2009.
55. den Heyer and Briand 1986; Koechlin, Naccache, Block, and Dehaene 1999; Reynvoet and Brysbaert 1999; Reynvoet, Brysbaert, and Fias 2002; Reynvoet and Brysbaert 2004; Reynvoet, Gevers, and Caessens 2005.
56. Van den Bussche and Reynvoet 2007; Van den Bussche, Notebaert, and Reynvoet 2009.
57. Naccache, Gaillard, Adam, Hasboun, Clémenceau, Baulac, Dehaene, and Cohen 2005.
58. Morris, Ohman, and Dolan 1999; Morris, Ohman, and Dolan 1998.

59. Kiefer and Spitzer 2000; Kiefer 2002; Kiefer and Brendel 2006.
60. Vogel, Luck, and Shapiro 1998; Luck, Vogel, and Shapiro 1996.
61. van Gaal, Naccache, Meeuwese, van Loon, Cohen, and Dehaene 2013.
62. For a demonstration of syntactic processing without awareness, see Batterink and Neville 2013.
63. Sergent, Baillet, and Dehaene 2005.
64. Cohen, Cavanagh, Chun, and Nakayama 2012; Posner and Rothbart 1998; Posner 1994.
65. For a review of dissociations between attention and consciousness, see Koch and Tsuchiya 2007.
66. McCormick 1997.
67. Bressan and Pizzighello 2008; Tsushima, Seitz, and Watanabe 2008; Tsushima, Sasaki, and Watanabe 2006.
68. Posner and Snyder 1975.
69. Naccache, Blandín, and Dehaene 2002; see also Lachter, Forster, and Ruthruff 2004; Kentridge, Nijboer, and Heywood 2008; Kiefer and Brendel 2006.
70. Woodman and Luck 2003.
71. Marti, Sigman, and Dehaene 2012.
72. Pessiglione, Schmidt, Draganski, Kalisch, Lau, Dolan, and Frith 2007.
73. Pessiglione, Petrovic, Daunizeau, Palminteri, Dolan, and Frith 2008.
74. Jaynes 1976, 23.
75. Hadamard 1945.
76. Bechara, Damasio, Tranel, and Damasio 1997. The findings were questioned by Maia and McClelland 2004, then later clarified by Persaud, Davidson, Maniscalco, Mobbs, Passingham, Cowey, and Lau 2011.
77. Lawrence, Jollant, O'Daly, Zelaya, and Phillips 2009.
78. Dijksterhuis, Bos, Nordgren, and van Baaren 2006.
79. Yang and Shadlen 2007.
80. de Lange, van Gaal, Lamme, and Dehaene 2011.
81. Van Opstal, de Lange, and Dehaene 2011.
82. Wagner, Gais, Haider, Verleger, and Born 2004.
83. Ji and Wilson 2007; Louie and Wilson 2001.
84. van Gaal, Ridderinkhof, Fahrenfort, Scholte, and Lamme 2008.
85. van Gaal, Ridderinkhof, Scholte, and Lamme 2010.
86. Nieuwenhuis, Ridderinkhof, Blom, Band, and Kok 2001.
87. Lau and Passingham 2007; see also Reuss, Kiesel, Kunde, and Hommel 2011.
88. Lau and Rosenthal 2011; Rosenthal 2008; Bargh and Morsella 2008; Velmans 1991.

3 WHAT IS CONSCIOUSNESS GOOD FOR?

1. Turing 1952.
2. Gould 1974.
3. Gould and Lewontin 1979.
4. Velmans 1991.
5. Nørretranders 1999.
6. Lau and Rosenthal 2011; Velmans 1991; Wegner 2003. Benjamin Libet expresses a more nuanced opinion, arguing that consciousness plays no role in initiating voluntary actions but may still veto them; see Libet 2004; Libet, Gleason, Wright, and Pearl 1983.
7. Peirce 1901.
8. Pack and Born 2001.
9. Pack, Berezovskii, and Born 2001.

10. Moreno-Bote, Knill, and Pouget 2011.
11. As discussed in Chapter 1. See Brascamp and Blake 2012; Zhang, Jamison, Engel, He, and He 2011.
12. Norris 2009; Norris 2006.
13. Schvaneveldt and Meyer 1976.
14. Vul, Hanus, and Kanwisher 2009; Vul, Nieuwenstein, and Kanwisher 2008.
15. Vul and Pashler 2008.
16. Fuster 1973; Fuster 2008; Funahashi, Bruce, and Goldman-Rakic 1989; Goldman-Rakic 1995.
17. Rounis, Maniscalco, Rothwell, Passingham, and Lau 2010; Del Cul, Dehaene, Reyes, Bravo, and Slachevsky 2009.
18. Clark, Manns, and Squire 2002; Clark and Squire 1998.
19. Carter, O'Doherty, Seymour, Koch, and Dolan 2006. See also Carter, Hofstotter, Tsuchiya, and Koch 2003. The value of the memory-trace conditioning test remains debated, however, because some vegetative-state patients seem to pass the test. See Bekinschtein, Shalom, Forcato, Herrera, Coleman, Manes, and Sigman 2009; Bekinschtein, Peeters, Shalom, and Sigman 2011.
20. Edelman 1989.
21. Han, O'Tuathaigh, van Trigt, Quinn, Fanselow, Mongeau, Koch, and Anderson 2003.
22. Mattler 2005; Greenwald, Draine, and Abrams 1996; Dupoux, de Gardelle, and Kouider 2008.
23. Naccache 2006b.
24. Soto, Mantyla, and Silvanto 2011.
25. Siegler 1987; Siegler 1988; Siegler 1989; Siegler and Jenkins 1989.
26. A recent and controversial report claims that human subjects can solve even complex subtraction problems, such as $9 - 4 - 3$, even when they are made invisible, by flashing a series of shapes to the other eye (Sklar, Levy, Goldstein, Mandel, Maril, and Hassin 2012). The design of that study, however, did not exclude the possibility that the subjects performed only part of the calculation (e.g., only $9 - 4$). Even if further research supported the capacity to combine several numbers into a calculation, I would still predict that this combination would be performed very differently under conscious and unconscious conditions. Sophisticated computations, such as the averaging of up to eight distinct numbers, may occur in parallel without consciousness (De Lange, van Gaal, Lamme, and Dehaene 2011; Van Opstal, de Lange, and Dehaene 2011). However, slow, serial, flexible, and controlled processing seems to be the prerogative of consciousness.
27. Zylberberg, Fernandez Slezak, Roelfsema, Dehaene, and Sigman 2010.
28. Zylberberg, Dehaene, Roelfsema, and Sigman 2011; Zylberberg, Fernandez Slezak, Roelfsema, Dehaene, and Sigman 2010; Zylberberg, Dehaene, Mindlin, and Sigman 2009; Dehaene and Sigman 2012. See also Shanahan and Baars 2005.
29. Turing 1936.
30. Anderson 1983; Anderson and Lebiere 1998.
31. Ashcraft and Stazyk 1981; Widaman, Geary, Cormier, and Little 1989.
32. Tombu and Jolicoeur 2003; Logan and Schulkind 2000; Moro, Tolboom, Khayat, and Roelfsema 2010.
33. Sackur and Dehaene 2009.
34. Dehaene and Cohen 2007; Dehaene 2009.
35. Calculating prodigies may seem to violate this prediction. I would, however, object that we do not know to what extent their calculating strategies do, in fact, rely on conscious and effortful strategies. After all, their calculations typically require several seconds of focused attention, during which time they cannot be distracted. They miss the verbal resources needed to explain their strategies (or refuse to do so), but this does not imply that

they draw a blank mind. For instance, some calculators report moving through vivid visual images of digit arrays or calendars (Howe and Smith 1988).

36. Sakur and Dehaene 2009.
37. de Lange, van Gaal, Lamme, and Dehaene 2011.
38. Van Opstal, de Lange, and Dehaene 2011.
39. Dijksterhuis, Bos, Nordgren, and van Baaren 2006.
40. de Lange, van Gaal, Lamme, and Dehaene 2011.
41. Levelt 1989.
42. Reed and Durlach 1998.
43. Dunbar 1996.
44. Bahrami, Olsen, Latham, Roepstorff, Rees, and Frith 2010.
45. Buckner, Andrews-Hanna, and Schacter 2008.
46. Yokoyama, Miura, Watanabe, Takemoto, Uchida, Sugiura, Horie, et al. 2010; Kikyo, Ohki, and Miyashita 2002; see also Rounis, Maniscalco, Rothwell, Passingham, and Lau 2010; Del Cul, Dehaene, Reyes, Bravo, and Slachevsky 2009; Fleming, Weil, Nagy, Dolan, and Rees 2010.
47. Saxe and Powell 2006; Perner and Aichhorn 2008.
48. Ochsner, Knierim, Ludlow, Hanelin, Ramachandran, Glover, and Mackey 2004; Vogeley, Bussfeld, Newen, Herrmann, Happe, Falkai, Maier, et al. 2001.
49. Jenkins, Macrae, and Mitchell 2008.
50. Ricoeur 1990.
51. Frith 2007.
52. Marti, Sackur, Sigman, and Dehaene 2010; Corallo, Sackur, Dehaene, and Sigman 2008.

4 THE SIGNATURES OF A CONSCIOUS THOUGHT

1. Ogawa, Lee, Kay, and Tank 1990.
2. Grill-Spector, Kushnir, Hendler, and Malach 2000.
3. Dehaene, Naccache, Cohen, Le Bihan, Mangin, Poline, and Rivière 2001.
4. Naccache and Dehaene 2001a.
5. Dehaene, Naccache, Cohen, Le Bihan, Mangin, Poline, and Rivière 2001. Nikos Logothetis and his colleagues had made similar observations using the technique of single-neuron recordings in the awake monkey; see Leopold and Logothetis 1996; Logothetis, Leopold, and Sheinberg 1996; Logothetis 1998.
6. Dehaene, Naccache, Cohen, Le Bihan, Mangin, Poline, and Rivière 2001. See also Rodriguez, George, Lachaux, Martinerie, Renault, and Varela 1999; Varela, Lachaux, Rodriguez, and Martinerie 2001 for similar suggestions but without contrasting seen and unseen stimuli.
7. Sadaghiani, Hesselmann, and Kleinschmidt 2009.
8. van Gaal, Ridderinkhof, Scholte, and Lamme 2010.
9. For more examples of prefrontal and parietal activity in relation to conscious effortful processing, see, e.g., Marois, Yi, and Chun 2004; Kouider, Dehaene, Jobert, and Le Bihan 2007; Stephan, Thaut, Wunderlich, Schicks, Tian, Tellmann, Schmitz, et al. 2002; McIntosh, Rajah, and Lobaugh 1999; Petersen, van Mier, Fiez, and Raichle 1998.
10. Sergent, Baillet, and Dehaene 2005.
11. Ibid.; Sergent and Dehaene 2004.
12. Williams, Baker, Op de Beeck, Shim, Dang, Triantafyllou, and Kanwisher 2008; Roelfsema, Lamme, and Spekreijse 1998; Roelfsema, Khayat, and Spekreijse 2003; Supèr, Spekreijse, and Lamme 2001a; Supèr, Spekreijse, and Lamme 2001b; Haynes, Driver, and Rees 2005; see also Williams, Visser, Cunnington, and Mattingley 2008.
13. Luck, Vogel, and Shapiro 1996.
14. Neuroscientists distinguish a P3a wave, which is automatically generated from a subset of regions in the mesial frontal lobe when a surprising or unexpected event occurs, and a

P3b wave, which indexes a very distributed pattern of neuronal activity spread through-out the cortex. The P3a wave may still be evoked under unconscious conditions, but the P3b wave seems to specifically index conscious states.

15. See, e.g., Lamy, Salti, and Bar-Haim 2009; Del Cul, Baillet, and Dehaene 2007; Donchin and Coles 1988; Bekinschtein, Dehaene, Rouhaut, Tadel, Cohen, and Naccache 2009; Picton 1992; Melloni, Molina, Pena, Torres, Singer, and Rodriguez 2007. For a review, see Dehaene 2011.

16. Marti, Sackur, Sigman, and Dehaene 2010; Sigman and Dehaene 2008; Marti, Sigman, and Dehaene 2012.

17. Dehaene 2008.

18. Levy, Pashler, and Boer 2006; Strayer, Drews, and Johnston 2003.

19. Pisella, Grea, Tilikete, Vighetto, Desmurget, Rode, Boisson, and Rossetti 2000.

20. The exact mechanism for this effect is still heavily debated. For glimpses of this fascinating debate, see Kanai, Carlson, Verstraten, and Walsh 2009; Eagleman and Sejnowski 2007; Krekelberg and Lappe 2001; Eagleman and Sejnowski 2000.

21. Nieuwenhuis, Ridderinkhof, Blom, Band, and Kok 2001.

22. Dehaene, Posner, and Tucker 1994; Gehring, Goss, Coles, Meyer, and Donchin 1993.

23. The idea that consciousness arises long after the fact was initially discussed by the California psychologist Benjamin Libet (see Libet 1991; Libet, Gleason, Wright, and Pearl 1983; Libet, Wright, Feinstein, and Pearl 1979; Libet, Alberts, Wright, and Feinstein 1967; Libet, Alberts, Wright, Delattre, Levin, and Feinstein 1964). His clever experiments were well in advance of their time (no pun intended). For instance, in 1967 he already noted that early event-related potentials remain present on unconsciously perceived trials and that later brain responses are a better correlate of consciousness. See Libet, Alberts, Wright, and Feinstein 1967; see also Libet 1965; Schiller and Chorover 1966. Unfortunately, his interpretations were excessive. He did not strive to identify the minimal interpretation of his findings and instead appealed to nonmaterial "mental fields" and backward time mechanisms; see Libet 2004. As a consequence, his work remained controversial; only recently have new neurophysiological interpretations of his findings been proposed (e.g., Schurger, Sitt, and Dehaene 2012).

24. Sergent, Baillet, and Dehaene 2005.

25. Lau and Passingham 2006.

26. Persaud, Davidson, Maniscalco, Mobbs, Passingham, Cowey, and Lau 2011.

27. Lamy, Salti, and Bar-Haim 2009.

28. Dehaene and Naccache 2001.

29. Hebb 1949.

30. Dehaene, Sergent, and Changeux 2003.

31. Dehaene and Naccache 2001.

32. Del Cul, Baillet, and Dehaene 2007.

33. Ibid.; Del Cul, Dehaene, and Leboyer 2006. We made similar observations in other paradigms: Sergent, Baillet, and Dehaene 2005; Sergent and Dehaene 2004. The discontinuity of conscious perception remains debated; see Overgaard, Rote, Mouridsen, and Ramsøy 2006. Part of the confusion may arise from the failure to distinguish our claim of all-or-none access to a fixed content (e.g., a digit) from the fact that the contents of consciousness may gradually change (one may see a bar, then a letter, then the entire word); see Kouider, de Gardelle, Sackur, and Dupoux 2010; Kouider and Dupoux 2004.

34. Gaillard, Dehaene, Adam, Clemenceau, Hasboun, Baulac, Cohen, and Naccache 2009; Gaillard, Del Cul, Naccache, Vinckier, Cohen, and Dehaene 2006; Gaillard, Naccache, Pinel, Clemenceau, Volle, Hasboun, Dupont, et al., 2006.

35. Fisch, Privman, Ramot, Harel, Nir, Kipervasser, Andelman, et al. 2009; Quiroga, Mukamel, Isham, Malach, and Fried 2008; Kreiman, Fried, and Koch 2002.

36. Gaillard, Dehaene, Adam, Clemenceau, Hasboun, Baulac, Cohen, and Naccache 2009.
37. Fisch, Privman, Ramot, Harel, Nir, Kipervasser, Andelman, et al. 2009.
38. Gaillard, Dehaene, Adam, Clemenceau, Hasboun, Baulac, Cohen, and Naccache 2009; Fisch, Privman, Ramot, Harel, Nir, Kipervasser, Andelman, et al. 2009; Aru, Axmacher, Do Lam, Fell, Elger, Singer, and Melloni 2012.
39. Whittingstall and Logothetis 2009; Fries, Nikolic, and Singer 2007; Cardin, Carlen, Meletis, Knoblich, Zhang, Deisseroth, Tsai, and Moore 2009; Buzsaki 2006.
40. Fries 2005.
41. Womelsdorf, Schoffelen, Oostenveld, Singer, Desimone, Engel, and Fries 2007; Fries 2005; Varela, Lachaux, Rodriguez, and Martinerie 2001.
42. Rodriguez, George, Lachaux, Martinerie, Renault, and Varela 1999; Gaillard, Dehaene, Adam, Clemenceau, Hasboun, Baulac, Cohen, and Naccache 2009; Gross, Schmitz, Schnitzler, Kessler, Shapiro, Hommel, and Schnitzler 2004; Melloni, Molina, Pena, Torres, Singer, and Rodriguez 2007.
43. Varela, Lachaux, Rodriguez, and Martinerie 2001.
44. He, Snyder, Zempel, Smyth, and Raichle 2008; He, Zempel, Snyder, and Raichle 2010; Canolty, Edwards, Dalal, Soltani, Nagarajan, Kirsch, Berger, et al. 2006.
45. Gaillard, Dehaene, Adam, Clemenceau, Hasboun, Baulac, Cohen, and Naccache 2009.
46. Pins and Ffytche 2003; Palva, Linkenkaer-Hansen, Naatanen, and Palva 2005; Fahrenfort, Scholte, and Lamme 2007; Railo and Koivisto 2009; Koivisto, Lahteenmaki, Sorensen, Vangkilde, Overgaard, and Revonsuo 2008.
47. van Aalderen-Smeets, Oosstenveld, and Schwarzbach 2006; Lamy, Salti, and Bar-Haim 2009.
48. Wyart, Dehaene, and Tallon-Baudry 2012.
49. Palva, Linkenkaer-Hansen, Naatanen, and Palva 2005; Wyart and Tallon-Baudry 2009; Boly, Balteau, Schnakers, Degueldre, Moonen, Luxen, Phillips, et al. 2007; Supèr, van der Togt, Spekreijse, and Lamme 2003; Sadaghiani, Hesselmann, Friston, and Kleinschmidt 2010.
50. Nieuwenhuis, Gilzenrat, Holmes, and Cohen 2005.
51. Lesions to the brain stem nuclei in the vicinity of the locus coeruleus may induce coma; see Parvizi and Damasio 2003.
52. Haynes 2009.
53. Shady, MacLeod, and Fisher 2004; Krolak-Salmon, Henaff, Tallon-Baudry, Yvert, Guenot, Vighetto, Mauguiere, and Bertrand 2003.
54. MacLeod and He 1993; He and MacLeod 2001.
55. Quiroga, Kreiman, Koch, and Fried 2008; Quiroga, Mukamel, Isham, Malach, and Fried 2008.
56. Wyler, Ojemann, and Ward 1982; Heit, Smith, and Halgren 1988.
57. Fried, MacDonald, and Wilson 1997.
58. Quiroga, Kreiman, Koch, and Fried 2008; Quiroga, Mukamel, Isham, Malach, and Fried 2008; Quiroga, Reddy, Kreiman, Koch, and Fried 2005; Kreiman, Fried, and Koch 2002; Kreiman, Koch, and Fried 2000a; Kreiman, Koch, and Fried 2000b.
59. Quiroga, Reddy, Kreiman, Koch, and Fried 2007.
60. Quiroga, Mukamel, Isham, Malach, and Fried 2008.
61. Kreiman, Fried, and Koch 2002. This research builds upon pioneering research by Nikos Logothetis and David Leopold in the macaque monkey, where animals were trained to report their conscious perception while neuronal discharges were being recorded. See Leopold and Logothetis 1996; Logothetis, Leopold, and Sheinberg 1996; Leopold and Logothetis 1999.
62. Kreiman, Koch, and Fried 2000b.
63. Fisch, Privman, Ramot, Harel, Nir, Kipervasser, Andelman, et al. 2009.

64. Vogel, McCollough, and Machizawa 2005; Vogel and Machizawa 2004.
65. Schurger, Pereira, Treisman, and Cohen 2009.
66. Dean and Platt 2006.
67. Derdikman and Moser 2010.
68. Jezek, Henriksen, Treves, Moser, and Moser 2011.
69. Peyrache, Khamassi, Benchenane, Wiener, and Battaglia 2009; Ji and Wilson 2007; Louie and Wilson 2001.
70. Horikawa, Tamaki, Miyawaki, and Kamitani 2013.
71. Thompson 1910; Magnusson and Stevens 1911.
72. Barker, Jalinous, and Freeston 1985; Pascual-Leone, Walsh, and Rothwell 2000; Hallett 2000.
73. Selimbeyoglu and Parvizi 2010; Parvizi, Jacques, Foster, Withoft, Rangarajan, Weiner, and Grill-Spector 2012.
74. Selimbeyoglu and Parvizi 2010.
75. Blanke, Ortigue, Landis, and Seeck 2002.
76. Desmurget, Reilly, Richard, Szathmari, Mottolese, and Sirigu 2009.
77. Taylor, Walsh, and Eimer 2010.
78. Silvanto, Lavie, and Walsh 2005; Silvanto, Cowey, Lavie, and Walsh 2005.
79. Halelamien, Wu, and Shimojo 2007.
80. Silvanto and Cattaneo 2010.
81. Lamme and Roelfsema 2000.
82. Lamme 2006.
83. Zeki 2003 actually defends the hypothesis of a "disunity of consciousness" and speculates that each brain region encodes a distinct form of "micro-consciousness."
84. Edelman 1987; Sporns, Tononi, and Edelman 1991.
85. Lamme and Roelfsema 2000; Roelfsema 2005.
86. Lamme, Zipser, and Spekreijse 1998; Pack and Born 2001.
87. Koivisto, Railo, and Salminen-Vaparanta 2010; Koivisto, Mantyla, and Silvanto 2010.
88. On change blindness, see Beck, Muggleton, Walsh, and Lavie 2006. On binocular rivalry, Carmel, Walsh, Lavie, and Rees 2010. On inattentional blindness, Babiloni, Vecchio, Rossi, De Capua, Bartalini, Ulivelli, and Rossini 2007. On attentional blink, Kihara, Ikeda, Matsuyoshi, Hirose, Mima, Fukuyama, and Osaka 2010.
89. Kanai, Muggleton, and Walsh 2008.
90. Rounis, Maniscalco, Rothwell, Passingham, and Lau 2010. My opinion is that—contrary to focal single-pulse stimulation, which seems safe—repeated, intense, and bilateral stimulation, as used by Rounis, Maniscalco, Rothwell, Passingham, and Lau 2010, should be avoided. Although the effect of such stimulation is reputed to wear out within the following hour, psychiatrists routinely apply repeated transcranial stimulation over longer periods in order to induce a month-long remission from depression, with detectable long-term changes in brain anatomy (e.g., May, Hajak, Ganssbauer, Steffens, Langguth, Kleinjung and Eichhammer 2007). In the current state of knowledge, I would not let them do it to my brain.
91. Carlen, Meletis, Siegle, Cardin, Futai, Vierling-Claassen, Ruhlmann, et al. 2011; Cardin, Carlen, Meletis, Knoblich, Zhang, Deisseroth, Tsai, and Moore 2009.
92. Adamantidis, Zhang, Aravanis, Deisseroth, and de Lecea 2007.

5 THEORIZING CONSCIOUSNESS

1. Dehaene, Kerszberg, and Changeux 1998; Dehaene, Changeux, Naccache, Sackur, and Sergent 2006; Dehaene and Naccache 2001. Global neuronal workspace theory relates directly to an earlier theory of a "global workspace," first presented by Bernard Baars in a

seminal book: Baars 1989. My colleagues and I fleshed it out in neuronal terms, specifically proposing that long-distance cortical networks play an essential role in its implementation: Dehaene, Kerszberg, and Changeux 1998.

2. Taine 1870.
3. Dennett 1991.
4. Dennett 1978.
5. Broadbent 1958.
6. Pashler 1994.
7. Chun and Potter 1995.
8. Shallice 1972; Shallice 1979; Posner and Snyder 1975; Posner and Rothbart 1998.
9. James 1890.
10. This hierarchical organization, emphasized by the British neurologist John Hughling Jackson in the nineteenth century, has become textbook knowledge in neurology.
11. van Gaal, Ridderinkhof, Fahrenfort, Scholte, and Lamme 2008; van Gaal, Ridderinkhof, Scholte, and Lamme 2010.
12. Tsao, Freiwald, Tootell, and Livingstone 2006.
13. Dehaene and Naccache 2001.
14. Denton, Shade, Zamarippa, Egan, Blair-West, McKinley, Lancaster, and Fox 1999.
15. Hagmann, Cammoun, Gigandet, Meuli, Honey, Wedeen, and Sporns 2008; Parvizi, Van Hoesen, Buckwalter, and Damasio 2006.
16. Goldman-Rakic 1988.
17. Sherman 2012.
18. Rigas and Castro-Alamancos 2007.
19. Elston 2003; Elston 2000.
20. Elston, Benavides-Piccione, and DeFelipe 2001.
21. Konopka, Wexler, Rosen, Mukamel, Osborn, Chen, Lu, et al. 2012.
22. Enard, Przeworski, Fisher, Lai, Wiebe, Kitano, Monaco, and Paabo 2002.
23. Pinel, Fauchereau, Moreno, Barbot, Lathrop, Zelenika, Le Bihan, et al. 2012.
24. Lai, Fisher, Hurst, Vargha-Khadem, and Monaco 2001.
25. Enard, Gehre, Hammerschmidt, Holter, Blass, Somel, Bruckner, et al. 2009; Vernes, Oliver, Spiteri, Lockstone, Puliyadi, Taylor, and Ho, et al. 2011.
26. Di Virgilio and Clarke 1997.
27. Tononi and Edelman 1998.
28. Hebb 1949.
29. Tsunoda, Yamane, Nishizaki, and Tanifuji 2001.
30. Selfridge 1959.
31. Felleman and Van Essen 1991; Salin and Bullier 1995.
32. Perin, Berger, and Markram 2011.
33. Hopfield 1982; Ackley, Hinton, and Sejnowski 1985; Amit 1989.
34. Crick 2003; Koch and Crick 2001.
35. Tononi 2008. Giulio Tononi has introduced a mathematical formalism for differentiation and integration that yields a quantitative measure of information integration called Φ. High values of this quantity would be necessary and sufficient for a conscious system: "consciousness is integrated information." I am reticent to accept this conclusion, however, because it leads to panpsychism, the view that any connected system, be it a colony of bacteria or a galaxy, has a certain degree of consciousness. It also fails to explain why complex yet unconscious visual and semantic processing occurs quite routinely in the human brain.
36. Meyer and Damasio 2009; Damasio 1989.
37. Edelman 1987.
38. Friston 2005; Kersten, Mamassian, and Yuille 2004.

39. Beck, Ma, Kiani, Hanks, Churchland, Roitman, Shadlen, et al. 2008.
40. Dehaene, Kerszberg, and Changeux 1998; Dehaene, Changeux, Naccache, Sackur, and Sergent 2006; Dehaene and Naccache 2001; Dehaene 2011.
41. Fries 2005; Womelsdorf, Schoffelen, Oostenveld, Singer, Desimone, Engel, and Fries 2007; Buschman and Miller 2007; Engel and Singer 2001.
42. He and Raichle 2009.
43. Rockstroh, Müller, Cohen, and Elbert 1992.
44. Vogel, McCollough, and Machizawa 2005; Vogel and Machizawa 2004.
45. Dehaene and Changeux 2005; Dehaene, Sergent, and Changeux 2003; Dehaene, Kerszberg, and Changeux 1998. Our simulations were inspired by a previous model (Lumer, Edelman, and Tononi 1997a; Lumer, Edelman, and Tononi 1997b), which was, however, limited to early visual cortex. Much more extensive and realistic simulations of the same ideas were later implemented by Ariel Zylberberg and Mariano Sigman at the University of Buenos Aires: Zylberberg, Fernandez Slezak, Roelfsema, Dehaene, and Sigman 2010; Zylberberg, Dehaene, Mindlin, and Sigman 2009. Along similar lines, Nancy Kopell and her colleagues at Boston University have developed detailed neurophysiological models of cortical dynamics, capable of simulating sleep and anesthesia: Ching, Cimenser, Purdon, Brown, and Kopell 2010; McCarthy, Brown, and Kopell 2008.
46. Ariel Zylberberg later extended the simulations to much broader networks. See Zylberberg, Fernandez Slezak, Roelfsema, Dehaene, and Sigman 2010; Zylberberg, Dehaene, Mindlin, and Sigman 2009.
47. The scientific literature contains several detailed proposals of phase transitions corresponding to anesthesia, vigilance, and conscious access. See Steyn-Ross, Steyn-Ross, and Sleigh 2004; Breshears, Roland, Sharma, Gaona, Freudenburg, Tempelhoff, Avidan, and Leuthardt 2010; Jordan, Stockmanns, Kochs, Pilge, and Schneider 2008; Ching, Cimenser, Purdon, Brown, and Kopell 2010; Dehaene and Changeux 2005.
48. Portas, Krakow, Allen, Josephs, Armony, and Frith 2000; Davis, Coleman, Absalom, Rodd, Johnsrude, Matta, Owen, and Menon 2007; Supp, Siegel, Hipp, and Engel 2011.
49. Tsodyks, Kenet, Grinvald, and Arieli 1999; Kenet, Bibitchkov, Tsodyks, Grinvald, and Arieli 2003.
50. He, Snyder, Zempel, Smyth, and Raichle 2008; Raichle, MacLeod, Snyder, Powers, Gusnard, and Shulman 2001; Raichle 2010; Greicius, Krasnow, Reiss, and Menon 2003.
51. He, Snyder, Zempel, Smyth, and Raichle 2008; Boly, Tshibanda, Vanhaudenhuyse, Noirhomme, Schnakers, Ledoux, Boveroux, et al. 2009; Greicius, Kiviniemi, Tervonen, Vainionpaa, Alahuhta, Reiss, and Menon 2008; Vincent, Patel, Fox, Snyder, Baker, Van Essen, Zempel, et al. 2007.
52. Buckner, Andrews-Hanna, and Schacter 2008.
53. Mason, Norton, Van Horn, Wegner, Grafton, and Macrae 2007; Christoff, Gordon, Smallwood, Smith, and Schooler 2009.
54. Smallwood, Beach, Schooler, and Handy 2008.
55. Dehaene and Changeux 2005.
56. Sadaghiani, Hesselmann, Friston, and Kleinschmidt 2010.
57. Raichle 2010.
58. Berkes, Orban, Lengyel, and Fiser 2011.
59. Changeux, Heidmann, and Patte 1984; Changeux and Danchin 1976; Edelman 1987; Changeux and Dehaene 1989.
60. Dehaene and Changeux 1997; Dehaene, Kerszberg, and Changeux 1998; Dehaene and Changeux 1991.
61. Rougier, Noelle, Braver, Cohen, and O'Reilly 2005.
62. Dehaene, Changeux, Naccache, Sackur, and Sergent 2006.
63. Ibid.

64. Sergent, Baillet, and Dehaene 2005; Dehaene, Sergent, and Changeux 2003; Zylberberg, Fernandez Slezak, Roelfsema, Dehaene, and Sigman 2010; Zylberberg, Dehaene, Mindlin, and Sigman 2009.
65. Sergent, Wyart, Babo-Rebelo, Cohen, Naccache, and Tallon-Baudry 2013; Marti, Sigman, and Dehaene 2012.
66. See also Enns and Di Lollo 2000; Di Lollo, Enns, and Rensink 2000.
67. Shady, MacLeod, and Fisher 2004; He and MacLeod 2001.
68. Gilbert, Sigman, and Crist 2001.
69. Haynes and Rees 2005a; Haynes and Rees 2005b; Haynes, Sakai, Rees, Gilbert, Frith, and Passingham 2007.
70. Stettler, Das, Bennett, and Gilbert 2002.
71. Gaser and Schlaug 2003; Bengtsson, Nagy, Skare, Forsman, Forssberg, and Ullen 2005.
72. Buckner and Koutstaal 1998; Buckner, Andrews-Hanna, and Schacter 2008.
73. Sigala, Kusunoki, Nimmo-Smith, Gaffan, and Duncan 2008; Saga, Iba, Tanji, and Hoshi 2011; Shima, Isoda, Mushiake, and Tanji 2007; Fujii and Graybiel 2003. For review, see Dehaene and Sigman 2012.
74. Tyler and Marslen-Wilson 2008; Griffiths, Marslen-Wilson, Stamatakis, and Tyler 2013; Pallier, Devauchelle, and Dehaene 2011; Saur, Schelter, Schnell, Kratochvil, Kupper, Kellmeyer, Kummerer, et al. 2010; Fedorenko, Duncan, and Kanwisher 2012.
75. Davis, Coleman, Absalom, Rodd, Johnsrude, Matta, Owen, and Menon 2007.
76. Beck, Ma, Kiani, Hanks, Churchland, Roitman, Shadlen, et al. 2008; Friston 2005; Deneve, Latham, and Pouget 2001.
77. Yang and Shadlen 2007.
78. Izhikevich and Edelman 2008.

6 THE ULTIMATE TEST

1. Laureys 2005.
2. Leon-Carrion, van Eeckhout, Dominguez-Morales Mdcl, and Perez-Santamaria 2002.
3. Schnakers, Vanhaudenhuyse, Giacino, Ventura, Boly, Majerus, Moonen, and Laureys 2009.
4. Smedira, Evans, Grais, Cohen, Lo, Cooke, Schecter, et al. 1990.
5. Laureys, Owen, and Schiff 2004.
6. Pontifical Academy of Sciences 2008.
7. Alving, Moller, Sindrup, and Nielsen 1979; Grindal, Suter, and Martinez 1977; Westmoreland, Klass, Sharbrough, and Reagan 1975.
8. Hanslmayr, Gross, Klimesch, and Shapiro 2011; Capotosto, Babiloni, Romani, and Corbetta 2009.
9. Supp, Siegel, Hipp, and Engel 2011.
10. Jennett and Plum 1972.
11. Jennett 2002.
12. Giacino 2005.
13. Giacino, Kezmarsky, DeLuca, and Cicerone 1991. Neurologists now use the Coma Recovery Scale Revised (CRS-R), as described by Giacino, Kalmar, and Whyte 2004. This battery of tests continues to be debated and improved. See for instance Schnakers, Vanhaudenhuyse, Giacino, Ventura, Boly, Majerus, Moonen, and Laureys 2009.
14. Giacino, Kalmar, and Whyte 2004; Schnakers, Vanhaudenhuyse, Giacino, Ventura, Boly, Majerus, Moonen, and Laureys 2009.
15. Bruno, Bernheim, Ledoux, Pellas, Demertzi, and Laureys 2011. See also Laureys 2005.
16. Owen, Coleman, Boly, Davis, Laureys, and Pickard 2006. Because this patient showed fluctuating behavioral responses to stimulation, there is an ongoing discussion among

clinicians as to whether she should have been classified as minimally conscious in the first place. Even then the contrast with her extensive and largely normal brain activation patterns would remain striking.

17. See, e.g., Davis, Coleman, Absalom, Rodd, Johnsrude, Matta, Owen, and Menon 2007; Portas, Krakow, Allen, Josephs, Armony, and Frith 2000.

18. Naccache 2006a; Nachev and Husain 2007; Greenberg 2007.

19. Ropper 2010.

20. Owen, Coleman, Boly, Davis, Laureys, Jolles, and Pickard 2007.

21. Monti, Vanhaudenhuyse, Coleman, Boly, Pickard, Tshibanda, Owen, and Laureys 2010.

22. Cyranoski 2012.

23. The undisputed pioneer of the field of EEG decoding and brain-computer interfaces is Neils Birbaumer from the University of Tübingen. For a review, see Birbaumer, Murguialday, and Cohen 2008.

24. Cruse, Chennu, Chatelle, Bekinschtein, Fernandez-Espejo, Pickard, Laureys, and Owen 2011.

25. Goldfine, Victor, Conte, Bardin, and Schiff 2012.

26. Goldfine, Victor, Conte, Bardin, and Schiff 2011.

27. Chatelle, Chennu, Noirhomme, Cruse, Owen, and Laureys 2012.

28. Hochberg, Bacher, Jarosiewicz, Masse, Simeral, Vogel, Haddadin, et al. 2012.

29. Brumberg, Nieto-Castanon, Kennedy, and Guenther 2010.

30. Squires, Squires, and Hillyard 1975; Squires, Wickens, Squires, and Donchin 1976.

31. Naatanen, Paavilainen, Rinne, and Alho 2007.

32. Wacongne, Changeux, and Dehaene 2012.

33. Although the mismatch response does not index consciousness, it is a useful clinical sign: coma patients with a clear mismatch response have a greater probability of later recovering than those who do not; see Fischer, Luaute, Adeleine, and Morlet 2004; Kane, Curry, Butler, and Cummins 1993; Naccache, Puybasset, Gaillard, Serve, and Willer 2005.

34. Bekinschtein, Dehaene, Rohaut, Tadel, Cohen, and Naccache 2009.

35. Ibid.

36. Faugeras, Rohaut, Weiss, Bekinschtein, Galanaud, Puybasset, Bolgert, et al. 2012; Faugeras, Rohaut, Weiss, Bekinschtein, Galanaud, Puybasset, Bolgert, et al. 2011.

37. Friston 2005; Wacongne, Labyt, van Wassenhove, Bekinschtein, Naccache, and Dehaene 2011.

38. King, Faugeras, Gramfort, Schurger, El Karoui, Sitt, Wacongne, et al. 2013. See also Tzovara, Rossetti, Spierer, Grivel, Murray, Oddo, and De Lucia 2012 for a similar approach.

39. Massimini, Ferrarelli, Huber, Esser, Singh, and Tononi 2005; Massimini, Boly, Casali, Rosanova, and Tononi 2009; Ferrarelli, Massimini, Sarasso, Casali, Riedner, Angelini, Tononi, and Pearce 2010.

40. Casali, Gosseries, Rosanova, Boly, Sarasso, Casali, Casarotto, et al. 2013.

41. Rosanova, Gosseries, Casarotto, Boly, Casali, Bruno, Mariotti, et al. 2012.

42. Laureys 2005; Laureys, Lemaire, Maquet, Phillips, and Franck 1999.

43. Schiff, Ribary, Moreno, Beattie, Kronberg, Blasberg, Giacino, et al. 2002; Schiff, Ribary, Plum, and Llinas 1999.

44. Galanaud, Perlbarg, Gupta, Stevens, Sanchez, Tollard, de Champfleur, et al. 2012; Tshibanda, Vanhaudenhuyse, Galanaud, Boly, Laureys, and Puybasset 2009; Galanaud, Naccache, and Puybasset 2007.

45. King, Faugeras, Gramfort, Schurger, El Karoui, Sitt, Wacongne, et al. 2013.

46. Our measure of "weighted symbolic mutual information" was inspired by an earlier proposal called "symbolic transfer entropy"; see Staniek and Lehnertz 2008.

47. Sitt, King, El Karoui, Rohaut, Faugeras, Gramfort, Cohen, et al. 2013.

48. The trade-off between high and low frequencies enters heavily into the computation of the bispectral index, a commercial system that purports to measure the depth of

unconsciousness during anesthesia. For a critical assessment, see for instance Miller, Sleigh, Barnard, and Steyn-Ross 2004; Schnakers, Ledoux, Majerus, Damas, Damas, Lambermont, Lamy, et al. 2008.

49. Schiff, Giacino, Kalmar, Victor, Baker, Gerber, Fritz, et al. 2007. The priority of this research has been questioned (Staunton 2008), as deep brain stimulation was frequently attempted in coma and vegetative-state patients starting in the 1960s. See, for instance, Tsubokawa, Yamamoto, Katayama, Hirayama, Maejima, and Moriya 1990. For a reply, see Schiff, Giacino, Kalmar, Victor, Baker, Gerber, Fritz, et al. 2008.
50. Moruzzi and Magoun 1949.
51. Shirvalkar, Seth, Schiff, and Herrera 2006.
52. Giacino, Fins, Machado, and Schiff 2012.
53. Schiff, Giacino, Kalmar, Victor, Baker, Gerber, Fritz, et al. 2007.
54. Voss, Uluc, Dyke, Watts, Kobylarz, McCandliss, Heier, et al. 2006. See also Sidaros, Engberg, Sidaros, Liptrot, Herning, Petersen, Paulson, et al. 2008.
55. Laureys, Faymonville, Luxen, Lamy, Franck, and Maquet 2000.
56. Matsuda, Matsumura, Komatsu, Yanaka, and Nose 2003.
57. Giacino, Fins, Machado, and Schiff 2012.
58. Brefel-Courbon, Payoux, Ory, Sommet, Slaoui, Raboyeau, Lemesle, et al. 2007.
59. Cohen, Chaaban, and Habert 2004.
60. Schiff 2010.
61. Striem-Amit, Cohen, Dehaene, and Amedi 2012.

7 THE FUTURE OF CONSCIOUSNESS

1. Tooley 1983.
2. Tooley 1972.
3. Singer 1993.
4. Diamond and Doar 1989; Diamond and Gilbert 1989; Diamond and Goldman-Rakic 1989.
5. Dubois, Dehaene-Lambertz, Perrin, Mangin, Cointepas, Duchesnay, Le Bihan, and Hertz-Pannier 2007; Jessica Dubois and Ghislaine Dehaene-Lambertz, ongoing research at Unicog lab, NeuroSpin Center, Gif-sur-Yvette, France.
6. Fransson, Skiold, Horsch, Nordell, Blennow, Lagercrantz, and Aden 2007; Doria, Beckmann, Arichi, Merchant, Groppo, Turkheimer, Counsell, et al. 2010; Lagercrantz and Changeux 2010.
7. Mehler, Jusczyk, Lambertz, Halsted, Bertoncini, and Amiel-Tison 1988.
8. Dehaene-Lambertz, Dehaene, and Hertz-Pannier 2002; Dehaene-Lambertz, Hertz-Pannier, and Dubois 2006; Dehaene-Lambertz, Hertz-Pannier, Dubois, Meriaux, Roche, Sigman, and Dehaene 2006; Dehaene-Lambertz, Montavont, Jobert, Allirol, Dubois, Hertz-Pannier, and Dehaene 2009.
9. Dehaene-Lambertz, Montavont, Jobert, Allirol, Dubois, Hertz-Pannier, and Dehaene 2009.
10. Leroy, Glasel, Dubois, Hertz-Pannier, Thirion, Mangin, and Dehaene-Lambertz 2011.
11. Dehaene-Lambertz, Hertz-Pannier, Dubois, Meriaux, Roche, Sigman, and Dehaene 2006.
12. Davis, Coleman, Absalom, Rodd, Johnsrude, Matta, Owen, and Menon 2007.
13. Dehaene-Lambertz, Hertz-Pannier, Dubois, Meriaux, Roche, Sigman, and Dehaene 2006.
14. Basirat, Dehaene, and Dehaene-Lambertz 2012.
15. Johnson, Dziurawiec, Ellis, and Morton 1991.
16. On infant experiments, see Gelskov and Kouider 2010; Kouider, Stahlhut, Gelskov, Barbosa, Dutat, de Gardelle, Christophe, et al. 2013. The adult paradigm, which I described in Chapter 4, was published in Del Cul, Baillet, and Dehaene 2007.
17. Diamond and Doar 1989.

18. de Haan and Nelson 1999; Csibra, Kushnerenko, and Grossman 2008.
19. Nelson, Thomas, de Haan, and Wewerka 1998.
20. Dehaene-Lambertz and Dehaene 1994.
21. Friederici, Friedrich, and Weber 2002.
22. Dubois, Dehaene-Lambertz, Perrin, Mangin, Cointepas, Duchesnay, Le Bihan, and Hertz-Pannier 2007.
23. Izard, Sann, Spelke, and Streri 2009.
24. Lagercrantz and Changeux 2009.
25. Han, O'Tuathaigh, van Trigt, Quinn, Fanselow, Mongeau, Koch, and Anderson 2003; Dos Santos Coura and Granon 2012.
26. Bolhuis and Gahr 2006.
27. Leopold and Logothetis 1996.
28. Kovacs, Vogels, and Orban 1995; Macknik and Haglund 1999.
29. Cowey and Stoerig 1995.
30. Fuster 2008.
31. Denys, Vanduffel, Fize, Nelissen, Sawamura, Georgieva, Vogels, et al. 2004.
32. Hasson, Nir, Levy, Fuhrmann, and Malach 2004.
33. Hayden, Smith, and Platt 2009.
34. Buckner, Andrews-Hanna, and Schacter 2008.
35. My colleagues and I are currently pursuing explorations of the local-global paradigm in monkeys (in collaboration with Lynn Uhrig and Bechir Jarraya) and in mice (with Karim Benchenane and Catherine Wacongne).
36. Smith, Schull, Strote, McGee, Egnor, and Erb 1995.
37. Terrace and Son 2009.
38. Hampton 2001; Kornell, Son, and Terrace 2007; Kiani and Shadlen 2009.
39. Kornell, Son, and Terrace 2007.
40. Nieuwenhuis, Ridderinkhof, Blom, Band, and Kok 2001; Logan and Crump 2010; Charles, Van Opstal, Marti, and Dehaene 2013.
41. Kiani and Shadlen 2009; Fleming, Weil, Nagy, Dolan, and Rees 2010. A specific part of the thalamus called the pulvinar, which is tightly interconnected to the prefrontal and parietal areas, also plays a key role in metacognitive judgments. See Komura, Nikkuni, Hirashima, Uetake, and Miyamoto 2013.
42. Meltzoff and Brooks 2008; Kovacs, Teglas, and Endress 2010.
43. Herrmann, Call, Hernandez-Lloreda, Hare, and Tomasello 2007.
44. Marticorena, Ruiz, Mukerji, Goddu, and Santos 2011.
45. Fuster 2008.
46. Elston, Benavides-Piccione, and DeFelipe 2001; Elston 2003.
47. Ochsner, Knierim, Ludlow, Hanelin, Ramachandran, Glover, and Mackey 2004; Saxe and Powell 2006; Fleming, Weil, Nagy, Dolan, and Rees 2010.
48. Schoenemann, Sheehan, and Glotzer 2005.
49. Schenker, Buxhoeveden, Blackmon, Amunts, Zilles, and Semendeferi 2008; Schenker, Hopkins, Spocter, Garrison, Stimpson, Erwin, Hof, and Sherwood 2009.
50. Nimchinsky, Gilissen, Allman, Perl, Erwin, and Hof 1999; Allman, Hakeem, and Watson 2002; Allman, Watson, Tetreault, and Hakeem 2005.
51. Dehaene and Changeux 2011.
52. Frith 1979; Frith 1996; Stephan, Friston, and Frith 2009.
53. Huron, Danion, Giacomoni, Grange, Robert, and Rizzo 1995; Danion, Meulemans, Kauffmann-Muller, and Vermaat 2001; Danion, Cuervo, Piolino, Huron, Riutort, Peretti, and Eustache 2005.
54. Dehaene, Artiges, Naccache, Martelli, Viard, Schurhoff, Recasens, et al. 2003; Del Cul, Dehaene, and Leboyer 2006. Our work specifically focused on the dissociation between

impaired conscious access and intact subliminal processing. For a review of earlier research into the masking deficit in schizophrenia, see McClure 2001.

55. Reuter, Del Cul, Audoin, Malikova, Naccache, Ranjeva, Lyon-Caen, et al. 2007.
56. Reuter, Del Cul, Malikova, Naccache, Confort-Gouny, Cohen, Cherif, et al. 2009.
57. Luck, Fuller, Braun, Robinson, Summerfelt, and Gold 2006; Luck, Kappenman, Fuller, Robinson, Summerfelt, and Gold 2009; Antoine Del Cul, Stanislas Dehaene, Marion Leboyer et al., unpublished experiments.
58. Uhlhaas, Linden, Singer, Haenschel, Lindner, Maurer, and Rodriguez 2006; Uhlhaas and Singer 2010.
59. Kubicki, Park, Westin, Nestor, Mulkern, Maier, Niznikiewicz, et al. 2005; Karlsgodt, Sun, Jimenez, Lutkenhoff, Willhite, van Erp, and Cannon 2008; Knochel, Oertel-Knochel, Schonmeyer, Rotarska-Jagiela, van de Ven, Prvulovic, Haenschel, et al. 2012.
60. Bassett, Bullmore, Verchinski, Mattay, Weinberger, and Meyer-Lindenberg 2008; Liu, Liang, Zhou, He, Hao, Song, Yu, et al. 2008; Bassett, Bullmore, Meyer-Lindenberg, Apud, Weinberger, and Coppola 2009; Lynall, Bassett, Kerwin, McKenna, Kitzbichler, Muller, and Bullmore 2010.
61. Ross, Margolis, Reading, Pletnikov, and Coyle 2006; Dickman and Davis 2009; Tang, Yang, Chen, Lu, Ji, Roche, and Lu 2009; Shao, Shuai, Wang, Feng, Lu, Li, Zhao, et al. 2011.
62. Self, Kooijmans, Supèr, Lamme, and Roelfsema 2012.
63. Dehaene, Sergent, and Changeux 2003; Dehaene and Changeux 2005.
64. Wong and Wang 2006.
65. Fletcher and Frith 2009; see also Stephan, Friston, and Frith 2009.
66. Friston 2005.
67. Dalmau, Tuzun, Wu, Masjuan, Rossi, Voloschin, Baehring, et al. 2007; Dalmau, Gleichman, Hughes, Rossi, Peng, Lai, Dessain, et al. 2008.
68. Block 2001; Block 2007.
69. Chalmers 1996.
70. Chalmers 1995, 81.
71. Weiss, Simoncelli, and Adelson 2002.
72. Lucretius, *De Rerum Natura* (On the Nature of Things), book 2.
73. Eccles 1994.
74. Penrose and Hameroff 1998.
75. Dennett 1984.
76. Edelman 1989.

BIBLIOGRAPHY

Abrams, R. L., and A. G. Greenwald. 2000. "Parts Outweigh the Whole (Word) in Unconscious Analysis of Meaning." *Psychological Science* 11 (2): 118–24.

Abrams, R. L., M. R. Klinger, and A. G. Greenwald. 2002. "Subliminal Words Activate Semantic Categories (Not Automated Motor Responses)." *Psychonomic Bulletin and Review* 9 (1): 100–6.

Ackley, D. H., G. E. Hinton, and T. J. Sejnowski. 1985. "A Learning Algorithm for Boltzmann Machines." *Cognitive Science* 9 (1): 147–69.

Adamantidis, A. R., F. Zhang, A. M. Aravanis, K. Deisseroth, and L. de Lecea. 2007. "Neural Substrates of Awakening Probed with Optogenetic Control of Hypocretin Neurons." *Nature* 450 (7168): 420–24.

Allman, J., A. Hakeem, and K. Watson. 2002. "Two Phylogenetic Specializations in the Human Brain." *Neuroscientist* 8 (4): 335–46.

Allman, J. M., K. K. Watson, N. A. Tetreault, and A. Y. Hakeem. 2005. "Intuition and Autism: A Possible Role for Von Economo Neurons." *Trends in Cognitive Sciences* 9 (8): 367–73.

Almeida, J., B. Z. Mahon, K. Nakayama, and A. Caramazza. 2008. "Unconscious Processing Dissociates Along Categorical Lines." *Proceedings of the National Academy of Sciences* 105 (39): 15214–18.

Alving, J., M. Moller, E. Sindrup, and B. L. Nielsen. 1979. "'Alpha Pattern Coma' Following Cerebral Anoxia." *Electroencephalography and Clinical Neurophysiology* 47 (1): 95–101.

Amit, D. 1989. *Modeling Brain Function: The World of Attractor Neural Networks.* New York: Cambridge University Press.

Anderson, J. R. 1983. *The Architecture of Cognition.* Cambridge, Mass.: Harvard University Press.

Anderson, J. R., and C. Lebiere. 1998. *The Atomic Components of Thought.* Mahwah, N.J.: Lawrence Erlbaum.

Aru, J., N. Axmacher, A. T. Do Lam, J. Fell, C. E. Elger, W. Singer, and L. Melloni. 2012. "Local Category-Specific Gamma Band Responses in the Visual Cortex Do Not Reflect Conscious Perception." *Journal of Neuroscience* 32 (43): 14909–14.

Ashcraft, M. H., and E. H. Stazyk. 1981. "Mental Addition : A Test of Three Verification Models." *Memory and Cognition* 9: 185–96.

Baars, B. J. 1989. *A Cognitive Theory of Consciousness.* Cambridge, U.K.: Cambridge University Press.

Babiloni, C., F. Vecchio, S. Rossi, A. De Capua, S. Bartalini, M. Ulivelli, and P. M. Rossini. 2007. "Human Ventral Parietal Cortex Plays a Functional Role on Visuospatial Attention and Primary Consciousness: A Repetitive Transcranial Magnetic Stimulation Study." *Cerebral Cortex* 17 (6): 1486–92.

Bahrami, B., K. Olsen, P. E. Latham, A. Roepstorff, G. Rees, and C. D. Frith. 2010. "Optimally Interacting Minds." *Science* 329 (5995): 1081–85.

Baker, C., M. Behrmann, and C. Olson. 2002. "Impact of Learning on Representation of Parts and Wholes in Monkey Inferotemporal Cortex." *Nature Neuroscience* 5 (11): 1210–16.

Bargh, J. A., and E. Morsella. 2008. "The Unconscious Mind." *Perspectives on Psychological Science* 3 (1): 73–79.

Barker, A. T., R. Jalinous, and I. L. Freeston. 1985. "Non-invasive Magnetic Stimulation of Human Motor Cortex." *Lancet* 1 (8437): 1106–7.

Basirat, A., S. Dehaene, and G. Dehaene-Lambertz. 2012. "A Hierarchy of Cortical Responses to Sequence Violations in Two-Month-Old Infants." *Cognition*, submitted.

Bassett, D. S., E. Bullmore, B. A. Verchinski, V. S. Mattay, D. R. Weinberger, and A. Meyer-Lindenberg. 2008. "Hierarchical Organization of Human Cortical Networks in Health and Schizophrenia." *Journal of Neuroscience* 28 (37): 9239–48.

Bassett, D. S., E. T. Bullmore, A. Meyer-Lindenberg, J. A. Apud, D. R. Weinberger, and R. Coppola. 2009. "Cognitive Fitness of Cost-Efficient Brain Functional Networks." *Proceedings of the National Academy of Sciences* 106 (28): 11747–52.

Batterink, L., and H. J. Neville. 2013. "The Human Brain Processes Syntax in the Absence of Conscious Awareness." *Journal of Neuroscience* 33 (19): 8528–33.

Bechara, A., H. Damasio, D. Tranel, and A. R. Damasio. 1997. "Deciding Advantageously Before Knowing the Advantageous Strategy." *Science* 275 (5304): 1293–95.

Beck, D. M., N. Muggleton, V. Walsh, and N. Lavie. 2006. "Right Parietal Cortex Plays a Critical Role in Change Blindness." *Cerebral Cortex* 16 (5): 712–17.

Beck, D. M., G. Rees, C. D. Frith, and N. Lavie. 2001. "Neural Correlates of Change Detection and Change Blindness." *Nature Neuroscience* 4: 645–50.

Beck, J. M., W. J. Ma, R. Kiani, T. Hanks, A. K. Churchland, J. Roitman, M. N. Shadlen, et al. 2008. "Probabilistic Population Codes for Bayesian Decision Making." *Neuron* 60 (6): 1142–52.

Bekinschtein, T. A., S. Dehaene, B. Rohaut, F. Tadel, L. Cohen, and L. Naccache. 2009. "Neural Signature of the Conscious Processing of Auditory Regularities." *Proceedings of the National Academy of Sciences* 106 (5): 1672–77.

Bekinschtein, T. A., M. Peeters, D. Shalom, and M. Sigman. 2011. "Sea Slugs, Subliminal Pictures, and Vegetative State Patients: Boundaries of Consciousness in Classical Conditioning." *Frontiers in Psychology* 2: 337.

Bekinschtein, T. A., D. E. Shalom, C. Forcato, M. Herrera, M. R. Coleman, F. F. Manes, and M. Sigman. 2009. "Classical Conditioning in the Vegetative and Minimally Conscious State." *Nature Neuroscience* 12 (10): 1343–49.

Bengtsson, S. L., Z. Nagy, S. Skare, L. Forsman, H. Forssberg, and F. Ullen. 2005. "Extensive Piano Practicing Has Regionally Specific Effects on White Matter Development." *Nature Neuroscience* 8 (9): 1148–50.

Berkes, P., G. Orban, M. Lengyel, and J. Fiser. 2011. "Spontaneous Cortical Activity Reveals Hallmarks of an Optimal Internal Model of the Environment." *Science* 331 (6013): 83–87.

Birbaumer, N., A. R. Murguialday, and L. Cohen. 2008. "Brain-Computer Interface in Paralysis." *Current Opinion in Neurology* 21 (6): 634–38.

Bisiach, E., C. Luzzatti, and D. Perani. 1979. "Unilateral Neglect, Representational Schema and Consciousness." *Brain* 102 (3): 609–18.

Blanke, O., T. Landis, L. Spinelli, and M. Seeck. 2004. "Out-of-Body Experience and Autoscopy of Neurological Origin." *Brain* 127 (Pt 2): 243–58.

Blanke, O., S. Ortigue, T. Landis, and M. Seeck. 2002. "Stimulating Illusory Own-Body Perceptions." *Nature* 419 (6904): 269–70.

Block, N. 2001. "Paradox and Cross Purposes in Recent Work on Consciousness." *Cognition* 79 (1–2): 197–219.

———. 2007. "Consciousness, Accessibility, and the Mesh Between Psychology and Neuroscience." *Behavioral and Brain Sciences* 30 (5–6): 481–99; discussion 499–548.

Bolhuis, J. J., and M. Gahr. 2006. "Neural Mechanisms of Birdsong Memory." *Nature Reviews Neuroscience* 7 (5): 347–57.

Boly, M., E. Balteau, C. Schnakers, C. Degueldre, G. Moonen, A. Luxen, C. Phillips, et al. 2007. "Baseline Brain Activity Fluctuations Predict Somatosensory Perception in Humans." *Proceedings of the National Academy of Sciences* 104 (29): 12187–92.

Boly, M., L. Tshibanda, A. Vanhaudenhuyse, Q. Noirhomme, C. Schnakers, D. Ledoux, P.

Boveroux, et al. 2009. "Functional Connectivity in the Default Network During Resting State Is Preserved in a Vegetative but Not in a Brain Dead Patient." *Human Brain Mapping* 30 (8): 239–400.

Botvinick, M., and J. Cohen. 1998. "Rubber Hands 'Feel' Touch That Eyes See." *Nature* 391 (6669): 756.

Bowers, J. S., G. Vigliocco, and R. Haan, 1998. "Orthographic, Phonological, and Articulatory Contributions to Masked Letter and Word Priming." *Journal of Experimental Psychology: Human Perception and Performance* 24 (6): 1705–19.

Brascamp, J. W., and R. Blake. 2012. "Inattention Abolishes Binocular Rivalry: Perceptual Evidence." *Psychological Science* 23 (10): 1159–67.

Brefel-Courbon, C., P. Payoux, F. Ory, A. Sommet, T. Slaoui, G. Raboyeau, B. Lemesle, et al. 2007. "Clinical and Imaging Evidence of Zolpidem Effect in Hypoxic Encephalopathy." *Annals of Neurology* 62 (1): 102–5.

Breitmeyer, B. G., A. Koc, H. Ogmen, and R. Ziegler. 2008. "Functional Hierarchies of Nonconscious Visual Processing." *Vision Research* 48 (14): 1509–13.

Breshears, J. D., J. L. Roland, M. Sharma, C. M. Gaona, Z. V. Freudenburg, R. Tempelhoff, M. S. Avidan, and E. C. Leuthardt. 2010. "Stable and Dynamic Cortical Electrophysiology of Induction and Emergence with Propofol Anesthesia." *Proceedings of the National Academy of Sciences* 107 (49): 21170–75.

Bressan, P., and S. Pizzighello. 2008. "The Attentional Cost of Inattentional Blindness." *Cognition* 106 (1): 370–83.

Brincat, S. L., and C. E. Connor. 2004. "Underlying Principles of Visual Shape Selectivity in Posterior Inferotemporal Cortex." *Nature Neuroscience* 7 (8): 880–86.

Broadbent, D. E. 1958. *Perception and Communication.* London: Pergamon.

———. 1962. "Attention and the Perception of Speech." *Scientific American* 206 (4): 143–51.

Brumberg, J. S., A. Nieto-Castanon, P. R. Kennedy, and F. H. Guenther. 2010. "Brain-Computer Interfaces for Speech Communication." *Speech Communication* 52 (4): 367–79.

Bruno, M. A., J. L. Bernheim, D. Ledoux, F. Pellas, A. Demertzi, and S. Laureys. 2011. "A Survey on Self-Assessed Well-Being in a Cohort of Chronic Locked-In Syndrome Patients: Happy Majority, Miserable Minority." *BMJ Open* 1 (1): e000039.

Buckner, R. L., J. R. Andrews-Hanna, and D. L. Schacter. 2008. "The Brain's Default Network: Anatomy, Function, and Relevance to Disease." *Annals of the New York Academy of Sciences* 1124: 1–38.

Buckner, R. L., and W. Koutstaal. 1998. "Functional Neuroimaging Studies of Encoding, Priming, and Explicit Memory Retrieval." *Proceedings of the National Academy of Sciences* 95 (3): 891–98.

Buschman, T. J., and E. K. Miller. 2007. "Top-Down Versus Bottom-Up Control of Attention in the Prefrontal and Posterior Parietal Cortices." *Science* 315 (5820): 1860–62.

Buzsaki, G. 2006. *Rhythms of the Brain.* New York: Oxford University Press.

Canolty, R. T., E. Edwards, S. S. Dalal, M. Soltani, S. S. Nagarajan, H. E. Kirsch, M. S. Berger, et al. 2006. "High Gamma Power Is Phase-Locked to Theta Oscillations in Human Neocortex." *Science* 313 (5793): 1626–28.

Capotosto, P., C. Babiloni, G. L. Romani, and M. Corbetta. 2009. "Frontoparietal Cortex Controls Spatial Attention Through Modulation of Anticipatory Alpha Rhythms." *Journal of Neuroscience* 29 (18): 5863–72.

Cardin, J. A., M. Carlen, K. Meletis, U. Knoblich, F. Zhang, K. Deisseroth, L. H. Tsai, and C. I. Moore. 2009. "Driving Fast-Spiking Cells Induces Gamma Rhythm and Controls Sensory Responses." *Nature* 459 (7247): 663–67.

Carlen, M., K. Meletis, J. H. Siegle, J. A. Cardin, K. Futai, D. Vierling-Claassen, C. Ruhlmann, et al. 2011. "A Critical Role for NMDA Receptors in Parvalbumin Interneurons for Gamma Rhythm Induction and Behavior." *Molecular Psychiatry* 17 (5): 537–48.

Carmel, D., V. Walsh, N. Lavie, and G. Rees. 2010. "Right Parietal TMS Shortens Dominance Durations in Binocular Rivalry." *Current Biology* 20 (18): R799–800.

Carter, R. M., C. Hofstotter, N. Tsuchiya, and C. Koch. 2003. "Working Memory and Fear Conditioning." *Proceedings of the National Academy of Sciences* 100 (3): 1399–404.

Carter, R. M., J. P. O'Doherty, B. Seymour, C. Koch, and R. J. Dolan. 2006. "Contingency Awareness in Human Aversive Conditioning Involves the Middle Frontal Gyrus." *NeuroImage* 29 (3): 1007–12.

Casali, A., O. Gosseries, M. Rosanova, M. Boly, S. Sarasso, K. R. Casali, S. Casarotto, et al. 2013. "A Theoretically Based Index of Consciousness Independent of Sensory Processing and Behavior." *Science Translational Medicine*, in press.

Chalmers, D. 1996. *The Conscious Mind*. New York: Oxford University Press.

Chalmers, D. J. 1995. "The Puzzle of Conscious Experience." *Scientific American* 273 (6): 80–86.

Changeux, J. P. 1983. *L'homme neuronal*. Paris: Fayard.

Changeux, J. P., and A. Danchin. 1976. "Selective Stabilization of Developing Synapses as a Mechanism for the Specification of Neuronal Networks." *Nature* 264: 705–12.

Changeux, J. P., and S. Dehaene. 1989. "Neuronal Models of Cognitive Functions." *Cognition* 33 (1–2): 63–109.

Changeux, J. P., T. Heidmann, and P. Patte. 1984. "Learning by Selection." In *The Biology of Learning*, edited by P. Marler and H. S. Terrace, 115–39. Springer: Berlin.

Charles, L., F. Van Opstal, S. Marti, and S. Dehaene. 2013. "Distinct Brain Mechanisms for Conscious Versus Subliminal Error Detection." *NeuroImage* 73: 80–94.

Chatelle, C., S. Chennu, Q. Noirhomme, D. Cruse, A. M. Owen, and S. Laureys. 2012. "Brain-Computer Interfacing in Disorders of Consciousness." *Brain Injury* 26 (12): 1510–22.

Chein, J. M., and W. Schneider. 2005. "Neuroimaging Studies of Practice-Related Change: fMRI and Meta-analytic Evidence of a Domain-General Control Network for Learning." *Brain Research: Cognitive Brain Research* 25 (3): 607–23.

Ching, S., A. Cimenser, P. L. Purdon, E. N. Brown, and N. J. Kopell. 2010. "Thalamocortical Model for a Propofol-Induced Alpha-Rhythm Associated with Loss of Consciousness." *Proceedings of the National Academy of Sciences* 107 (52): 22665–70.

Chong, S. C., and R. Blake. 2006. "Exogenous Attention and Endogenous Attention Influence Initial Dominance in Binocular Rivalry." *Vision Research* 46 (11): 1794–803.

Chong, S. C., D. Tadin, and R. Blake. 2005. "Endogenous Attention Prolongs Dominance Durations in Binocular Rivalry." *Journal of Vision* 5 (11): 1004–12.

Christoff, K., A. M. Gordon, J. Smallwood, R. Smith, and J. W. Schooler. 2009. "Experience Sampling During fMRI Reveals Default Network and Executive System Contributions to Mind Wandering." *Proceedings of the National Academy of Sciences* 106 (21): 8719–24.

Chun, M. M., and M. C. Potter. 1995. "A Two-Stage Model for Multiple Target Detection in Rapid Serial Visual Presentation." *Journal of Experimental Psychology: Human Perception and Performance* 21 (1): 109–27.

Churchland, P. S. 1986. *Neurophilosophy: Toward a Unified Understanding of the Mind/Brain*. Cambridge, Mass.: MIT Press.

Clark, R. E., J. R. Manns, and L. R. Squire. 2002. "Classical Conditioning, Awareness, and Brain Systems." *Trends in Cognitive Sciences* 6 (12): 524–31.

Clark, R. E., and L. R. Squire. 1998. "Classical Conditioning and Brain Systems: The Role of Awareness." *Science* 280 (5360): 77–81.

Cohen, L., B. Chaaban, and M. O. Habert. 2004. "Transient Improvement of Aphasia with Zolpidem." *New England Journal of Medicine* 350 (9): 949–50.

Cohen, M. A., P. Cavanagh, M. M. Chun, and K. Nakayama. 2012. "The Attentional Requirements of Consciousness." *Trends in Cognitive Sciences* 16 (8): 411–17.

Comte, A. 1830–42. *Cours de philosophie positive*. Paris: Bachelier.

Corallo, G., J. Sackur, S. Dehaene, and M. Sigman. 2008. "Limits on Introspection: Distorted Subjective Time During the Dual-Task Bottleneck." *Psychological Science* 19 (11): 1110–17.

Cowey, A., and P. Stoerig. 1995. "Blindsight in Monkeys." *Nature* 373 (6511): 247–49.

Crick, F., and C. Koch. 1990a. "Some Reflections on Visual Awareness." *Cold Spring Harbor Symposia on Quantitative Biology* 55: 953–62.

———. 1990b. "Toward a Neurobiological Theory of Consciousness." *Seminars in Neuroscience* 2: 263–75.

———. 2003. "A Framework for Consciousness." *Nature Neuroscience* 6 (2): 119–26.

Cruse, D., S. Chennu, C. Chatelle, T. A. Bekinschtein, D. Fernandez-Espejo, J. D. Pickard, S. Laureys, and A. M. Owen. 2011. "Bedside Detection of Awareness in the Vegetative State: A Cohort Study." *Lancet* 378 (9809): 2088–94.

Csibra, G., E. Kushnerenko, and T. Grossman. 2008. "Electrophysiological Methods in Studying Infant Cognitive Development." In *Handbook of Developmental Cognitive Neuroscience*, 2nd ed., edited by C. A. Nelson and M. Luciana. Cambridge, Mass.: MIT Press.

Cyranoski, D. 2012. "Neuroscience: The Mind Reader." *Nature* 486 (7402): 178–80.

Dalmau, J., A. J. Gleichman, E. G. Hughes, J. E. Rossi, X. Peng, M. Lai, S. K. Dessain, et al. 2008. "Anti-NMDA-Receptor Encephalitis: Case Series and Analysis of the Effects of Antibodies." *Lancet Neurology* 7 (12): 1091–98.

Dalmau, J., E. Tuzun, H. Y. Wu, J. Masjuan, J. E. Rossi, A. Voloschin, J. M. Baehring, et al. 2007. "Paraneoplastic Anti-N-Methyl-D-Aspartate Receptor Encephalitis Associated with Ovarian Teratoma." *Annals of Neurology* 61 (1): 25–36.

Damasio, A. R. 1989. "The Brain Binds Entities and Events by Multiregional Activation from Convergence Zones." *Neural Computation* 1: 123–32.

———. 1994. *Descartes' Error: Emotion, Reason, and the Human Brain.* New York: G. P. Putnam.

Danion, J. M., C. Cuervo, P. Piolino, C. Huron, M. Riutort, C. S. Peretti, and F. Eustache. 2005. "Conscious Recollection in Autobiographical Memory: An Investigation in Schizophrenia." *Consciousness and Cognition* 14 (3): 535–47.

Danion, J. M., T. Meulemans, F. Kauffmann-Muller, and H. Vermaat. 2001. "Intact Implicit Learning in Schizophrenia." *American Journal of Psychiatry* 158 (6): 944–48.

Davis, M. H., M. R. Coleman, A. R. Absalom, J. M. Rodd, I. S. Johnsrude, B. F. Matta, A. M. Owen, and D. K. Menon. 2007. "Dissociating Speech Perception and Comprehension at Reduced Levels of Awareness." *Proceedings of the National Academy of Sciences* 104 (41): 16032–37.

de Groot, A. D., and F. Gobet. 1996. *Perception and Memory in Chess.* Assen, Netherlands: Van Gorcum.

de Haan, M., and C. A. Nelson. 1999. "Brain Activity Differentiates Face and Object Processing in 6-Month-Old Infants." *Developmental Psychology* 35 (4): 1113–21.

de Lange, F. P., S. van Gaal, V. A. Lamme, and S. Dehaene. 2011. "How Awareness Changes the Relative Weights of Evidence During Human Decision-Making." *PLOS Biology* 9 (11): e1001203.

Dean, H. L., and M. L. Platt. 2006. "Allocentric Spatial Referencing of Neuronal Activity in Macaque Posterior Cingulate Cortex." *Journal of Neuroscience* 26 (4): 1117–27.

Dehaene, S. 2008. "Conscious and Nonconscious Processes: Distinct Forms of Evidence Accumulation?" In *Better Than Conscious? Decision Making, the Human Mind, and Implications for Institutions. Strüngmann Forum Report*, edited by C. Engel and W. Singer. Cambridge, Mass.: MIT Press.

———. 2009. *Reading in the Brain.* New York: Viking.

———. 2011. *The Number Sense*, 2nd ed. New York: Oxford University Press.

Dehaene, S., E. Artiges, L. Naccache, C. Martelli, A. Viard, F. Schurhoff, C. Recasens, et al. 2003. "Conscious and Subliminal Conflicts in Normal Subjects and Patients with Schizophrenia: The Role of the Anterior Cingulate." *Proceedings of the National Academy of Sciences* 100 (23): 13722–27.

292   *Bibliography*

Dehaene, S., and J. P. Changeux. 1991. "The Wisconsin Card Sorting Test: Theoretical Analysis and Modelling in a Neuronal Network." *Cerebral Cortex* 1: 62–79.
———. 1997. "A Hierarchical Neuronal Network for Planning Behavior." *Proceedings of the National Academy of Sciences* 94 (24): 13293–98.
———. 2005. "Ongoing Spontaneous Activity Controls Access to Consciousness: A Neuronal Model for Inattentional Blindness." *PLOS Biology* 3 (5): e141.
———. 2011. "Experimental and Theoretical Approaches to Conscious Processing." *Neuron* 70 (2): 200–27.
Dehaene, S., J. P. Changeux, L. Naccache, Sackur, J., and C. Sergent. 2006. "Conscious, Preconscious, and Subliminal Processing: A Testable Taxonomy." *Trends in Cognitive Sciences* 10 (5): 204–11.
Dehaene, S., and L. Cohen. 2007. "Cultural Recycling of Cortical Maps." *Neuron* 56 (2): 384–98.
Dehaene, S., A. Jobert, L. Naccache, P. Ciuciu, J. B. Poline, D. Le Bihan, and L. Cohen. 2004. "Letter Binding and Invariant Recognition of Masked Words: Behavioral and Neuroimaging Evidence." *Psychological Science* 15 (5): 307–13.
Dehaene, S., M. Kerszberg, and J. P. Changeux. 1998. "A Neuronal Model of a Global Workspace in Effortful Cognitive Tasks." *Proceedings of the National Academy of Sciences* 95 (24): 14529–34.
Dehaene, S., and L. Naccache. 2001. "Towards a Cognitive Neuroscience of Consciousness: Basic Evidence and a Workspace Framework." *Cognition* 79 (1–2): 1–37.
Dehaene, S., L. Naccache, L. Cohen, D. Le Bihan, J. F. Mangin, J. B. Poline, and D. Rivière. 2001. "Cerebral Mechanisms of Word Masking and Unconscious Repetition Priming." *Nature Neuroscience* 4 (7): 752–58.
Dehaene, S., L. Naccache, G. Le Clec'H, E. Koechlin, M. Mueller, G. Dehaene-Lambertz, P. F. van de Moortele, and D. Le Bihan. 1998. "Imaging Unconscious Semantic Priming." *Nature* 395 (6702): 597–600.
Dehaene, S., F. Pegado, L. W. Braga, P. Ventura, G. Nunes Filho, A. Jobert, G. Dehaene-Lambertz, et al. 2010. "How Learning to Read Changes the Cortical Networks for Vision and Language." *Science* 330 (6009): 1359–64.
Dehaene, S., M. I. Posner, and D. M. Tucker. 1994. "Localization of a Neural System for Error Detection and Compensation." *Psychological Science* 5: 303–5.
Dehaene, S., C. Sergent, and J. P. Changeux. 2003. "A Neuronal Network Model Linking Subjective Reports and Objective Physiological Data During Conscious Perception." *Proceedings of the National Academy of Sciences* 100: 8520–25.
Dehaene, S., and M. Sigman. 2012. "From a Single Decision to a Multi-step Algorithm." *Current Opinion in Neurobiology* 22 (6): 937–45.
Dehaene-Lambertz, G., and S. Dehaene. 1994. "Speed and Cerebral Correlates of Syllable Discrimination in Infants." *Nature* 370: 292–95.
Dehaene-Lambertz, G., S. Dehaene, and L. Hertz-Pannier. 2002. "Functional Neuroimaging of Speech Perception in Infants." *Science* 298 (5600): 2013–15.
Dehaene-Lambertz, G., L. Hertz-Pannier, and J. Dubois. 2006. "Nature and Nurture in Language Acquisition: Anatomical and Functional Brain-Imaging Studies in Infants." *Trends in Neurosciences* 29 (7): 367–73.
Dehaene-Lambertz, G., L. Hertz-Pannier, J. Dubois, S. Meriaux, A. Roche, M. Sigman, and S. Dehaene. 2006. "Functional Organization of Perisylvian Activation During Presentation of Sentences in Preverbal Infants." *Proceedings of the National Academy of Sciences* 103 (38): 14240–45.
Dehaene-Lambertz, G., A. Montavont, A. Jobert, L. Allirol, J. Dubois, L. Hertz-Pannier, and S. Dehaene. 2009. "Language or Music, Mother or Mozart? Structural and Environmental Influences on Infants' Language Networks." *Brain Language* 114 (2): 53–65.

Del Cul, A., S. Baillet, and S. Dehaene. 2007. "Brain Dynamics Underlying the Nonlinear Threshold for Access to Consciousness." *PLOS Biology* 5 (10): e260.

Del Cul, A., S. Dehaene, and M. Leboyer. 2006. "Preserved Subliminal Processing and Impaired Conscious Access in Schizophrenia." *Archives of General Psychiatry* 63 (12): 1313–23.

Del Cul, A., S. Dehaene, P. Reyes, E. Bravo, and A. Slachevsky. 2009. "Causal Role of Prefrontal Cortex in the Threshold for Access to Consciousness." *Brain* 132 (9): 2531–40.

Dell'Acqua, R., and J. Grainger. 1999. "Unconscious Semantic Priming from Pictures." *Cognition* 73 (1): B1–B15.

den Heyer, K., and K. Briand. 1986. "Priming Single Digit Numbers: Automatic Spreading Activation Dissipates as a Function of Semantic Distance." *American Journal of Psychology* 99 (3): 315–40.

Deneve, S., P. E. Latham, and A. Pouget. 2001. "Efficient Computation and Cue Integration with Noisy Population Codes." *Nature Neuroscience* 4 (8): 826–31.

Dennett, D. 1978. *Brainstorms.* Cambridge, Mass.: MIT Press.

———. 1984. *Elbow Room: The Varieties of Free Will Worth Wanting.* Cambridge, Mass.: MIT Press.

———. 1991. *Consciousness Explained.* London: Penguin.

Denton, D., R. Shade, F. Zamarippa, G. Egan, J. Blair-West, M. McKinley, J. Lancaster, and P. Fox. 1999. "Neuroimaging of Genesis and Satiation of Thirst and an Interoceptor-Driven Theory of Origins of Primary Consciousness." *Proceedings of the National Academy of Sciences* 96 (9): 5304–9.

Denys, K., W. Vanduffel, D. Fize, K. Nelissen, H. Sawamura, S. Georgieva, R. Vogels, et al. 2004. "Visual Activation in Prefrontal Cortex Is Stronger in Monkeys Than in Humans." *Journal of Cognitive Neuroscience* 16 (9): 1505–16.

Derdikman, D., and E. I. Moser. 2010. "A Manifold of Spatial Maps in the Brain." *Trends in Cognitive Sciences* 14 (12): 561–69.

Descartes, R. 1985. *The Philosophical Writings of Descartes.* Translated by J. Cottingham, R. Stoothoff, and D. Murdoch. New York: Cambridge University Press.

Desmurget, M., K. T. Reilly, N. Richard, A. Szathmari, C. Mottolese, and A. Sirigu. 2009. "Movement Intention After Parietal Cortex Stimulation in Humans." *Science* 324 (5928): 811–13.

Di Lollo, V., J. T. Enns, and R. A. Rensink. 2000. "Competition for Consciousness Among Visual Events: The Psychophysics of Reentrant Visual Processes." *Journal of Experimental Psychology: General* 129 (4): 481–507.

Di Virgilio, G., and S. Clarke. 1997. "Direct Interhemispheric Visual Input to Human Speech Areas." *Human Brain Mapping* 5 (5): 347–54.

Diamond, A., and B. Doar. 1989. "The Performance of Human Infants on a Measure of Frontal Cortex Function, the Delayed Response Task." *Developmental Psychobiology* 22 (3): 271–94.

Diamond, A., and J. Gilbert. 1989. "Development as Progressive Inhibitory Control of Action: Retrieval of a Contiguous Object." *Cognitive Development* 4 (3): 223–50.

Diamond, A., and P. S. Goldman-Rakic. 1989. "Comparison of Human Infants and Rhesus Monkeys on Piaget's A-not-B Task: Evidence for Dependence on Dorsolateral Prefrontal Cortex." *Experimental Brain Research* 74 (1): 24–40.

Dickman, D. K., and G. W. Davis. 2009. "The Schizophrenia Susceptibility Gene Dysbindin Controls Synaptic Homeostasis." *Science* 326 (5956): 1127–30.

Dijksterhuis, A., M. W. Bos, L. F. Nordgren, and R. B. van Baaren. 2006. "On Making the Right Choice: The Deliberation-Without-Attention Effect." *Science* 311 (5763): 1005–7.

Donchin, E., and M. G. H. Coles. 1988. "Is the P300 Component a Manifestation of Context Updating?" *Behavioral and Brain Sciences* 11 (3): 357–427.

Doria, V., C. F. Beckmann, T. Arichi, N. Merchant, M. Groppo, F. E. Turkheimer, S. J. Counsell, et al. 2010. "Emergence of Resting State Networks in the Preterm Human Brain." *Proceedings of the National Academy of Sciences* 107 (46): 20015–20.

Dos Santos Coura, R., and S. Granon. 2012. "Prefrontal Neuromodulation by Nicotinic Receptors for Cognitive Processes." *Psychopharmacology (Berlin)* 221 (1): 1–18.

Driver, J., and P. Vuilleumier. 2001. "Perceptual Awareness and Its Loss in Unilateral Neglect and Extinction." *Cognition* 79 (1–2): 39–88.

Dubois, J., G. Dehaene-Lambertz, M. Perrin, J. F. Mangin, Y. Cointepas, F. Duchesnay, D. Le Bihan, and L. Hertz-Pannier. 2007. "Asynchrony of the Early Maturation of White Matter Bundles in Healthy Infants: Quantitative Landmarks Revealed Noninvasively by Diffusion Tensor Imaging." *Human Brain Mapping* 29 (1): 14–27.

Dunbar, R. 1996. *Grooming, Gossip and the Evolution of Language*. London: Faber and Faber.

Dupoux, E., V. de Gardelle, and S. Kouider. 2008. "Subliminal Speech Perception and Auditory Streaming." *Cognition* 109 (2): 267–73.

Eagleman, D. M., and T. J. Sejnowski. 2000. "Motion Integration and Postdiction in Visual Awareness." *Science* 287 (5460): 2036–38.

———. 2007. "Motion Signals Bias Localization Judgments: A Unified Explanation for the Flash-Lag, Flash-Drag, Flash-Jump, and Frohlich Illusions." *Journal of Vision* 7 (4): 3.

Eccles, J. C. 1994. *How the Self Controls Its Brain*. New York: Springer Verlag.

Edelman, G. 1987. *Neural Darwinism*. New York: Basic Books.

———. 1989. *The Remembered Present*. New York: Basic Books.

Ehrsson, H. H. 2007. "The Experimental Induction of Out-of-Body Experiences." *Science* 317 (5841): 1048.

Ehrsson, H. H., C. Spence, and R. E. Passingham. 2004. "That's My Hand! Activity in Premotor Cortex Reflects Feeling of Ownership of a Limb." *Science* 305 (5685): 875–77.

Eliasmith, C., T. C. Stewart, X. Choo, T. Bekolay, T. DeWolf, Y. Tang, and D. Rasmussen. 2012. "A Large-Scale Model of the Functioning Brain." *Science* 338 (6111): 1202–5.

Ellenberger, H. F. 1970. *The Discovery of the Unconscious: The History and Evolution of Dynamic Psychiatry*. New York: Basic Books.

Elston, G. N. 2000. "Pyramidal Cells of the Frontal Lobe: All the More Spinous to Think With." *Journal of Neuroscience* 20 (18): RC95.

———. 2003. "Cortex, Cognition and the Cell: New Insights into the Pyramidal Neuron and Prefrontal Function." *Cerebral Cortex* 13 (11): 1124–38.

Elston, G. N., R. Benavides-Piccione, and J. DeFelipe. 2001. "The Pyramidal Cell in Cognition: A Comparative Study in Human and Monkey." *Journal of Neuroscience* 21 (17): RC163.

Enard, W., S. Gehre, K. Hammerschmidt, S. M. Holter, T. Blass, M. Somel, M. K. Bruckner, et al. 2009. "A Humanized Version of Foxp2 Affects Cortico-Basal Ganglia Circuits in Mice." *Cell* 137 (5): 961–71.

Enard, W., M. Przeworski, S. E. Fisher, C. S. Lai, V. Wiebe, T. Kitano, A. P. Monaco, and S. Paabo. 2002. "Molecular Evolution of FOXP2, a Gene Involved in Speech and Language." *Nature* 418 (6900): 869–72.

Engel, A. K., and W. Singer. 2001. "Temporal Binding and the Neural Correlates of Sensory Awareness." *Trends in Cognitive Sciences* 5 (1): 16–25.

Enns, J. T., and V. Di Lollo. 2000. "What's New in Visual Masking." *Trends in Cognitive Sciences* 4 (9): 345–52.

Epstein, R., R. P. Lanza, and B. F. Skinner. 1981. "'Self-Awareness' in the Pigeon." *Science* 212 (4495): 695–96.

Fahrenfort, J. J., H. S. Scholte, and V. A. Lamme. 2007. "Masking Disrupts Reentrant Processing in Human Visual Cortex." *Journal of Cognitive Neuroscience* 19 (9): 1488–97.

Faugeras, F., B. Rohaut, N. Weiss, T. A. Bekinschtein, D. Galanaud, L. Puybasset, F. Bolgert, et al. 2011. "Probing Consciousness with Event-Related Potentials in the Vegetative State." *Neurology* 77 (3): 264–68.

———. 2012. "Event Related Potentials Elicited by Violations of Auditory Regularities in Patients with Impaired Consciousness." *Neuropsychologia* 50 (3): 403–18.

Fedorenko, E., J. Duncan, and N. Kanwisher. 2012. "Language-Selective and Domain-General Regions Lie Side by Side Within Broca's Area." *Current Biology* 22 (21): 2059–62.

Felleman, D. J., and D. C. Van Essen. 1991. "Distributed Hierarchical Processing in the Primate Cerebral Cortex." *Cerebral Cortex* 1 (1): 1–47.

Ferrarelli, F., M. Massimini, S. Sarasso, A. Casali, B. A. Riedner, G. Angelini, G. Tononi, and R. A. Pearce. 2010. "Breakdown in Cortical Effective Connectivity During Midazolam-Induced Loss of Consciousness." *Proceedings of the National Academy of Sciences* 107 (6): 2681–86.

Ffytche, D. H., R. J. Howard, M. J. Brammer, A. David, P. Woodruff, and S. Williams. 1998. "The Anatomy of Conscious Vision: An fMRI Study of Visual Hallucinations." *Nature Neuroscience* 1 (8): 738–42.

Finger, S. 2001. *Origins of Neuroscience: A History of Explorations into Brain Function.* Oxford: Oxford University Press.

Finkel, L. H., and G. M. Edelman. 1989. "Integration of Distributed Cortical Systems by Reentry: A Computer Simulation of Interactive Functionally Segregated Visual Areas." *Journal of Neuroscience* 9 (9): 3188–208.

Fisch, L., E. Privman, M. Ramot, M. Harel, Y. Nir, S. Kipervasser, F. Andelman, et al. 2009. "Neural 'Ignition': Enhanced Activation Linked to Perceptual Awareness in Human Ventral Stream Visual Cortex." *Neuron* 64 (4): 562–74.

Fischer, C., J. Luaute, P. Adeleine, and D. Morlet. 2004. "Predictive Value of Sensory and Cognitive Evoked Potentials for Awakening from Coma." *Neurology* 63 (4): 669–73.

Fleming, S. M., R. S. Weil, Z. Nagy, R. J. Dolan, and G. Rees. 2010. "Relating Introspective Accuracy to Individual Differences in Brain Structure." *Science* 329 (5998): 1541–43.

Fletcher, P. C., and C. D. Frith. 2009. "Perceiving Is Believing: A Bayesian Approach to Explaining the Positive Symptoms of Schizophrenia." *Nature Reviews Neuroscience* 10 (1): 48–58.

Forster, K. I. 1998. "The Pros and Cons of Masked Priming." *Journal of Psycholinguistic Research* 27 (2): 203–33.

Forster, K. I., and C. Davis. 1984. "Repetition Priming and Frequency Attenuation in Lexical Access." *Journal of Experimental Psychology: Learning, Memory, and Cognition* 10 (4): 680–98.

Fransson, P., B. Skiold, S. Horsch, A. Nordell, M. Blennow, H. Lagercrantz, and U. Aden. 2007. "Resting-State Networks in the Infant Brain." *Proceedings of the National Academy of Sciences* 104 (39): 15531–36.

Fried, I., K. A. MacDonald, and C. L. Wilson. 1997. "Single Neuron Activity in Human Hippocampus and Amygdala During Recognition of Faces and Objects." *Neuron* 18 (5): 753–65.

Friederici, A. D., M. Friedrich, and C. Weber. 2002. "Neural Manifestation of Cognitive and Precognitive Mismatch Detection in Early Infancy." *NeuroReport* 13 (10): 1251–54.

Fries, P. 2005. "A Mechanism for Cognitive Dynamics: Neuronal Communication Through Neuronal Coherence." *Trends in Cognitive Sciences* 9 (10): 474–80.

Fries, P., D. Nikolic, and W. Singer. 2007. "The Gamma Cycle." *Trends in Neurosciences* 30 (7): 309–16.

Fries, P., J. H. Schroder, P. R. Roelfsema, W. Singer, and A. K. Engel. 2002. "Oscillatory Neuronal Synchronization in Primary Visual Cortex as a Correlate of Stimulus Selection." *Journal of Neuroscience* 22 (9): 3739–54.

Friston, K. 2005. "A Theory of Cortical Responses." *Philosophical Transactions of the Royal Society B: Biological Sciences* 360 (1456): 815–36.

Frith, C. 1996. "The Role of the Prefrontal Cortex in Self-Consciousness: The Case of Auditory Hallucinations." *Philosophical Transactions of the Royal Society B: Biological Sciences* 351 (1346): 1505–12.

———. 1979. "Consciousness, Information Processing and Schizophrenia." *British Journal of Psychiatry* 134 (3): 225–35.

———. 2007. *Making Up the Mind: How the Brain Creates Our Mental World.* London: Blackwell.

Fujii, N., and A. M. Graybiel. 2003. "Representation of Action Sequence Boundaries by Macaque Prefrontal Cortical Neurons." *Science* 301 (5637): 1246–49.

Funahashi, S., C. J. Bruce, and P. S. Goldman-Rakic. 1989. "Mnemonic Coding of Visual Space in the Monkey's Dorsolateral Prefrontal Cortex." *Journal of Neurophysiology* 61 (2): 331–49.

Fuster, J. M. 1973. "Unit Activity in Prefrontal Cortex During Delayed-Response Performance: Neuronal Correlates of Transient Memory." *Journal of Neurophysiology* 36 (1): 61–78.

———. 2008. *The Prefrontal Cortex,* 4th ed. London: Academic Press.

Gaillard, R., S. Dehaene, C. Adam, S. Clemenceau, D. Hasboun, M. Baulac, L. Cohen, and L. Naccache. 2009. "Converging Intracranial Markers of Conscious Access." *PLOS Biology* 7 (3): e61.

Gaillard, R., A. Del Cul, L. Naccache, F. Vinckier, L. Cohen, and S. Dehaene. 2006. "Nonconscious Semantic Processing of Emotional Words Modulates Conscious Access." *Proceedings of the National Academy of Sciences* 103 (19): 7524–29.

Gaillard, R., L. Naccache, P. Pinel, S. Clemenceau, E. Volle, D. Hasboun, S. Dupont, et al. 2006. "Direct Intracranial, fMRI, and Lesion Evidence for the Causal Role of Left Inferotemporal Cortex in Reading." *Neuron* 50 (2): 191–204.

Galanaud, D., L. Naccache, and L. Puybasset. 2007. "Exploring Impaired Consciousness: The MRI Approach." *Current Opinion in Neurology* 20 (6): 627–31.

Galanaud, D., V. Perlbarg, R. Gupta, R. D. Stevens, P. Sanchez, E. Tollard, N. M. de Champfleur, et al. 2012. "Assessment of White Matter Injury and Outcome in Severe Brain Trauma: A Prospective Multicenter Cohort." *Anesthesiology* 117 (6): 1300–10.

Gallup, G. G. 1970. "Chimpanzees: Self-Recognition." *Science* 167: 86–87.

Gaser, C., and G. Schlaug. 2003. "Brain Structures Differ Between Musicians and Nonmusicians." *Journal of Neuroscience* 23 (27): 9240–45.

Gauchet, M. 1992. *L'inconscient cérébral.* Paris: Le Seuil.

Gehring, W. J., B. Goss, M. G. H. Coles, D. E. Meyer, and E. Donchin. 1993. "A Neural System for Error Detection and Compensation." *Psychological Science* 4 (6): 385–90.

Gelskov, S. V., and S. Kouider. 2010. "Psychophysical Thresholds of Face Visibility During Infancy." *Cognition* 114 (2): 285–92.

Giacino, J., J. J. Fins, A. Machado, and N. D. Schiff. 2012. "Central Thalamic Deep Brain Stimulation to Promote Recovery from Chronic Posttraumatic Minimally Conscious State: Challenges and Opportunities." *Neuromodulation* 15 (4): 339–49.

Giacino, J. T. 2005. "The Minimally Conscious State: Defining the Borders of Consciousness." *Progress in Brain Research* 150: 381–95.

Giacino, J. T., K. Kalmar, and J. Whyte. 2004. "The JFK Coma Recovery Scale–Revised: Measurement Characteristics and Diagnostic Utility." *Archives of Physical Medicine and Rehabilitation* 85 (12): 2020–29.

Giacino, J. T., M. A. Kezmarsky, J. DeLuca, and K. D. Cicerone. 1991. "Monitoring Rate of Recovery to Predict Outcome in Minimally Responsive Patients." *Archives of Physical Medicine and Rehabilitation* 72 (11): 897–901.

Giacino, J. T., J. Whyte, E. Bagiella, K. Kalmar, N. Childs, A. Khademi, B. Eifert, et al. 2012. "Placebo-Controlled Trial of Amantadine for Severe Traumatic Brain Injury." *New England Journal of Medicine* 366 (9): 819–26.

Giesbrecht, B., and V. Di Lollo. 1998. "Beyond the Attentional Blink: Visual Masking by Object Substitution." *Journal of Experimental Psychology: Human Perception and Performance* 24 (5): 1454–66.

Gilbert, C. D., M. Sigman, and R. E. Crist. 2001. "The Neural Basis of Perceptual Learning." *Neuron* 31 (5): 681–97.

Gobet, F., and H. A. Simon. 1998. "Expert Chess Memory: Revisiting the Chunking Hypothesis." *Memory* 6 (3): 225–55.

Goebel, R., L. Muckli, F. E. Zanella, W. Singer, and P. Stoerig. 2001. "Sustained Extrastriate Cortical Activation Without Visual Awareness Revealed by fMRI Studies of Hemianopic Patients." *Vision Research* 41 (10–11): 1459–74.

Goldfine, A. M., J. D. Victor, M. M. Conte, J. C. Bardin, and N. D. Schiff. 2011. "Determination of Awareness in Patients with Severe Brain Injury Using EEG Power Spectral Analysis." *Clinical Neurophysiology* 122 (11): 2157–68.

———. 2012. "Bedside Detection of Awareness in the Vegetative State." *Lancet* 379 (9827): 1701–2.

Goldman-Rakic, P. S. 1988. "Topography of Cognition: Parallel Distributed Networks in Primate Association Cortex." *Annual Review of Neuroscience* 11: 137–56.

———. 1995. "Cellular Basis of Working Memory." *Neuron* 14 (3): 477–85.

Goodale, M. A., A. D. Milner, L. S. Jakobson, and D. P. Carey. 1991. "A Neurological Dissociation Between Perceiving Objects and Grasping Them." *Nature* 349 (6305): 154–56.

Gould, S. J. 1974. "The Origin and Function of 'Bizarre' Structures: Antler Size and Skull Size in the 'Irish Elk,' *Megaloceros giganteus.*" *Evolution* 28 (2): 191–220.

Gould, S. J., and R. C. Lewontin. 1979. "The Spandrels of San Marco and the Panglossian Paradigm: A Critique of the Adaptationist Programme." *Proceedings of the Royal Society B: Biological Sciences* 205 (1161): 581–98.

Greenberg, D. L. 2007. Comment on "Detecting Awareness in the Vegetative State." *Science* 315 (5816): 1221; author reply 1221.

Greenwald, A. G., R. L. Abrams, L. Naccache, and S. Dehaene. 2003. "Long-Term Semantic Memory Versus Contextual Memory in Unconscious Number Processing." *Journal of Experimental Psychology: Learning, Memory, Cognition* 29 (2): 235–47.

Greenwald, A. G., S. C. Draine, and R. L. Abrams. 1996. "Three Cognitive Markers of Unconscious Semantic Activation." *Science* 273 (5282): 1699–702.

Greicius, M. D., V. Kiviniemi, O. Tervonen, V. Vainionpaa, S. Alahuhta, A. L. Reiss, , and V. Menon. 2008. "Persistent Default-Mode Network Connectivity During Light Sedation." *Human Brain Mapping* 29 (7): 839–47.

Greicius, M. D., B. Krasnow, A. L. Reiss, and V. Menon. 2003. "Functional Connectivity in the Resting Brain: A Network Analysis of the Default Mode Hypothesis." *Proceedings of the National Academy of Sciences* 100 (1): 253–58.

Griffiths, J. D., W. D. Marslen-Wilson, E. A. Stamatakis, and L. K. Tyler. 2013. "Functional Organization of the Neural Language System: Dorsal and Ventral Pathways Are Critical for Syntax." *Cerebral Cortex* 23 (1): 139–47.

Grill-Spector, K., T. Kushnir, T. Hendler, and R. Malach. 2000. "The Dynamics of Object-Selective Activation Correlate with Recognition Performance in Humans." *Nature Neuroscience* 3 (8): 837–43.

Grindal, A. B., C. Suter, and A. J. Martinez. 1977. "Alpha-Pattern Coma: 24 Cases with 9 Survivors." *Annals of Neurology* 1 (4): 371–77.

Gross, J., F. Schmitz, I. Schnitzler, K. Kessler, K. Shapiro, B. Hommel, and A. Schnitzler. 2004. "Modulation of Long-Range Neural Synchrony Reflects Temporal Limitations of Visual Attention in Humans." *Proceedings of the National Academy of Sciences* 101 (35): 13050–55.

Hadamard, J. 1945. *An Essay on the Psychology of Invention in the Mathematical Field.* Princeton, N.J.: Princeton University Press.

Hagmann, P., L. Cammoun, X. Gigandet, R. Meuli, C. J. Honey, V. J. Wedeen, and O. Sporns. 2008. "Mapping the Structural Core of Human Cerebral Cortex." *PLOS Biology* 6 (7): e159.

Halelamien, N., D.-A. Wu, and S. Shimojo. 2007. "TMS Induces Detail-Rich 'Instant Replays' of Natural Images." *Journal of Vision* 7 (9).

Hallett, M. 2000. "Transcranial Magnetic Stimulation and the Human Brain." *Nature* 406 (6792): 147–50.

Hampton, R. R. 2001. "Rhesus Monkeys Know When They Remember." *Proceedings of the National Academy of Sciences* 98 (9): 5359–62.

Han, C. J., C. M. O'Tuathaigh, L. van Trigt, J. J. Quinn, M. S. Fanselow, R. Mongeau, C. Koch, and D. J. Anderson. 2003. "Trace but Not Delay Fear Conditioning Requires Attention and the Anterior Cingulate Cortex." *Proceedings of the National Academy of Sciences* 100 (22): 13087–92.

Hanslmayr, S., J. Gross, W. Klimesch, and K. L. Shapiro. 2011. "The Role of Alpha Oscillations in Temporal Attention." *Brain Research Reviews* 67 (1–2): 331–43.

Hasson, U., Y. Nir, I. Levy, G. Fuhrmann, and R. Malach. 2004. "Intersubject Synchronization of Cortical Activity During Natural Vision." *Science* 303 (5664): 1634–40.

Hasson, U., J. I. Skipper, H. C. Nusbaum, and S. L. Small. 2007. "Abstract Coding of Audiovisual Speech: Beyond Sensory Representation." *Neuron* 56 (6): 1116–26.

Hayden, B. Y., D. V. Smith, and M. L. Platt. 2009. "Electrophysiological Correlates of Default-Mode Processing in Macaque Posterior Cingulate Cortex." *Proceedings of the National Academy of Sciences* 106 (14): 5948–53.

Haynes, J. D. 2009. "Decoding Visual Consciousness from Human Brain Signals." *Trends in Cognitive Sciences* 13: 194–202.

Haynes, J. D., R. Deichmann, and G. Rees. 2005. "Eye-Specific Effects of Binocular Rivalry in the Human Lateral Geniculate Nucleus." *Nature* 438 (7067): 496–99.

Haynes, J. D., J. Driver, and G. Rees. 2005. "Visibility Reflects Dynamic Changes of Effective Connectivity Between V1 and Fusiform Cortex." *Neuron* 46 (5): 811–21.

Haynes, J. D., and G. Rees. 2005a. "Predicting the Orientation of Invisible Stimuli from Activity in Human Primary Visual Cortex." *Nature Neuroscience* 8 (5): 686–91.

———. 2005b. "Predicting the Stream of Consciousness from Activity in Human Visual Cortex." *Current Biology* 15 (14): 1301–7.

Haynes, J. D., K. Sakai, G. Rees, S. Gilbert, C. Frith, and R. E. Passingham. 2007. "Reading Hidden Intentions in the Human Brain." *Current Biology* 17 (4): 323–28.

He, B. J., and M. E. Raichle. 2009. "The fMRI Signal, Slow Cortical Potential and Consciousness." *Trends in Cognitive Sciences* 13 (7): 302–9.

He, B. J., A. Z. Snyder, J. M. Zempel, M. D. Smyth, and M. E. Raichle. 2008. "Electrophysiological Correlates of the Brain's Intrinsic Large-Scale Functional Architecture." *Proceedings of the National Academy of Sciences* 105 (41): 16039–44.

He, B. J., J. M. Zempel, A. Z. Snyder, and M. E. Raichle. 2010. "The Temporal Structures and Functional Significance of Scale-Free Brain Activity." *Neuron* 66 (3): 353–69.

He, S., and D. I. MacLeod. 2001. "Orientation-Selective Adaptation and Tilt After-Effect from Invisible Patterns." *Nature* 411 (6836): 473–76.

Hebb, D. O. 1949. *The Organization of Behavior.* New York: Wiley.

Heit, G., M. E. Smith, and E. Halgren. 1988. "Neural Encoding of Individual Words and Faces by the Human Hippocampus and Amygdala." *Nature* 333 (6175): 773–75.

Henson, R. N., E. Mouchlianitis, W. J. Matthews, and S. Kouider. 2008. "Electrophysiological Correlates of Masked Face Priming." *NeuroImage* 40 (2): 884–95.

Herrmann, E., J. Call, M. V. Hernandez-Lloreda, B. Hare, and M. Tomasello. 2007. "Humans Have Evolved Specialized Skills of Social Cognition: The Cultural Intelligence Hypothesis." *Science* 317 (5843): 1360–66.

Hochberg, L. R., D. Bacher, B. Jarosiewicz, N. Y. Masse, J. D. Simeral, J. Vogel, S. Haddadin, et al. 2012. "Reach and Grasp by People with Tetraplegia Using a Neurally Controlled Robotic Arm." *Nature* 485 (7398): 372–75.

Hofstadter, D. 2007. *I Am a Strange Loop*. New York: Basic Books.

Holender, D. 1986. "Semantic Activation Without Conscious Identification in Dichotic Listening Parafoveal Vision and Visual Masking: A Survey and Appraisal." *Behavioral and Brain Sciences* 9 (1): 1–23.

Holender, D., and K. Duscherer. 2004. "Unconscious Perception: The Need for a Paradigm Shift." *Perception and Psychophysics* 66 (5): 872–81; discussion 888–95.

Hopfield, J. J. 1982. "Neural Networks and Physical Systems with Emergent Collective Computational Abilities." *Proceedings of the National Academy of Sciences* 79 (8): 2554–58.

Horikawa, T., M. Tamaki, Y. Miyawaki, and Y. Kamitani. 2013. "Neural Decoding of Visual Imagery During Sleep." *Science* 340 (6132): 639–42.

Howard, I. P. 1996. "Alhazen's Neglected Discoveries of Visual Phenomena." *Perception* 25 (10): 1203–17.

Howe, M. J. A., and J. Smith. 1988. "Calendar Calculating in 'Idiots Savants': How Do They Do It?" *British Journal of Psychology* 79 (3): 371–86.

Huron, C., J. M. Danion, F. Giacomoni, D. Grange, P. Robert, and L. Rizzo. 1995. "Impairment of Recognition Memory With, but Not Without, Conscious Recollection in Schizophrenia." *American Journal of Psychiatry* 152 (12): 1737–42.

Izard, V., C. Sann, E. S. Spelke, and A. Streri. 2009. "Newborn Infants Perceive Abstract Numbers." *Proceedings of the National Academy of Sciences* 106 (25): 10382–85.

Izhikevich, E. M., and G. M. Edelman. 2008. "Large-Scale Model of Mammalian Thalamocortical Systems." *Proceedings of the National Academy of Sciences* 105 (9): 3593–98.

James, W. 1890. *The Principles of Psychology*. New York: Holt.

Jaynes, J. 1976. *The Origin of Consciousness in the Breakdown of the Bicameral Mind*. New York: Houghton Mifflin.

Jenkins, A. C., C. N. Macrae, and J. P. Mitchell. 2008. "Repetition Suppression of Ventromedial Prefrontal Activity During Judgments of Self and Others." *Proceedings of the National Academy of Sciences* 105 (11): 4507–12.

Jennett, B. 2002. *The Vegetative State: Medical Facts, Ethical and Legal Dilemmas*. New York: Cambridge University Press.

Jennett, B., and F. Plum. 1972. "Persistent Vegetative State After Brain Damage: A Syndrome in Search of a Name." *Lancet* 1 (7753): 734–37.

Jezek, K., E. J. Henriksen, A. Treves, E. I. Moser, and M. B. Moser. 2011. "Theta-Paced Flickering Between Place-Cell Maps in the Hippocampus." *Nature* 478 (7368): 246–49.

Ji, D., and M. A. Wilson. 2007. "Coordinated Memory Replay in the Visual Cortex and Hippocampus During Sleep." *Nature Neuroscience* 10 (1): 100–7.

Johansson, P., L. Hall, S. Sikstrom, and A. Olsson. 2005. "Failure to Detect Mismatches Between Intention and Outcome in a Simple Decision Task." *Science* 310 (5745): 116–19.

Johnson, M. H., S. Dziurawiec, H. Ellis, and J. Morton. 1991. "Newborns' Preferential Tracking of Face-Like Stimuli and Its Subsequent Decline." *Cognition* 40 (1–2): 1–19.

Jolicoeur, P. 1999. "Concurrent Response-Selection Demands Modulate the Attentional Blink." *Journal of Experimental Psychology: Human Perception and Performance* 25 (4): 1097–113.

Jordan, D., G. Stockmanns, E. F. Kochs, S. Pilge, and G. Schneider. 2008. "Electroencephalographic Order Pattern Analysis for the Separation of Consciousness and Unconsciousness: An Analysis of Approximate Entropy, Permutation Entropy, Recurrence Rate, and Phase Coupling of Order Recurrence Plots." *Anesthesiology* 109 (6): 1014–22.

Jouvet, M. 1999. *The Paradox of Sleep*. Cambridge, Mass.: MIT Press.

Kahneman, D., and A. Treisman. 1984. "Changing Views of Attention and Automaticity." In *Varieties of Attention*, edited by R. Parasuraman, R. Davies, and J. Beatty, 29–61. New York: Academic Press.

Kanai, R., T. A. Carlson, F. A. Verstraten, and V. Walsh. 2009. "Perceived Timing of New Objects and Feature Changes." *Journal of Vision* 9 (7): 5.

Kanai, R., N. G. Muggleton, and V. Walsh. 2008. "TMS over the Intraparietal Sulcus Induces Perceptual Fading." *Journal of Neurophysiology* 100 (6): 3343–50.

Kane, N. M., S. H. Curry, S. R. Butler, and B. H. Cummins. 1993. "Electrophysiological Indicator of Awakening from Coma." *Lancet* 341 (8846): 688.

Kanwisher, N. 2001. "Neural Events and Perceptual Awareness." *Cognition* 79 (1–2): 89–113.

Karlsgodt, K. H., D. Sun, A. M. Jimenez, E. S. Lutkenhoff, R. Willhite, T. G. van Erp, and T. D. Cannon. 2008. "Developmental Disruptions in Neural Connectivity in the Pathophysiology of Schizophrenia." *Development and Psychopathology* 20 (4): 1297–327.

Kenet, T., D. Bibitchkov, M. Tsodyks, A. Grinvald, and A. Arieli. 2003. "Spontaneously Emerging Cortical Representations of Visual Attributes." *Nature* 425 (6961): 954–56.

Kentridge, R. W., T. C. Nijboer, and C. A. Heywood. 2008. "Attended but Unseen: Visual Attention Is Not Sufficient for Visual Awareness." *Neuropsychologia* 46 (3): 864–69.

Kersten, D., P. Mamassian, and A. Yuille. 2004. "Object Perception as Bayesian Inference." *Annual Review of Psychology* 55: 271–304.

Kiani, R., and M. N. Shadlen. 2009. "Representation of Confidence Associated with a Decision by Neurons in the Parietal Cortex." *Science* 324 (5928): 759–64.

Kiefer, M. 2002. "The N400 Is Modulated by Unconsciously Perceived Masked Words: Further Evidence for an Automatic Spreading Activation Account of N400 Priming Effects." *Brain Research: Cognitive Brain Research* 13 (1): 27–39.

Kiefer, M., and D. Brendel. 2006. "Attentional Modulation of Unconscious 'Automatic' Processes: Evidence from Event-Related Potentials in a Masked Priming Paradigm." *Journal of Cognitive Neuroscience* 18 (2): 184–98.

Kiefer, M., and M. Spitzer. 2000. "Time Course of Conscious and Unconscious Semantic Brain Activation." *NeuroReport* 11 (11): 2401–7.

Kiesel, A., W. Kunde, C. Pohl, M. P. Berner, and J. Hoffmann. 2009. "Playing Chess Unconsciously." *Journal of Experimental Psychology: Learning, Memory, Cognition* 35 (1): 292–98.

Kihara, K., T. Ikeda, D. Matsuyoshi, N. Hirose, T. Mima, H. Fukuyama, and N. Osaka. 2010. "Differential Contributions of the Intraparietal Sulcus and the Inferior Parietal Lobe to Attentional Blink: Evidence from Transcranial Magnetic Stimulation." *Journal of Cognitive Neuroscience* 23 (1): 247–56.

Kikyo, H., K. Ohki, and Y. Miyashita. 2002. "Neural Correlates for Feeling-of-Knowing: An fMRI Parametric Analysis." *Neuron* 36 (1): 177–86.

Kim, C. Y., and R. Blake. 2005. "Psychophysical Magic: Rendering the Visible 'Invisible.'" *Trends in Cognitive Sciences* 9 (8): 381–88.

King, J. R., F. Faugeras, A. Gramfort, A. Schurger, I. El Karoui, J. D. Sitt, C. Wacongne, et al. 2013. "Single-Trial Decoding of Auditory Novelty Responses Facilitates the Detection of Residual Consciousness." *NeuroImage*, in press.

King, J. R., J. D. Sitt, F. Faugeras, B. Rohaut, I. El Karoui, L. Cohen, L. Naccache, and S. Dehaene. 2013. "Long-Distance Information Sharing Indexes the State of Consciousness of Unresponsive Patients." Submitted.

Knochel, C., V. Oertel-Knochel, R. Schonmeyer, A. Rotarska-Jagiela, V. van de Ven, D. Prvulovic, C. Haenschel, et al. 2012. "Interhemispheric Hypoconnectivity in Schizophrenia: Fiber Integrity and Volume Differences of the Corpus Callosum in Patients and Unaffected Relatives." *NeuroImage* 59 (2): 926–34.

Koch, C., and F. Crick. 2001. "The Zombie Within." *Nature* 411 (6840): 893.

Koch, C., and N. Tsuchiya. 2007. "Attention and Consciousness: Two Distinct Brain Processes." *Trends in Cognitive Sciences* 11 (1): 16–22.

Koechlin, E., L. Naccache, E. Block, and S. Dehaene. 1999. "Primed Numbers: Exploring the

Modularity of Numerical Representations with Masked and Unmasked Semantic Priming." *Journal of Experimental Psychology: Human Perception and Performance* 25 (6): 1882–905.

Koivisto, M., M. Lahteenmaki, T. A. Sorensen, S. Vangkilde, M. Overgaard, and A. Revonsuo. 2008. "The Earliest Electrophysiological Correlate of Visual Awareness?" *Brain and Cognition* 66 (1): 91–103.

Koivisto, M., T. Mantyla, and J. Silvanto. 2010. "The Role of Early Visual Cortex (V1/V2) in Conscious and Unconscious Visual Perception." *NeuroImage* 51 (2): 828–34.

Koivisto, M., H. Railo, and N. Salminen-Vaparanta. 2010. "Transcranial Magnetic Stimulation of Early Visual Cortex Interferes with Subjective Visual Awareness and Objective Forced-Choice Performance." *Consciousness and Cognition* 20 (2): 288–98.

Komura, Y., A. Nikkuni, N. Hirashima, T. Uetake, and A. Miyamoto. 2013. "Responses of Pulvinar Neurons Reflect a Subject's Confidence in Visual Categorization." *Nature Neuroscience* 16: 749–55.

Konopka, G., E. Wexler, E. Rosen, Z. Mukamel, G. E. Osborn, L. Chen, D. Lu, et al. 2012. "Modeling the Functional Genomics of Autism Using Human Neurons." *Molecular Psychiatry* 17 (2): 202–14.

Kornell, N., L. K. Son, and H. S. Terrace. 2007. "Transfer of Metacognitive Skills and Hint Seeking in Monkeys." *Psychological Science* 18 (1): 64–71.

Kouider, S., V. de Gardelle, J. Sackur, and E. Dupoux. 2010. "How Rich Is Consciousness? The Partial Awareness Hypothesis." *Trends in Cognitive Sciences* 14 (7): 301–7.

Kouider, S., and S. Dehaene. 2007. "Levels of Processing During Non-conscious Perception: A Critical Review of Visual Masking." *Philosophical Transactions of the Royal Society B: Biological Sciences* 362 (1481): 857–75.

———. 2009. "Subliminal Number Priming Within and Across the Visual and Auditory Modalities." *Experimental Psychology*, in press.

Kouider, S., S. Dehaene, A. Jobert, and D. Le Bihan. 2007. "Cerebral Bases of Subliminal and Supraliminal Priming During Reading." *Cerebral Cortex* 17 (9): 2019–29.

Kouider, S., and E. Dupoux. 2004. "Partial Awareness Creates the 'Illusion' of Subliminal Semantic Priming." *Psychological Science* 15 (2): 75–81.

Kouider, S., E. Eger, R. Dolan, and R. N. Henson. 2009. "Activity in Face-Responsive Brain Regions Is Modulated by Invisible, Attended Faces: Evidence from Masked Priming." *Cerebral Cortex* 19 (1): 13–23.

Kouider, S., C. Stahlhut, S. V. Gelskov, L. Barbosa, M. Dutat, V. de Gardelle, A. Christophe, et al. 2013. "A Neural Marker of Perceptual Consciousness in Infants." *Science* 340 (6130): 376–80.

Kovacs, A. M., E. Teglas, and A. D. Endress. 2010. "The Social Sense: Susceptibility to Others' Beliefs in Human Infants and Adults." *Science* 330 (6012): 1830–34.

Kovacs, G., R. Vogels, and G. A. Orban. 1995. "Cortical Correlate of Pattern Backward Masking." *Proceedings of the National Academy of Sciences* 92 (12): 5587–91.

Kreiman, G., I. Fried, and C. Koch. 2002. "Single-Neuron Correlates of Subjective Vision in the Human Medial Temporal Lobe." *Proceedings of the National Academy of Sciences* 99 (12): 8378–83.

Kreiman, G., C. Koch, and I. Fried. 2000a. "Category-Specific Visual Responses of Single Neurons in the Human Medial Temporal Lobe." *Nature Neuroscience* 3 (9): 946–53.

———. 2000b. "Imagery Neurons in the Human Brain." *Nature* 408 (6810): 357–61.

Krekelberg, B., and M. Lappe. 2001. "Neuronal Latencies and the Position of Moving Objects." *Trends in Neurosciences* 24 (6): 335–39.

Krolak-Salmon, P., M. A. Henaff, C. Tallon-Baudry, B. Yvert, M. Guenot, A. Vighetto, F. Mauguiere, and O. Bertrand. 2003. "Human Lateral Geniculate Nucleus and Visual Cortex Respond to Screen Flicker." *Annals of Neurology* 53 (1): 73–80.

Kruger, J., and D. Dunning. 1999. "Unskilled and Unaware of It: How Difficulties in Recognizing One's Own Incompetence Lead to Inflated Self-Assessments." *Journal of Personality and Social Psychology* 77 (6): 1121–34.

Kubicki, M., H. Park, C. F. Westin, P. G. Nestor, R. V. Mulkern, S. E. Maier, M. Niznikiewicz, et al. 2005. "DTI and MTR Abnormalities in Schizophrenia: Analysis of White Matter Integrity." *NeuroImage* 26 (4): 1109–18.

Lachter, J., K. I. Forster, and E. Ruthruff. 2004. "Forty-Five Years After Broadbent (1958): Still No Identification Without Attention." *Psychology Review* 111 (4): 880–913.

Lagercrantz, H., and J. P. Changeux. 2009. "The Emergence of Human Consciousness: From Fetal to Neonatal Life." *Pediatric Research* 65 (3): 255–60.

———. 2010. "Basic Consciousness of the Newborn." *Seminars in Perinatology* 34 (3): 201–6.

Lai, C. S., S. E. Fisher, J. A. Hurst, F. Vargha-Khadem, and A. P. Monaco. 2001. "A Forkhead-Domain Gene Is Mutated in a Severe Speech and Language Disorder." *Nature* 413 (6855): 519–23.

Lamme, V. A. 2006. "Towards a True Neural Stance on Consciousness." *Trends in Cognitive Sciences* 10 (11): 494–501.

Lamme, V. A., and P. R. Roelfsema. 2000. "The Distinct Modes of Vision Offered by Feedforward and Recurrent Processing." *Trends in Neurosciences* 23 (11): 571–79.

Lamme, V. A., K. Zipser, and H. Spekreijse. 1998. "Figure-Ground Activity in Primary Visual Cortex Is Suppressed by Anesthesia." *Proceedings of the National Academy of Sciences* 95 (6): 3263–68.

Lamy, D., M. Salti, and Y. Bar-Haim. 2009. "Neural Correlates of Subjective Awareness and Unconscious Processing: An ERP Study." *Journal of Cognitive Neuroscience* 21 (7): 1435–46.

Landman, R., H. Spekreijse, and V. A. Lamme. 2003. "Large Capacity Storage of Integrated Objects Before Change Blindness." *Vision Research* 43 (2): 149–64.

Lau, H., and D. Rosenthal. 2011. "Empirical Support for Higher-Order Theories of Conscious Awareness." *Trends in Cognitive Sciences* 15 (8): 365–73.

Lau, H. C., and R. E. Passingham. 2006. "Relative Blindsight in Normal Observers and the Neural Correlate of Visual Consciousness." *Proceedings of the National Academy of Sciences* 103 (49): 18763–68.

———. 2007. "Unconscious Activation of the Cognitive Control System in the Human Prefrontal Cortex." *Journal of Neuroscience* 27 (21): 5805–11.

Laureys, S. 2005. "The Neural Correlate of (Un)Awareness: Lessons from the Vegetative State." *Trends in Cognitive Sciences* 9 (12): 556–59.

Laureys, S., M. E. Faymonville, A. Luxen, M. Lamy, G. Franck, and P. Maquet. 2000. "Restoration of Thalamocortical Connectivity After Recovery from Persistent Vegetative State." *Lancet* 355 (9217): 1790–91.

Laureys, S., C. Lemaire, P. Maquet, C. Phillips, and G. Franck. 1999. "Cerebral Metabolism During Vegetative State and After Recovery to Consciousness." *Journal of Neurology, Neurosurgery and Psychiatry* 67 (1): 121.

Laureys, S., A. M. Owen, and N. D. Schiff. 2004. "Brain Function in Coma, Vegetative State, and Related Disorders." *Lancet Neurology* 3 (9): 537–46.

Laureys, S., F. Pellas, P. Van Eeckhout, S. Ghorbel, C. Schnakers, F. Perrin, J. Berre, et al. 2005. "The Locked-In Syndrome: What Is It Like to Be Conscious but Paralyzed and Voiceless?" *Progress in Brain Research* 150: 495–511.

Lawrence, N. S., F. Jollant, O. O'Daly, F. Zelaya, and M. L. Phillips. 2009. "Distinct Roles of Prefrontal Cortical Subregions in the Iowa Gambling Task." *Cerebral Cortex* 19 (5): 1134–43.

Ledoux, J. 1996. *The Emotional Brain*. New York: Simon and Schuster.

Lenggenhager, B., M. Mouthon, and O. Blanke. 2009. "Spatial Aspects of Bodily Self-Consciousness." *Consciousness and Cognition* 18 (1): 110–17.

Lenggenhager, B., T. Tadi, T. Metzinger, and O. Blanke. 2007. "Video Ergo Sum: Manipulating Bodily Self-Consciousness." *Science* 317 (5841): 1096–99.

Leon-Carrion, J., P. van Eeckhout, R. Dominguez-Morales Mdel, and F. J. Perez-Santamaria. 2002. "The Locked-In Syndrome: A Syndrome Looking for a Therapy." *Brain Injury* 16 (7): 571–82.

Leopold, D. A., and N. K. Logothetis. 1996. "Activity Changes in Early Visual Cortex Reflect Monkeys' Percepts During Binocular Rivalry." *Nature* 379 (6565): 549–53.

———. 1999. "Multistable Phenomena: Changing Views in Perception." *Trends in Cognitive Sciences* 3 (7): 254–64.

Leroy, F., H. Glasel, J. Dubois, L. Hertz-Pannier, B. Thirion, J. F. Mangin, and G. Dehaene-Lambertz. 2011. "Early Maturation of the Linguistic Dorsal Pathway in Human Infants." *Journal of Neuroscience* 31 (4): 1500–6.

Levelt, W. J. M. 1989. *Speaking: From Intention to Articulation.* Cambridge, Mass.: MIT Press.

Levy, J., H. Pashler, and E. Boer. 2006. "Central Interference in Driving: Is There Any Stopping the Psychological Refractory Period?" *Psychological Science* 17 (3): 228–35.

Lewis, J. L. 1970. "Semantic Processing of Unattended Messages Using Dichotic Listening." *Journal of Experimental Psychology* 85 (2): 225–28.

Libet, B. 1965. "Cortical Activation in Conscious and Unconscious Experience." *Perspectives in Biology and Medicine* 9 (1): 77–86.

———. 1991. "Conscious vs Neural Time." *Nature* 352 (6330): 27–28.

———. 2004. *Mind Time: The Temporal Factor in Consciousness.* Cambridge, Mass.: Harvard University Press.

Libet, B., W. W. Alberts, E. W. Wright, Jr., L. D. Delattre, G. Levin, and B. Feinstein. 1964. "Production of Threshold Levels of Conscious Sensation by Electrical Stimulation of Human Somatosensory Cortex." *Journal of Neurophysiology* 27: 546–78.

Libet, B., W. W. Alberts, E. W. Wright, Jr., and B. Feinstein. 1967. "Responses of Human Somatosensory Cortex to Stimuli Below Threshold for Conscious Sensation." *Science* 158 (808): 1597–600.

Libet, B., C. A. Gleason, E. W. Wright, and D. K. Pearl. 1983. "Time of Conscious Intention to Act in Relation to Onset of Cerebral Activity (Readiness-Potential). The Unconscious Initiation of a Freely Voluntary Act." *Brain* 106 (3): 623–42.

Libet, B., E. W. Wright, Jr., B. Feinstein, and D. K. Pearl. 1979. "Subjective Referral of the Timing for a Conscious Sensory Experience: A Functional Role for the Somatosensory Specific Projection System in Man." *Brain* 102 (1): 193–224.

Liu, Y., M. Liang, Y. Zhou, Y. He, Y. Hao, M. Song, C. Yu, et al. 2008. "Disrupted Small-World Networks in Schizophrenia." *Brain* 131 (4): 945–61.

Logan, G. D., and M. J. Crump. 2010. "Cognitive Illusions of Authorship Reveal Hierarchical Error Detection in Skilled Typists." *Science* 330 (6004): 683–86.

Logan, G. D., and M. D. Schulkind. 2000. "Parallel Memory Retrieval in Dual-Task Situations: I. Semantic Memory." *Journal of Experimental Psychology: Human Perception and Performance* 26 (3): 1072–90.

Logothetis, N. K. 1998. "Single Units and Conscious Vision." *Philosophical Transactions of the Royal Society B: Biological Sciences* 353 (1377): 1801–18.

Logothetis, N. K., D. A. Leopold, and D. L. Sheinberg. 1996. "What Is Rivalling During Binocular Rivalry?" *Nature* 380 (6575): 621–24.

Louie, K., and M. A. Wilson. 2001. "Temporally Structured Replay of Awake Hippocampal Ensemble Activity During Rapid Eye Movement Sleep." *Neuron* 29 (1): 145–56.

Luck, S. J., R. L. Fuller, E. L. Braun, B. Robinson, A. Summerfelt, and J. M. Gold. 2006. "The Speed of Visual Attention in Schizophrenia: Electrophysiological and Behavioral Evidence." *Schizophrenia Research* 85 (1–3): 174–95.

Luck, S. J., E. S. Kappenman, R. L. Fuller, B. Robinson, A. Summerfelt, and J. M. Gold. 2009. "Impaired Response Selection in Schizophrenia: Evidence from the P3 Wave and the Lateralized Readiness Potential." *Psychophysiology* 46 (4): 776–86.

Luck, S. J., E. K. Vogel, and K. L. Shapiro. 1996. "Word Meanings Can Be Accessed but Not Reported During the Attentional Blink." *Nature* 383 (6601): 616–18.

Lumer, E. D., G. M. Edelman, and G. Tononi. 1997a. "Neural Dynamics in a Model of the Thalamocortical System. I. Layers, Loops and the Emergence of Fast Synchronous Rhythms." *Cerebral Cortex* 7 (3): 207–27.

———. 1997b. "Neural Dynamics in a Model of the Thalamocortical System. II. The Role of Neural Synchrony Tested Through Perturbations of Spike Timing." *Cerebral Cortex* 7 (3): 228–36.

Lumer, E. D., K. J. Friston, and G. Rees. 1998. "Neural Correlates of Perceptual Rivalry in the Human Brain." *Science* 280 (5371): 1930–34.

Lynall, M. E., D. S. Bassett, R. Kerwin, P. J. McKenna, M. Kitzbichler, U. Muller, and E. Bullmore. 2010. "Functional Connectivity and Brain Networks in Schizophrenia." *Journal of Neuroscience* 30 (28): 9477–87.

Mack, A., and I. Rock. 1998. *Inattentional Blindness*. Cambridge, Mass.: MIT Press.

Macknik, S. L., and M. M. Haglund. 1999. "Optical Images of Visible and Invisible Percepts in the Primary Visual Cortex of Primates." *Proceedings of the National Academy of Sciences* 96 (26): 15208–10.

MacLeod, D. I., and S. He. 1993. "Visible Flicker from Invisible Patterns." *Nature* 361 (6409): 256–58.

Magnusson, C. E., and H. C. Stevens. 1911. "Visual Sensations Created by a Magnetic Field." *American Journal of Physiology* 29: 124–36.

Maia, T. V., and J. L. McClelland. 2004. "A Reexamination of the Evidence for the Somatic Marker Hypothesis: What Participants Really Know in the Iowa Gambling Task." *Proceedings of the National Academy of Sciences* 101 (45): 16075–80.

Maier, A., M. Wilke, C. Aura, C. Zhu, F. Q. Ye, and D. A. Leopold. 2008. "Divergence of fMRI and Neural Signals in V1 During Perceptual Suppression in the Awake Monkey." *Nature Neuroscience* 11 (10): 1193–200.

Marcel, A. J. 1980. "Conscious and Preconscious Recognition of Polysemous Words: Locating the Selective Effect of Prior Verbal Context." In *Attention and Performance*, edited by R.S. Nickerson, vol. 8. Hillsdale, N.J.: Lawrence Erlbaum.

———. 1983. "Conscious and Unconscious Perception: Experiments on Visual Masking and Word Recognition." *Cognitive Psychology* 15: 197–237.

Marois, R., D. J. Yi, and M. M. Chun. 2004. "The Neural Fate of Consciously Perceived and Missed Events in the Attentional Blink." *Neuron* 41 (3): 465–72.

Marshall, J. C., and P. W. Halligan. 1988. "Blindsight and Insight in Visuo-Spatial Neglect." *Nature* 336 (6201): 766–67.

Marti, S., J. Sackur, M. Sigman, and S. Dehaene. 2010. "Mapping Introspection's Blind Spot: Reconstruction of Dual-Task Phenomenology Using Quantified Introspection." *Cognition* 115 (2): 303–13.

Marti, S., M. Sigman, and S. Dehaene. 2012. "A Shared Cortical Bottleneck Underlying Attentional Blink and Psychological Refractory Period." *NeuroImage* 59 (3): 2883–98.

Marticorena, D. C., A. M. Ruiz, C. Mukerji, A. Goddu, and L. R. Santos. 2011. "Monkeys Represent Others' Knowledge but Not Their Beliefs." *Developmental Science* 14 (6): 1406–16.

Mason, M. F., M. I. Norton, J. D. Van Horn, D. M. Wegner, S. T. Grafton, and C. N. Macrae. 2007. "Wandering Minds: The Default Network and Stimulus-Independent Thought." *Science* 315 (5810): 393–95.

Massimini, M., M. Boly, A. Casali, M. Rosanova, and G. Tononi. 2009. "A Perturbational

Approach for Evaluating the Brain's Capacity for Consciousness." *Progress in Brain Research* 177: 201–14.

Massimini, M., F. Ferrarelli, R. Huber, S. K. Esser, H. Singh, and G. Tononi. 2005. "Breakdown of Cortical Effective Connectivity During Sleep." *Science* 309 (5744): 2228–32.

Matsuda, W., A. Matsumura, Y. Komatsu, K. Yanaka, and T. Nose. 2003. "Awakenings from Persistent Vegetative State: Report of Three Cases with Parkinsonism and Brain Stem Lesions on MRI." *Journal of Neurology, Neurosurgery, and Psychiatry* 74 (11): 1571–73.

Mattler, U. 2005. "Inhibition and Decay of Motor and Nonmotor Priming." *Attention, Perception and Psychophysics* 67 (2): 285–300.

Maudsley, H. 1868. *The Physiology and Pathology of the Mind*. London: Macmillan.

May, A., G. Hajak, S. Ganssbauer, T. Steffens, B. Langguth, T. Kleinjung, and P. Eichhammer. 2007. "Structural Brain Alterations Following 5 Days of Intervention: Dynamic Aspects of Neuroplasticity." *Cerebral Cortex* 17 (1): 205–10.

McCarthy, M. M., E. N. Brown, and N. Kopell. 2008. "Potential Network Mechanisms Mediating Electroencephalographic Beta Rhythm Changes During Propofol-Induced Paradoxical Excitation." *Journal of Neuroscience* 28 (50): 13488–504.

McClure, R. K. 2001. "The Visual Backward Masking Deficit in Schizophrenia." *Progress in Neuro-psychopharmacology and Biological Psychiatry* 25 (2): 301–11.

McCormick, P. A. 1997. "Orienting Attention Without Awareness." *Journal of Experimental Psychology: Human Perception and Performance* 23 (1): 168–80.

McGlinchey-Berroth, R., W. P. Milberg, M. Verfaellie, M. Alexander, and P. Kilduff. 1993. "Semantic Priming in the Neglected Field: Evidence from a Lexical Decision Task." *Cognitive Neuropsychology* 10: 79–108.

McGurk, H., and J. MacDonald. 1976. "Hearing Lips and Seeing Voices." *Nature* 264 (5588): 746–48.

McIntosh, A. R., M. N. Rajah, and N. J. Lobaugh. 1999. "Interactions of Prefrontal Cortex in Relation to Awareness in Sensory Learning." *Science* 284 (5419): 1531–33.

Mehler, J., P. Jusczyk, G. Lambertz, N. Halsted, J. Bertoncini, and C. Amiel-Tison. 1988. "A Precursor of Language Acquisition in Young Infants." *Cognition* 29 (?): 143–78.

Melloni, L., C. Molina, M. Peña, D. Torres, W. Singer, and E. Rodriguez. 2007. "Synchronization of Neural Activity Across Cortical Areas Correlates with Conscious Perception." *Journal of Neuroscience* 27 (11): 2858–65.

Meltzoff, A. N., and R. Brooks. 2008. "Self-Experience as a Mechanism for Learning About Others: A Training Study in Social Cognition." *Developmental Psychology* 44 (5): 1257–65.

Merikle, P. M. 1992. "Perception Without Awareness: Critical Issues." *American Psychologist* 47: 792–96.

Merikle, P. M., and S. Joordens. 1997. "Parallels Between Perception Without Attention and Perception Without Awareness." *Consciousness and Cognition* 6 (2–3): 219–36.

Meyer, K., and A. Damasio. 2009. "Convergence and Divergence in a Neural Architecture for Recognition and Memory." *Trends in Neurosciences* 32 (7): 376–82.

Miller, A., J. W. Sleigh, J. Barnard, and D. A. Steyn-Ross. 2004. "Does Bispectral Analysis of the Electroencephalogram Add Anything but Complexity?" *British Journal of Anaesthesia* 92 (1): 8–13.

Milner, A. D., and M. A. Goodale. 1995. *The Visual Brain in Action*. New York: Oxford University Press.

Monti, M. M., A. Vanhaudenhuyse, M. R. Coleman, M. Boly, J. D. Pickard, L. Tshibanda, A. M. Owen, and S. Laureys. 2010. "Willful Modulation of Brain Activity in Disorders of Consciousness." *New England Journal of Medicine* 362 (7): 579–89.

Moray, N. 1959. "Attention in Dichotic Listening: Affective Cues and the Influence of Instructions." *Quarterly Journal of Experimental Psychology* 9: 56–60.

Moreno-Bote, R., D. C. Knill, and A. Pouget. 2011. "Bayesian Sampling in Visual Perception." *Proceedings of the National Academy of Sciences* 108 (30): 12491–96.

Morland, A. B., S. Le, E. Carroll, M. B. Hoffmann, and A. Pambakian. 2004. "The Role of Spared Calcarine Cortex and Lateral Occipital Cortex in the Responses of Human Hemianopes to Visual Motion." *Journal of Cognitive Neuroscience* 16 (2): 204–18.

Moro, S. I., M. Tolboom, P. S. Khayat, and P. R. Roelfsema. 2010. "Neuronal Activity in the Visual Cortex Reveals the Temporal Order of Cognitive Operations." *Journal of Neuroscience* 30 (48): 16293–303.

Morris, J. S., B. DeGelder, L. Weiskrantz, and R. J. Dolan. 2001. "Differential Extrageniculostriate and Amygdala Responses to Presentation of Emotional Faces in a Cortically Blind Field." *Brain* 124 (6): 1241–52.

Morris, J. S., A. Ohman, and R. J. Dolan. 1998. "Conscious and Unconscious Emotional Learning in the Human Amygdala." *Nature* 393 (6684): 467–70.

———. 1999. "A Subcortical Pathway to the Right Amygdala Mediating 'Unseen' Fear." *Proceedings of the National Academy of Sciences* 96 (4): 1680–85.

Moruzzi, G., and H. W. Magoun. 1949. "Brain Stem Reticular Formation and Activation of the EEG." *Electroencephalography and Clinical Neurophysiology* 1 (4): 455–73.

Naatanen, R., P. Paavilainen, T. Rinne, and K. Alho. 2007. "The Mismatch Negativity (MMN) in Basic Research of Central Auditory Processing: A Review." *Clinical Neurophysiology* 118 (12): 2544–90.

Naccache, L. 2006a. "Is She Conscious?" *Science* 313 (5792): 1395–96.

———. 2006b. *Le nouvel inconscient*. Paris: Editions Odile Jacob.

Naccache, L., E. Blandin, and S. Dehaene. 2002. "Unconscious Masked Priming Depends on Temporal Attention." *Psychological Science* 13: 416–24.

Naccache, L., and S. Dehaene. 2001a. "The Priming Method: Imaging Unconscious Repetition Priming Reveals an Abstract Representation of Number in the Parietal Lobes." *Cerebral Cortex* 11 (10): 966–74.

———. 2001b. "Unconscious Semantic Priming Extends to Novel Unseen Stimuli." *Cognition* 80 (3): 215–29.

Naccache, L., R. Gaillard, C. Adam, D. Hasboun, S. Clémenceau, M. Baulac, S. Dehaene, and L. Cohen. 2005. "A Direct Intracranial Record of Emotions Evoked by Subliminal Words." *Proceedings of the National Academy of Sciences* 102: 7713–17.

Naccache, L., L. Puybasset, R. Gaillard, E. Serve, and J. C. Willer. 2005. "Auditory Mismatch Negativity Is a Good Predictor of Awakening in Comatose Patients: A Fast and Reliable Procedure." *Clinical Neurophysiology* 116 (4): 988–89.

Nachev, P., and M. Husain. 2007. Comment on "Detecting Awareness in the Vegetative State." *Science* 315 (5816): 1221; author reply 1221.

Nelson, C. A., K. M. Thomas, M. de Haan, and S. S. Wewerka. 1998. "Delayed Recognition Memory in Infants and Adults as Revealed by Event-Related Potentials." *International Journal of Psychophysiology* 29 (2): 145–65.

New, J. J., and B. J. Scholl. 2008. "'Perceptual Scotomas': A Functional Account of Motion-Induced Blindness." *Psychological Science* 19 (7): 653–59.

Nieder, A., and S. Dehaene. 2009. "Representation of Number in the Brain." *Annual Review of Neuroscience* 32: 185–208.

Nieder, A., and E. K. Miller. 2004. "A Parieto-Frontal Network for Visual Numerical Information in the Monkey." *Proceedings of the National Academy of Sciences* 101 (19): 7457–62.

Nieuwenhuis, S., M. S. Gilzenrat, B. D. Holmes, and J. D. Cohen. 2005. "The Role of the Locus Coeruleus in Mediating the Attentional Blink: A Neurocomputational Theory." *Journal of Experimental Psychology: General* 134 (3): 291–307.

Nieuwenhuis, S., K. R. Ridderinkhof, J. Blom, G. P. Band, and A. Kok. 2001. "Error-Related Brain

Potentials Are Differentially Related to Awareness of Response Errors: Evidence from an Antisaccade Task." *Psychophysiology* 38 (5): 752–60.

Nimchinsky, E. A., E. Gilissen, J. M. Allman, D. P. Perl, J. M. Erwin, and P. R. Hof. 1999. "A Neuronal Morphologic Type Unique to Humans and Great Apes." *Proceedings of the National Academy of Sciences* 96 (9): 5268–73.

Nisbett, R. E., and T. D. Wilson. 1977. "Telling More Than We Can Know: Verbal Reports on Mental Processes." *Psychological Review* 84 (3): 231–59.

Nørretranders, T. 1999. *The User Illusion: Cutting Consciousness Down to Size.* London: Penguin.

Norris, D. 2006. "The Bayesian Reader: Explaining Word Recognition as an Optimal Bayesian Decision Process." *Psychological Review* 113 (2): 327–57.

———. 2009. "Putting It All Together: A Unified Account of Word Recognition and Reaction-Time Distributions." *Psychological Review* 116 (1): 207–19.

Ochsner, K. N., K. Knierim, D. H. Ludlow, J. Hanelin, T. Ramachandran, G. Glover, and S. C. Mackey. 2004. "Reflecting upon Feelings: An fMRI Study of Neural Systems Supporting the Attribution of Emotion to Self and Other." *Journal of Cognitive Neuroscience* 16 (10): 1746–72.

Ogawa, S., T. M. Lee, A. R. Kay, and D. W. Tank. 1990. "Brain Magnetic Resonance Imaging with Contrast Dependent on Blood Oxygenation." *Proceedings of the National Academy of Sciences* 87 (24): 9868–72.

Overgaard, M., J. Rote, K. Mouridsen, and T. Z. Ramsøy. 2006. "Is Conscious Perception Gradual or Dichotomous? A Comparison of Report Methodologies During a Visual Task." *Consciousness and Cognition* 15 (4): 700–8.

Owen, A., M. R. Coleman, M. Boly, M. H. Davis, S. Laureys, D. Jolles, and J. D. Pickard. 2007. "Response to Comments on 'Detecting Awareness in the Vegetative State.'" *Science* 315 (5816): 1221.

Owen, A. M., M. R. Coleman, M. Boly, M. H. Davis, S. Laureys, and J. D. Pickard. 2006. "Detecting Awareness in the Vegetative State." *Science* 313 (5792): 1402.

Pack, C. C., V. K. Berezovskii, and R. T. Born. 2001. "Dynamic Properties of Neurons in Cortical Area MT in Alert and Anaesthetized Macaque Monkeys." *Nature* 414 (6866): 905–8.

Pack, C. C., and R. T. Born. 2001. "Temporal Dynamics of a Neural Solution to the Aperture Problem in Visual Area MT of Macaque Brain." *Nature* 409 (6823): 1040–42.

Pallier, C., A. D. Devauchelle, and S. Dehaene. 2011. "Cortical Representation of the Constituent Structure of Sentences." *Proceedings of the National Academy of Sciences* 108 (6): 2522–27.

Palva, S., K. Linkenkaer-Hansen, R. Naatanen, and J. M. Palva. 2005. "Early Neural Correlates of Conscious Somatosensory Perception." *Journal of Neuroscience* 25 (21): 5248–58.

Parvizi, J., and A. R. Damasio. 2003. "Neuroanatomical Correlates of Brainstem Coma." *Brain* 126 (7): 1524–36.

Parvizi, J., C. Jacques, B. L. Foster, N. Withoft, V. Rangarajan, K. S. Weiner, and K. Grill-Spector. 2012. "Electrical Stimulation of Human Fusiform Face-Selective Regions Distorts Face Perception." *Journal of Neuroscience* 32 (43): 14915–20.

Parvizi, J., G. W. Van Hoesen, J. Buckwalter, and A. Damasio. 2006. "Neural Connections of the Posteromedial Cortex in the Macaque." *Proceedings of the National Academy of Sciences* 103 (5): 1563–68.

Pascual-Leone, A., V. Walsh, and J. Rothwell. 2000. "Transcranial Magnetic Stimulation in Cognitive Neuroscience—Virtual Lesion, Chronometry, and Functional Connectivity." *Current Opinion in Neurobiology* 10 (2): 232–37.

Pashler, H. 1984. "Processing Stages in Overlapping Tasks: Evidence for a Central Bottleneck." *Journal of Experimental Psychology: Human Perception and Performance* 10 (3): 358–77.

———. 1994. "Dual-Task Interference in Simple Tasks: Data and Theory." *Psychological Bulletin* 116 (2): 220–44.

Peirce, C. S. 1901. "The Proper Treatment of Hypotheses: A Preliminary Chapter, Toward an Examination of Hume's Argument Against Miracles, in Its Logic and in Its History." *Historical Perspectives* 2: 890–904.

Penrose, R., and S. Hameroff. 1998. "The Penrose-Hameroff 'Orch OR' Model of Consciousness." *Philosophical Transactions of the Royal Society London (A)* 356: 1869–96.

Perin, R., T. K. Berger, and H. Markram. 2011. "A Synaptic Organizing Principle for Cortical Neuronal Groups." *Proceedings of the National Academy of Sciences* 108 (13): 5419–24.

Perner, J., and M. Aichhorn. 2008. "Theory of Mind, Language and the Temporoparietal Junction Mystery." *Trends in Cognitive Sciences* 12 (4): 123–26.

Persaud, N., M. Davidson, B. Maniscalco, D. Mobbs, R. E. Passingham, A. Cowey, and H. Lau. 2011. "Awareness-Related Activity in Prefrontal and Parietal Cortices in Blindsight Reflects More Than Superior Visual Performance." *NeuroImage* 58 (2): 605–11.

Pessiglione, M., P. Petrovic, J. Daunizeau, S. Palminteri, R. J. Dolan, and C. D. Frith. 2008. "Subliminal Instrumental Conditioning Demonstrated in the Human Brain." *Neuron* 59 (4): 561–67.

Pessiglione, M., L. Schmidt, B. Draganski, R. Kalisch, H. Lau, R. J. Dolan, and C. D. Frith. 2007. "How the Brain Translates Money into Force: A Neuroimaging Study of Subliminal Motivation." *Science* 316 (5826): 904–6.

Petersen, S. E., H. van Mier, J. A. Fiez, and M. E. Raichle. 1998. "The Effects of Practice on the Functional Anatomy of Task Performance." *Proceedings of the National Academy of Sciences* 95 (3): 853–60.

Peyrache, A., M. Khamassi, K. Benchenane, S. I. Wiener, and F. P. Battaglia. 2009. "Replay of Rule-Learning Related Neural Patterns in the Prefrontal Cortex During Sleep." *Nature Neuroscience* 12 (7): 919–26.

Piazza, M., V. Izard, P. Pinel, D. Le Bihan, and S. Dehaene. 2004. "Tuning Curves for Approximate Numerosity in the Human Intraparietal Sulcus." *Neuron* 44 (3): 547–55.

Piazza, M., P. Pinel, D. Le Bihan, and S. Dehaene. 2007. "A Magnitude Code Common to Numerosities and Number Symbols in Human Intraparietal Cortex." *Neuron* 53: 293–305.

Picton, T. W. 1992. "The P300 Wave of the Human Event-Related Potential." *Journal of Clinical Neurophysiology* 9 (4): 456–79.

Pinel, P., F. Fauchereau, A. Moreno, A. Barbot, M. Lathrop, D. Zelenika, D. Le Bihan, et al. 2012. "Genetic Variants of FOXP2 and KIAA0319/TTRAP/THEM2 Locus Are Associated with Altered Brain Activation in Distinct Language-Related Regions." *Journal of Neuroscience* 32 (3): 817–25.

Pins, D., and D. Ffytche. 2003. "The Neural Correlates of Conscious Vision." *Cerebral Cortex* 13 (5): 461–74.

Pisella, L., H. Grea, C. Tilikete, A. Vighetto, M. Desmurget, G. Rode, D. Boisson, and Y. Rossetti. 2000. "An 'Automatic Pilot' for the Hand in Human Posterior Parietal Cortex: Toward Reinterpreting Optic Ataxia." *Nature Neuroscience* 3 (7): 729–36.

Plotnik, J. M., F. B. de Waal, and D. Reiss. 2006. "Self-Recognition in an Asian Elephant." *Proceedings of the National Academy of Sciences* 103 (45): 17053–57.

Pontifical Academy of Sciences. 2008. *Why the Concept of Death Is Valid as a Definition of Brain Death. Statement by the Pontifical Academy of Sciences and Responses to Objections.* http://www.pas.va/content/accademia/en/publications/extraseries/braindeath.html.

Portas, C. M., K. Krakow, P. Allen, O. Josephs, J. L. Armony, and C. D. Frith. 2000. "Auditory Processing Across the Sleep-Wake Cycle: Simultaneous EEG and fMRI Monitoring in Humans." *Neuron* 28 (3): 991–99.

Posner, M. I. 1994. "Attention: The Mechanisms of Consciousness." *Proceedings of the National Academy of Sciences* 91: 7398–403.

Posner, M. I., and M. K. Rothbart. 1998. "Attention, Self-Regulation and Consciousness." *Philosophical Transactions of the Royal Society B: Biological Sciences* 353 (1377): 1915–27.

Posner, M. I., and C. R. R. Snyder. 1975/2004. "Attention and Cognitive Control." In *Cognitive Psychology: Key Readings*, edited by D. A. Balota, and E. J. Marsh, 205–23. New York: Psychology Press.

———. 1975. "Attention and Cognitive Control." In *Information Processing and Cognition: The Loyola Symposium*, edited by R. L. Solso, 55–85. Hillsdale, N.J.: Lawrence Erlbaum.

Prior, H., A. Schwarz, and O. Gunturkun. 2008. "Mirror-Induced Behavior in the Magpie (Pica Pica): Evidence of Self-Recognition." *PLOS Biology* 6 (8): e202.

Quiroga, R. Q., G. Kreiman, C. Koch, and I. Fried. 2008. "Sparse but Not 'Grandmother-Cell' Coding in the Medial Temporal Lobe." *Trends in Cognitive Sciences* 12 (3): 87–91.

Quiroga, R. Q., R. Mukamel, E. A. Isham, R. Malach, and I. Fried. 2008. "Human Single-Neuron Responses at the Threshold of Conscious Recognition." *Proceedings of the National Academy of Sciences* 105 (9): 3599–604.

Quiroga, R. Q., L. Reddy, C. Koch, and I. Fried. 2007. "Decoding Visual Inputs from Multiple Neurons in the Human Temporal Lobe." *Journal of Neurophysiology* 98 (4): 1997–2007.

Quiroga, R. Q., L. Reddy, G. Kreiman, C. Koch, and I. Fried. 2005. "Invariant Visual Representation by Single Neurons in the Human Brain." *Nature* 435 (7045): 1102–7.

Raichle, M. E. 2010. "Two Views of Brain Function." *Trends in Cognitive Sciences* 14 (4): 180–90.

Raichle, M. E., J. A. Fiesz, T. O. Videen, and A. K. MacLeod. 1994. "Practice-Related Changes in Human Brain Functional Anatomy During Nonmotor Learning." *Cerebral Cortex* 4: 8–26.

Raichle, M. E., A. M. MacLeod, A. Z. Snyder, W. J. Powers, D. A. Gusnard, and G. L. Shulman. 2001. "A Default Mode of Brain Function." *Proceedings of the National Academy of Sciences* 98 (2): 676–82.

Railo, H., and M. Koivisto. 2009. "The Electrophysiological Correlates of Stimulus Visibility and Metacontrast Masking." *Consciousness and Cognition* 18 (3): 794–803.

Ramachandran, V. S., and R. L. Gregory. 1991. "Perceptual Filling In of Artificially Induced Scotomas in Human Vision." *Nature* 350 (6320): 699–702.

Raymond, J. E., K. L. Shapiro, and K. M. Arnell. 1992. "Temporary Suppression of Visual Processing in an RSVP Task: An Attentional Blink?" *Journal of Experimental Psychology: Human Perception and Performance* 18 (3): 849–60.

Reddy, L., R. Q. Quiroga, P. Wilken, C. Koch, and I. Fried. 2006. "A Single-Neuron Correlate of Change Detection and Change Blindness in the Human Medial Temporal Lobe." *Current Biology* 16 (20): 2066–72.

Reed, C. M., and N. I. Durlach. 1998. "Note on Information Transfer Rates in Human Communication." *Presence: Teleoperators and Virtual Environments* 7 (5): 509–18.

Reiss, D., and L. Marino. 2001. "Mirror Self-Recognition in the Bottlenose Dolphin: A Case of Cognitive Convergence." *Proceedings of the National Academy of Sciences* 98 (10): 5937–42.

Rensink, R. A., J. K. O'Regan, and J. Clark. 1997. "To See or Not to See: The Need for Attention to Perceive Changes in Scenes." *Psychological Science* 8: 368–73.

Reuss, H., A. Kiesel, W. Kunde, and B. Hommel. 2011. "Unconscious Activation of Task Sets." *Consciousness and Cognition* 20 (3): 556–67.

Reuter, F., A. Del Cul, B. Audoin, I. Malikova, L. Naccache, J. P. Ranjeva, O. Lyon-Caen, et al. 2007. "Intact Subliminal Processing and Delayed Conscious Access in Multiple Sclerosis." *Neuropsychologia* 45 (12): 2683–91.

Reuter, F., A. Del Cul, I. Malokova, L. Naccache, S. Confort-Gouny, L. Cohen, A. A. Cherif, et al. 2009. "White Matter Damage Impairs Access to Consciousness in Multiple Sclerosis." *NeuroImage* 44 (2): 590–99.

Reynvoet, B., and M. Brysbaert. 1999. "Single-Digit and Two-Digit Arabic Numerals Address the Same Semantic Number Line." *Cognition* 72 (2): 191–201.

———. 2004. "Cross-Notation Number Priming Investigated at Different Stimulus Onset Asynchronies in Parity and Naming Tasks." *Journal of Experimental Psychology* 51 (2): 81–90.

Reynvoet, B., M. Brysbaert, and W. Fias. 2002. "Semantic Priming in Number Naming." *Quarterly Journal of Experimental Psychology A* 55 (4): 1127–39.

Reynvoet, B., W. Gevers, and B. Caessens. 2005. "Unconscious Primes Activate Motor Codes Through Semantics." *Journal of Experimental Psychology: Learning, Memory, Cognition* 31 (5): 991–1000.

Ricoeur, P. 1990. *Soi-même comme un autre*. Paris: Le Seuil.

Rigas, P., and M. A. Castro-Alamancos. 2007. "Thalamocortical Up States: Differential Effects of Intrinsic and Extrinsic Cortical Inputs on Persistent Activity." *Journal of Neuroscience* 27 (16): 4261–72.

Rockstroh, B., M. Müller, R. Cohen, and T. Elbert. 1992. "Probing the Functional Brain State During P300 Evocation." *Journal of Psychophysiology* 6: 175–84.

Rodriguez, E., N. George, J. P. Lachaux, J. Martinerie, B. Renault, and F. J. Varela. 1999. "Perception's Shadow: Long-Distance Synchronization of Human Brain Activity." *Nature* 397 (6718): 430–33.

Roelfsema, P. R. 2005. "Elemental Operations in Vision." *Trends in Cognitive Sciences* 9 (5): 226–33.

Roelfsema, P. R., P. S. Khayat, and H. Spekreijse. 2003. "Subtask Sequencing in the Primary Visual Cortex." *Proceedings of the National Academy of Sciences* 100 (9): 5467–72.

Roelfsema, P. R., V. A. Lamme, and H. Spekreijse. 1998. "Object-Based Attention in the Primary Visual Cortex of the Macaque Monkey." *Nature* 395 (6700): 376–81.

Ropper, A. H. 2010. "*Cogito Ergo Sum* by MRI." *New England Journal of Medicine* 362 (7): 648–49.

Rosanova, M., O. Gosseries, S. Casarotto, M. Boly, A. G. Casali, M. A. Bruno, M. Mariotti, et al. 2012. "Recovery of Cortical Effective Connectivity and Recovery of Consciousness in Vegetative Patients." *Brain* 135 (4): 1308–20.

Rosenthal, D. M. 2008. "Consciousness and Its Function." *Neuropsychologia* 46 (3): 829–40.

Ross, C. A., R. L. Margolis, S. A. Reading, M. Pletnikov, and J. T. Coyle. 2006. "Neurobiology of Schizophrenia." *Neuron* 52 (1): 139–53.

Rougier, N. P., D. C. Noelle, T. S. Braver, J. D. Cohen, and R. C. O'Reilly. 2005. "Prefrontal Cortex and Flexible Cognitive Control: Rules Without Symbols." *Proceedings of the National Academy of Sciences* 10 (220): 7338–43.

Rounis, E., B. Maniscalco, J. C. Rothwell, R. Passingham, and H. Lau. 2010. "Theta-Burst Transcranial Magnetic Stimulation to the Prefrontal Cortex Impairs Metacognitive Visual Awareness." *Cognitive Neuroscience* 1 (3): 165–75.

Sackur, J., and S. Dehaene. 2009. "The Cognitive Architecture for Chaining of Two Mental Operations." *Cognition* 111 (2): 187–211.

Sackur, J., L. Naccache, P. Pradat-Diehl, P. Azouvi, D. Mazevet, R. Katz, L. Cohen, and S. Dehaene. 2008. "Semantic Processing of Neglected Numbers." *Cortex* 44 (6): 673–82.

Sadaghiani, S., G. Hesselmann, K. J. Friston, and A. Kleinschmidt. 2010. "The Relation of Ongoing Brain Activity, Evoked Neural Responses, and Cognition." *Frontiers in Systems Neuroscience* 4: 20.

Sadaghiani, S., G. Hesselmann, and A. Kleinschmidt. 2009. "Distributed and Antagonistic Contributions of Ongoing Activity Fluctuations to Auditory Stimulus Detection." *Journal of Neuroscience* 29 (42): 13410–17.

Saga, Y., M. Iba, J. Tanji, and E. Hoshi. 2011. "Development of Multidimensional Representations of Task Phases in the Lateral Prefrontal Cortex." *Journal of Neuroscience* 31 (29): 10648–65.

Sahraie, A., L. Weiskrantz, J. L. Barbur, A. Simmons, S. C. R. Williams, and M. J. Brammer.

1997. "Pattern of Neuronal Activity Associated with Conscious and Unconscious Processing of Visual Signals." *Proceedings of the National Academy of Sciences* 94: 9406–11.

Salin, P. A., and J. Bullier. 1995. "Corticocortical Connections in the Visual System: Structure and Function." *Physiological Reviews* 75 (1): 107–54.

Saur, D., B. Schelter, S. Schnell, D. Kratochvil, H. Kupper, P. Kellmeyer, D. Kummerer, et al. 2010. "Combining Functional and Anatomical Connectivity Reveals Brain Networks for Auditory Language Comprehension." *NeuroImage* 49 (4): 3187–97.

Saxe, R. 2006. "Uniquely Human Social Cognition." *Current Opinion in Neurobiology* 16 (2): 235–39.

Saxe, R., and L. J. Powell. 2006. "It's the Thought That Counts: Specific Brain Regions for One Component of Theory of Mind." *Psychological Science* 17 (8): 692–99.

Schenker, N. M., D. P. Buxhoeveden, W. L. Blackmon, K. Amunts, K. Zilles, and K. Semendeferi. 2008. "A Comparative Quantitative Analysis of Cytoarchitecture and Minicolumnar Organization in Broca's Area in Humans and Great Apes." *Journal of Comparative Neurology* 510 (1): 117–28.

Schenker, N. M., W. D. Hopkins, M. A. Spocter, A. R. Garrison, C. D. Stimpson, J. M. Erwin, P. R. Hof, and C. C. Sherwood. 2009. "Broca's Area Homologue in Chimpanzees (*Pan troglodytes*): Probabilistic Mapping, Asymmetry, and Comparison to Humans." *Cerebral Cortex* 20 (3): 730–42.

Schiff, N., U. Ribary, F. Plum, and R. Llinas. 1999. "Words Without Mind." *Journal of Cognitive Neuroscience* 11 (6): 650–56.

Schiff, N. D. 2010. "Recovery of Consciousness After Brain Injury: A Mesocircuit Hypothesis." *Trends in Neurosciences* 33 (1): 1–9.

Schiff, N. D., J. T. Giacino, K. Kalmar, J. D. Victor, K. Baker, M. Gerber, B. Fritz, et al. 2007. "Behavioural Improvements with Thalamic Stimulation After Severe Traumatic Brain Injury." *Nature* 448 (7153): 600–3.

———. 2008. "Behavioural Improvements with Thalamic Stimulation After Severe Traumatic Brain Injury." *Nature* 452 (7183): 120.

Schiff, N. D., U. Ribary, D. R. Moreno, B. Beattie, E. Kronberg, R. Blasberg, J. Giacino, et al. 2002. "Residual Cerebral Activity and Behavioural Fragments Can Remain in the Persistently Vegetative Brain." *Brain* 125 (6): 1210–34.

Schiller, P. H., and S. L. Chorover. 1966. "Metacontrast: Its Relation to Evoked Potentials." *Science* 153 (742): 1398–400.

Schmid, M. C., S. W. Mrowka, J. Turchi, R. C. Saunders, M. Wilke, A. J. Peters, F. Q. Ye, and D. A. Leopold. 2010. "Blindsight Depends on the Lateral Geniculate Nucleus." *Nature* 466 (7304): 373–77.

Schmid, M. C., T. Panagiotaropoulos, M. A. Augath, N. K. Logothetis, and S. M. Smirnakis. 2009. "Visually Driven Activation in Macaque Areas V2 and V3 Without Input from the Primary Visual Cortex." *PLOS One* 4 (5): e5527.

Schnakers, C., D. Ledoux, S. Majerus, P. Damas, F. Damas, B. Lambermont, M. Lamy, et al. 2008. "Diagnostic and Prognostic Use of Bispectral Index in Coma, Vegetative State and Related Disorders." *Brain Injury* 22 (12): 926–31.

Schnakers, C., A. Vanhaudenhuyse, J. Giacino, M. Ventura, M. Boly, S. Majerus, G. Moonen, and S. Laureys. 2009. "Diagnostic Accuracy of the Vegetative and Minimally Conscious State: Clinical Consensus Versus Standardized Neurobehavioral Assessment." *BMC Neurology* 9: 35.

Schneider, W., and R. M. Shiffrin. 1977. "Controlled and Automatic Human Information Processing. I. Detection, Search, and Attention." *Psychological Review* 84 (1): 1–66.

Schoenemann, P. T., M. J. Sheehan, and L. D. Glotzer. 2005. "Prefrontal White Matter Volume Is Disproportionately Larger in Humans Than in Other Primates." *Nature Neuroscience* 8 (2): 242–52.

Schurger, A., F. Pereira, A. Treisman, and J. D. Cohen. 2009. "Reproducibility Distinguishes Conscious from Nonconscious Neural Representations." *Science* 327 (5961): 97–99.

Schurger, A., J. D. Sitt, and S. Dehaene. 2012. "An Accumulator Model for Spontaneous Neural Activity Prior to Self-Initiated Movement." *Proceedings of the National Academy of Sciences* 109 (42): E2904–13.

Schvaneveldt, R. W., and D. E. Meyer. 1976. "Lexical Ambiguity, Semantic Context, and Visual Word Recognition." *Journal of Experimental Psychology: Human Perception and Performance* 2 (2): 243–56.

Self, M. W., R. N. Kooijmans, H. Supèr, V. A. Lamme, and P. R. Roelfsema. 2012. "Different Glutamate Receptors Convey Feedforward and Recurrent Processing in Macaque V1." *Proceedings of the National Academy of Sciences* 109 (27): 11031–36.

Selfridge, O. G. 1959. "Pandemonium: A Paradigm for Learning." In *Proceedings of the Symposium on Mechanisation of Thought Processes*, edited by D. V. Blake and A. M. Uttley, 511–29. London: H. M. Stationery Office.

Selimbeyoglu, A., and J. Parvizi. 2010. "Electrical Stimulation of the Human Brain: Perceptual and Behavioral Phenomena Reported in the Old and New Literature." *Frontiers in Human Neuroscience* 4: 46.

Sergent, C., S. Baillet, and S. Dehaene. 2005. "Timing of the Brain Events Underlying Access to Consciousness During the Attentional Blink." *Nature Neuroscience* 8 (10): 1391–400.

Sergent, C., and S. Dehaene. 2004. "Is Consciousness a Gradual Phenomenon? Evidence for an All-or-None Bifurcation During the Attentional Blink." *Psychological Science* 15 (11): 720–28.

Sergent, C., V. Wyart, M. Babo-Rebelo, L. Cohen, L. Naccache, and C. Tallon-Baudry. 2013. "Cueing Attention After the Stimulus Is Gone Can Retrospectively Trigger Conscious Perception." *Current Biology* 23 (2): 150–55.

Shady, S., D. I. MacLeod, and H. S. Fisher. 2004. "Adaptation from Invisible Flicker." *Proceedings of the National Academy of Sciences* 101 (14): 5170–73.

Shallice, T. 1972. "Dual Functions of Consciousness." *Psychological Review* 79 (5): 383–93.

———. 1979. "A Theory of Consciousness." *Science* 204 (4395): 827.

———. 1988. *From Neuropsychology to Mental Structure*. New York: Cambridge University Press.

Shanahan, M., and B. Baars. 2005. "Applying Global Workspace Theory to the Frame Problem." *Cognition* 98 (2): 157–76.

Shao, L., Y. Shuai, J. Wang, S. Feng, B. Lu, Z. Li, Y. Zhao, et al. 2011. "Schizophrenia Susceptibility Gene Dysbindin Regulates Glutamatergic and Dopaminergic Functions via Distinctive Mechanisms in Drosophila." *Proceedings of the National Academy of Sciences* 108 (46): 18831–836.

Sherman, S. M. 2012. "Thalamocortical Interactions." *Current Opinion in Neurobiology* 22 (4): 575–79.

Shiffrin, R. M., and W. Schneider. 1977. "Controlled and Automatic Human Information Processing. II. Perceptual Learning, Automatic Attending, and a General Theory." *Psychological Review* 84 (2): 127–90.

Shima, K., M. Isoda, H. Mushiake, and J. Tanji. 2007. "Categorization of Behavioural Sequences in the Prefrontal Cortex." *Nature* 445 (7125): 315–18.

Shirvalkar, P., M. Seth, N. D. Schiff, and D. G. Herrera. 2006. "Cognitive Enhancement with Central Thalamic Electrical Stimulation." *Proceedings of the National Academy of Sciences* 103 (45): 17007–12.

Sidaros, A., A. W. Engberg, K. Sidaros, M. G. Liptrot, M. Herning, P. Petersen, O. B. Paulson, et al. 2008. "Diffusion Tensor Imaging During Recovery from Severe Traumatic Brain Injury and Relation to Clinical Outcome: A Longitudinal Study." *Brain* 131 (2): 559–72.

Sidis, B. 1898. *The Psychology of Suggestion*. New York: D. Appleton.

Siegler, R. S. 1987. "Strategy Choices in Subtraction." In *Cognitive Processes in Mathematics*, edited by J. Sloboda and D. Rogers, 81–106. Oxford: Clarendon Press.

———. 1988. "Strategy Choice Procedures and the Development of Multiplication Skill." *Journal of Experimental Psychology: General* 117 (3): 258–75.

———. 1989. "Mechanisms of Cognitive Development." *Annual Review of Psychology* 40: 353–79.

Siegler, R. S., and E. A. Jenkins. 1989. *How Children Discover New Strategies.* Hillsdale, N.J.: Lawrence Erlbaum.

Sigala, N., M. Kusunoki, I. Nimmo-Smith, D. Gaffan, and J. Duncan. 2008. "Hierarchical Coding for Sequential Task Events in the Monkey Prefrontal Cortex." *Proceedings of the National Academy of Sciences* 105 (33): 11969–74.

Sigman, M., and S. Dehaene. 2005. "Parsing a Cognitive Task: A Characterization of the Mind's Bottleneck." *PLOS Biology* 3 (2): e37.

———. 2008. "Brain Mechanisms of Serial and Parallel Processing During Dual-Task Performance." *Journal of Neuroscience* 28 (30): 7585–98.

Silvanto, J., and Z. Cattaneo. 2010. "Transcranial Magnetic Stimulation Reveals the Content of Visual Short-Term Memory in the Visual Cortex." *NeuroImage* 50 (4): 1683–89.

Silvanto, J., A. Cowey, N. Lavie, and V. Walsh. 2005. "Striate Cortex (V1) Activity Gates Awareness of Motion." *Nature Neuroscience* 8 (2): 143–44.

Silvanto, J., N. Lavie, and V. Walsh. 2005. "Double Dissociation of V1 and V5/MT Activity in Visual Awareness." *Cerebral Cortex* 15 (11): 1736–41.

Simons, D. J., and M. S. Ambinder. 2005. "Change Blindness: Theory and Consequences." *Current Directions in Psychological Science* 14 (1): 44–48.

Simons, D. J., and C. F. Chabris. 1999. "Gorillas in Our Midst: Sustained Inattentional Blindness for Dynamic Events." *Perception* 28 (9): 1059–74.

Singer, P. 1993. *Practical Ethics.* 2nd ed. Cambridge: Cambridge University Press.

Singer, W. 1998. "Consciousness and the Structure of Neuronal Representations." *Philosophical Transactions of the Royal Society B: Biological Sciences* 353 (1377): 1829–40.

Sitt, J. D., J. R. King, I. El Karoui, B. Rohaut, F. Faugeras, A. Gramfort, L. Cohen, et al. 2013. "Signatures of Consciousness and Predictors of Recovery in Vegetative and Minimally Conscious Patients." Submitted.

Sklar, A. Y., N. Levy, A. Goldstein, R. Mandel, A. Maril, and R. R. Hassin. 2012. "Reading and Doing Arithmetic Nonconsciously." *Proceedings of the National Academy of Sciences* 109 (48): 19614–19.

Smallwood, J., E. Beach, J. W. Schooler, and T. C. Handy. 2008. "Going AWOL in the Brain: Mind Wandering Reduces Cortical Analysis of External Events." *Journal of Cognitive Neuroscience* 20 (3): 458–69.

Smedira, N. G., B. H. Evans, L. S. Grais, N. H. Cohen, B. Lo, M. Cooke, W. P. Schecter, et al. 1990. "Withholding and Withdrawal of Life Support from the Critically Ill." *New England Journal of Medicine* 322 (5): 309–15.

Smith, J. D., J. Schull, J. Strote, K. McGee, R. Egnor, and L. Erb. 1995. "The Uncertain Response in the Bottlenosed Dolphin (*Tursiops truncatus*)." *Journal of Experimental Psychology: General* 124 (4): 391–408.

Soto, D., T. Mantyla, and J. Silvanto. 2011. "Working Memory Without Consciousness." *Current Biology* 21 (22): R912–13.

Sporns, O., G. Tononi, and G. M. Edelman. 1991. "Modeling Perceptual Grouping and Figure-Ground Segregation by Means of Active Reentrant Connections." *Proceedings of the National Academy of Sciences* 88 (1): 129–33.

Squires, K. C., C. Wickens, N. K. Squires, and E. Donchin. 1976. "The Effect of Stimulus Sequence on the Waveform of the Cortical Event-Related Potential." *Science* 193 (4258): 1142–46.

Squires, N. K., K. C. Squires, and S. A. Hillyard. 1975. "Two Varieties of Long-Latency Positive Waves Evoked by Unpredictable Auditory Stimuli in Man." *Electroencephalography and Clinical Neurophysiology* 38 (4): 387–401.

Srinivasan, R., D. P. Russell, G. M. Edelman, and G. Tononi. 1999. "Increased Synchronization of Neuromagnetic Responses During Conscious Perception." *Journal of Neuroscience* 19 (13): 5435–48.

Staniek, M., and K. Lehnertz. 2008. "Symbolic Transfer Entropy." *Physical Review Letters* 100 (15): 158101.

Staunton, H. 2008. "Arousal by Stimulation of Deep-Brain Nuclei." *Nature* 452 (7183): E1; discussion E1–2.

Stephan, K. E., K. J. Friston, and C. D. Frith. 2009. "Dysconnection in Schizophrenia: From Abnormal Synaptic Plasticity to Failures of Self-Monitoring." *Schizophrenia Bulletin* 35 (3): 509–27.

Stephan, K. M., M. H. Thaut, G. Wunderlich, W. Schicks, B. Tian, L. Tellmann, T. Schmitz, et al. 2002. "Conscious and Subconscious Sensorimotor Synchronization—Prefrontal Cortex and the Influence of Awareness." *NeuroImage* 15 (2): 345–52.

Stettler, D. D., A. Das, J. Bennett, and C. D. Gilbert. 2002. "Lateral Connectivity and Contextual Interactions in Macaque Primary Visual Cortex." *Neuron* 36 (4): 739–50.

Steyn-Ross, M. L., D. A. Steyn-Ross, and J. W. Sleigh. 2004. "Modelling General Anaesthesia as a First-Order Phase Transition in the Cortex." *Progress in Biophysics and Molecular Biology* 85 (2–3): 369–85.

Strayer, D. L., F. A. Drews, and W. A. Johnston. 2003. "Cell Phone–Induced Failures of Visual Attention During Simulated Driving." *Journal of Experimental Psychology: Applied* 9 (1): 23–32.

Striem-Amit, E., L. Cohen, S. Dehaene, and A. Amedi. 2012. "Reading with Sounds: Sensory Substitution Selectively Activates the Visual Word Form Area in the Blind." *Neuron* 76 (3): 640–52.

Suddendorf, T., and D. L. Butler. 2013. "The Nature of Visual Self-Recognition." *Trends in Cognitive Sciences* 17 (3): 121–27.

Supèr, H., H. Spekreijse, and V. A. Lamme. 2001a. "Two Distinct Modes of Sensory Processing Observed in Monkey Primary Visual Cortex (V1)." *Nature Neuroscience* 4 (3): 304–10.

———. 2001b. "A Neural Correlate of Working Memory in the Monkey Primary Visual Cortex." *Science* 293 (5527): 120–24.

Supèr, H., C. van der Togt, H. Spekreijse, and V. A. Lamme. 2003. "Internal State of Monkey Primary Visual Cortex (V1) Predicts Figure-Ground Perception." *Journal of Neuroscience* 23 (8): 3407–14.

Supp, G. G., M. Siegel, J. F. Hipp, and A. K. Engel. 2011. "Cortical Hypersynchrony Predicts Breakdown of Sensory Processing During Loss of Consciousness." *Current Biology* 21 (23): 1988–93.

Taine, H. 1870. *De l'intelligence.* Paris: Hachette.

Tang, T. T., F. Yang, B. S. Chen, Y. Lu, Y. Ji, K. W. Roche, and B. Lu. 2009. "Dysbindin Regulates Hippocampal LTP by Controlling NMDA Receptor Surface Expression." *Proceedings of the National Academy of Sciences* 106 (50): 21395–400.

Taylor, P. C., V. Walsh, and M. Eimer. 2010. "The Neural Signature of Phosphene Perception." *Human Brain Mapping* 31 (9): 1408–17.

Telford, C. W. 1931. "The Refractory Phase of Voluntary and Associative Responses." *Journal of Experimental Psychology* 14 (1): 1–36.

Terrace, H. S., and L. K. Son. 2009. "Comparative Metacognition." *Current Opinion in Neurobiology* 19 (1): 67–74.

Thompson, S. P. 1910. "A Physiological Effect of an Alternating Magnetic Field." *Proceedings of the Royal Society B: Biological Sciences* B82: 396–99.

Tombu, M., and P. Jolicoeur. 2003. "A Central Capacity Sharing Model of Dual-Task Performance." *Journal of Experimental Psychology: Human Perception and Performance* 29 (1): 3–18.

Tononi, G. 2008. "Consciousness as Integrated Information: A Provisional Manifesto." *Biological Bulletin* 215 (3): 216–42.

Tononi, G., and G. M. Edelman. 1998. "Consciousness and Complexity." *Science* 282 (5395): 1846–51.

Tooley, M. 1972. "Abortion and Infanticide." *Philosophy and Public Affairs* 2 (1): 37–65.

———. 1983. *Abortion and Infanticide*. London: Clarendon Press.

Treisman, A., and G. Gelade. 1980. "A Feature-Integration Theory of Attention." *Cognitive Psychology* 12: 97–136.

Treisman, A., and J. Souther. 1986. "Illusory Words: The Roles of Attention and of Top-Down Constraints in Conjoining Letters to Form Words." *Journal of Experimental Psychology: Human Perception and Performance* 12: 3–17.

Tsao, D. Y., W. A. Freiwald, R. B. Tootell, and M. S. Livingstone. 2006. "A Cortical Region Consisting Entirely of Face-Selective Cells." *Science* 311 (5761): 670–74.

Tshibanda, L., A. Vanhaudenhuyse, D. Galanaud, M. Boly, S. Laureys, and L. Puybasset. 2009. "Magnetic Resonance Spectroscopy and Diffusion Tensor Imaging in Coma Survivors: Promises and Pitfalls." *Progress in Brain Research* 177: 215–29.

Tsodyks, M., T. Kenet, A. Grinvald, and A. Arieli. 1999. "Linking Spontaneous Activity of Single Cortical Neurons and the Underlying Functional Architecture." *Science* 286 (5446): 1943–46.

Tsubokawa, T., T. Yamamoto, Y. Katayama, T. Hirayama, S. Maejima, and T. Moriya. 1990. "Deep-Brain Stimulation in a Persistent Vegetative State: Follow-Up Results and Criteria for Selection of Candidates." *Brain Injury* 4 (4): 315–27.

Tsuchiya, N., and C. Koch. 2005. "Continuous Flash Suppression Reduces Negative Afterimages." *Nature Neuroscience* 8 (8): 1096–101.

Tsunoda, K., Y. Yamane, M. Nishizaki, and M. Tanifuji. 2001. "Complex Objects Are Represented in Macaque Inferotemporal Cortex by the Combination of Feature Columns." *Nature Neuroscience* 4 (8): 832–38.

Tsushima, Y., Y. Sasaki, and T. Watanabe. 2006. "Greater Disruption Due to Failure of Inhibitory Control on an Ambiguous Distractor." *Science* 314 (5806): 1786–88.

Tsushima, Y., A. R. Seitz, and T. Watanabe. 2008. "Task-Irrelevant Learning Occurs Only When the Irrelevant Feature Is Weak." *Current Biology* 18 (12): R516–517.

Turing, A. M. 1936. "On Computable Numbers, with an Application to the Entscheidungsproblem." *Proceedings of the London Mathematical Society* 42: 230–65.

———. 1952. "The Chemical Basis of Morphogenesis." *Philosophical Transactions of the Royal Society B: Biological Sciences* 237: 37–72.

Tyler, L. K., and W. Marslen-Wilson. 2008. "Fronto-Temporal Brain Systems Supporting Spoken Language Comprehension." *Philosophical Transactions of the Royal Society B: Biological Sciences* 363 (1493): 1037–54.

Tzovara, A., A. O. Rossetti, L. Spierer, J. Grivel, M. M. Murray, M. Oddo, and M. De Lucia. 2012. "Progression of Auditory Discrimination Based on Neural Decoding Predicts Awakening from Coma." *Brain* 136 (1): 81–89.

Uhlhaas, P. J., D. E. Linden, W. Singer, C. Haenschel, M. Lindner, K. Maurer, and E. Rodriguez. 2006. "Dysfunctional Long-Range Coordination of Neural Activity During Gestalt Perception in Schizophrenia." *Journal of Neuroscience* 26 (31): 8168–75.

Uhlhaas, P. J., and W. Singer. 2010. "Abnormal Neural Oscillations and Synchrony in Schizophrenia." *Nature Reviews Neuroscience* 11 (2): 100–13.

van Aalderen-Smeets, S. I., R. Oostenveld, and J. Schwarzbach. 2006. "Investigating Neurophysiological Correlates of Metacontrast Masking with Magnetoencephalography." *Advances in Cognitive Psychology* 2 (1): 21–35.

Van den Bussche, E., K. Notebaert, and B. Reynvoet. 2009. "Masked Primes Can Be Genuinely Semantically Processed." *Journal of Experimental Psychology* 56 (5): 295–300.

Van den Bussche, E., and B. Reynvoet. 2007. "Masked Priming Effects in Semantic Categorization Are Independent of Category Size." *Journal of Experimental Psychology* 54 (3): 225–35.

van Gaal, S., L. Naccache, J. D. I. Meeuwese, A. M. van Loon, L. Cohen, and S. Dehaene. 2013. "Can Multiple Words Be Integrated Unconsciously?" Submitted.

van Gaal, S., K. R. Ridderinkhof, J. J. Fahrenfort, H. S. Scholte, and V. A. Lamme. 2008. "Frontal Cortex Mediates Unconsciously Triggered Inhibitory Control." *Journal of Neuroscience* 28 (32): 8053–62.

van Gaal, S., K. R. Ridderinkhof, H. S. Scholte, and V. A. Lamme. 2010. "Unconscious Activation of the Prefrontal No-Go Network." *Journal of Neuroscience* 30 (11): 4143–50.

Van Opstal, F., F. P. de Lange, and S. Dehaene. 2011. "Rapid Parallel Semantic Processing of Numbers Without Awareness." *Cognition* 120 (1): 136–47.

Varela, F., J. P. Lachaux, E. Rodriguez, and J. Martinerie. 2001. "The Brainweb: Phase Synchronization and Large-Scale Integration." *Nature Reviews Neuroscience* 2 (4): 229–39.

Velmans, M. 1991. "Is Human Information Processing Conscious?" *Behavioral and Brain Sciences* 14: 651–726.

Vernes, S. C., P. L. Oliver, E. Spiteri, H. E. Lockstone, R. Puliyadi, J. M. Taylor, and J. Ho, et al. 2011. "Foxp2 Regulates Gene Networks Implicated in Neurite Outgrowth in the Developing Brain." *PLOS Genetics* 7 (7): e1002145.

Vincent, J. L., G. H. Patel, M. D. Fox, A. Z. Snyder, J. T. Baker, D. C. Van Essen, J. M. Zempel, et al. 2007. "Intrinsic Functional Architecture in the Anaesthetized Monkey Brain." *Nature* 447 (7140): 83–86.

Vogel, E. K., S. J. Luck, and K. L. Shapiro. 1998. "Electrophysiological Evidence for a Postperceptual Locus of Suppression During the Attentional Blink." *Journal of Experimental Psychology: Human Perception and Performance* 24 (6): 1656–74.

Vogel, E. K., and M. G. Machizawa. 2004. "Neural Activity Predicts Individual Differences in Visual Working Memory Capacity." *Nature* 428 (6984): 748–51.

Vogel, E. K., A. W. McCollough, and M. G. Machizawa. 2005. "Neural Measures Reveal Individual Differences in Controlling Access to Working Memory." *Nature* 438 (7067): 500–3.

Vogeley, K., P. Bussfeld, A. Newen, S. Herrmann, F. Happe, P. Falkai, W. Maier, et al. 2001. "Mind Reading: Neural Mechanisms of Theory of Mind and Self-Perspective." *NeuroImage* 14 (1 pt. 1): 170–81.

Voss, H. U., A. M. Uluc, J. P. Dyke, R. Watts, E. J. Kobylarz, B. D. McCandliss, L. A. Heier, et al. 2006. "Possible Axonal Regrowth in Late Recovery from the Minimally Conscious State." *Journal of Clinical Investigation* 116 (7): 2005–11.

Vuilleumier, P., N. Sagiv, E. Hazeltine, R. A. Poldrack, D. Swick, R. D. Rafal, and J. D. Gabrieli. 2001. "Neural Fate of Seen and Unseen Faces in Visuospatial Neglect: A Combined Event-Related Functional MRI and Event-Related Potential Study." *Proceedings of the National Academy of Sciences* 98 (6): 3495–500.

Vul, E., D. Hanus, and N. Kanwisher. 2009. "Attention as Inference: Selection Is Probabilistic; Responses Are All-or-None Samples." *Journal of Experimental Psychology: General* 138 (4): 546–60.

Vul, E., M. Nieuwenstein, and N. Kanwisher. 2008. "Temporal Selection Is Suppressed, Delayed, and Diffused During the Attentional Blink." *Psychological Science* 19 (1): 55–61.

Vul, E., and H. Pashler. 2008. "Measuring the Crowd Within: Probabilistic Representations Within Individuals." *Psychological Science (Wiley-Blackwell)* 19 (7): 645–47.

Wacongne, C., J. P. Changeux, and S. Dehaene. 2012. "A Neuronal Model of Predictive Coding Accounting for the Mismatch Negativity." *Journal of Neuroscience* 32 (11): 3665–78.

Wacongne, C., E. Labyt, V. van Wassenhove, T. Bekinschtein, L. Naccache, and S. Dehaene.

2011. "Evidence for a Hierarchy of Predictions and Prediction Errors in Human Cortex." *Proceedings of the National Academy of Sciences* 108 (51): 20754–59.

Wagner, U., S. Gais, H. Haider, R. Verleger, and J. Born. 2004. "Sleep Inspires Insight." *Nature* 427 (6972): 352–55.

Watson, J. B. 1913. "Psychology as the Behaviorist Views It." *Psychological Review* 20: 158–77.

Wegner, D. M. 2003. *The Illusion of Conscious Will*. Cambridge, Mass.: MIT Press.

Weinberger, J. 2000. "William James and the Unconscious: Redressing a Century-Old Misunderstanding." *Psychological Science* 11 (6): 439–45.

Weiskrantz, L. 1986. *Blindsight: A Case Study and Its Implications*. Oxford: Clarendon Press.

———. 1997. *Consciousness Lost and Found: A Neuropsychological Exploration*. New York: Oxford University Press.

Weiss, Y., E. P. Simoncelli, and E. H. Adelson. 2002. "Motion Illusions as Optimal Percepts." *Nature Neuroscience* 5 (6): 598–604.

Westmoreland, B. F., D. W. Klass, F. W. Sharbrough, and T. J. Reagan. 1975. "Alpha-Coma: Electroencephalographic, Clinical, Pathologic, and Etiologic Correlations." *Archives of Neurology* 32 (11): 713–18.

Whittingstall, K., and N. K. Logothetis. 2009. "Frequency-Band Coupling in Surface EEG Reflects Spiking Activity in Monkey Visual Cortex." *Neuron* 64 (2): 281–89.

Widaman, K. F., D. C. Geary, P. Cormier, and T. D. Little. 1989. "A Componential Model for Mental Addition." *Journal of Experimental Psychology: Learning, Memory, and Cognition* 15: 898–919.

Wilke, M., N. K. Logothetis, and D. A. Leopold. 2003. "Generalized Flash Suppression of Salient Visual Targets." *Neuron* 39 (6): 1043–52.

———. 2006. "Local Field Potential Reflects Perceptual Suppression in Monkey Visual Cortex." *Proceedings of the National Academy of Sciences* 103 (46): 17507–12.

Williams, M. A., C. I. Baker, H. P. Op de Beeck, W. M. Shim, S. Dang, C. Triantafyllou, and N. Kanwisher. 2008. "Feedback of Visual Object Information to Foveal Retinotopic Cortex." *Nature Neuroscience* 11 (12): 1439–45.

Williams, M. A., T. A. Visser, R. Cunnington, and J. B. Mattingley. 2008. "Attenuation of Neural Responses in Primary Visual Cortex During the Attentional Blink." *Journal of Neuroscience* 28 (39): 9890–94.

Womelsdorf, T., J. M. Schoffelen, R. Oostenveld, W. Singer, R. Desimone, A. K. Engel, and P. Fries. 2007. "Modulation of Neuronal Interactions Through Neuronal Synchronization." *Science* 316 (5831): 1609–12.

Wong, K. F. 2002. "The Relationship Between Attentional Blink and Psychological Refractory Period." *Journal of Experimental Psychology: Human Perception and Performance* 28 (1): 54–71.

Wong, K. F., and X. J. Wang. 2006. "A Recurrent Network Mechanism of Time Integration in Perceptual Decisions." *Journal of Neuroscience* 26 (4): 1314–28.

Woodman, G. F., and S. J. Luck. 2003. "Dissociations Among Attention, Perception, and Awareness During Object-Substitution Masking." *Psychological Science* 14 (6): 605–11.

Wyart, V., S. Dehaene, and C. Tallon-Baudry. 2012. "Early Dissociation Between Neural Signatures of Endogenous Spatial Attention and Perceptual Awareness During Visual Masking." *Frontiers in Human Neuroscience* 6: 16.

Wyart, V., and C. Tallon-Baudry. 2008. "Neural Dissociation Between Visual Awareness and Spatial Attention." *Journal of Neuroscience* 28 (10): 2667–79.

———. 2009. "How Ongoing Fluctuations in Human Visual Cortex Predict Perceptual Awareness: Baseline Shift Versus Decision Bias." *Journal of Neuroscience* 29 (27): 8715–25.

Wyler, A. R., G. A. Ojemann, and A. A. Ward, Jr. 1982. "Neurons in Human Epileptic Cortex: Correlation Between Unit and EEG Activity." *Annals of Neurology* 11 (3): 301–8.

Yang, T., and M. N. Shadlen. 2007. "Probabilistic Reasoning by Neurons." *Nature* 447 (7148): 1075–80.

Yokoyama, O., N. Miura, J. Watanabe, A. Takemoto, S. Uchida, M. Sugiura, K. Horie, et al. 2010. "Right Frontopolar Cortex Activity Correlates with Reliability of Retrospective Rating of Confidence in Short-Term Recognition Memory Performance." *Neuroscience Research* 68 (3): 199–206.

Zeki, S. 2003. "The Disunity of Consciousness." *Trends in Cognitive Sciences* 7 (5): 214–18.

Zhang, P., K. Jamison, S. Engel, B. He, and S. He. 2011. "Binocular Rivalry Requires Visual Attention." *Neuron* 71 (2): 362–69.

Zylberberg, A., S. Dehaene, G. B. Mindlin, and M. Sigman. 2009. "Neurophysiological Bases of Exponential Sensory Decay and Top-Down Memory Retrieval: A Model." *Frontiers in Computational Neuroscience* 3: 4.

Zylberberg, A., S. Dehaene, P. R. Roelfsema, and M. Sigman. 2011. "The Human Turing Machine: A Neural Framework for Mental Programs." *Trends in Cognitive Sciences* 15 (7): 293–300.

Zylberberg, A., D. Fernandez Slezak, P. R. Roelfsema, S. Dehaene, and M. Sigman. 2010. "The Brain's Router: A Cortical Network Model of Serial Processing in the Primate Brain." *PLOS Computational Biology* 6 (4): e1000765.

INDEX

Page numbers in *italics* refer to illustrations.

ILLUSTRATION CREDITS

Figure 1: © Ministère de la Culture—Médiathèque du Patrimoine, Dist. RMN-Grand Palais / image IGN.

Figure 4 (top right): By the author.

Figure 4 (bottom): Adapted by the author from D. A. Leopold and N. K. Logothetis. 1999. "Multistable Phenomena: Changing Views in Perception." *Trends in Cognitive Sciences* 3: 254–64. Copyright © 1999. With permission from Elsevier.

Figure 5: By the author.

Figure 6 (top): D. J. Simons and C. F. Chabris. 1999. "Gorillas in Our Midst: Sustained Inattentional Blindness for Dynamic Events." *Perception* 28: 1059–74.

Figure 7 (top and middle): Adapted by the author from S. Kouider and S. Dehaene. 2007. "Levels of Processing During Non-conscious Perception: A Critical Review of Visual Masking." *Philosophical Transactions of the Royal Society B: Biological Sciences* 362 (1481): 857–75. Figure 1, p. 859.

Figure 7 (bottom): By the author.

Figure 9 (top): Courtesy of Melvyn Goodale.

Figure 10: Courtesy of Edward Adelson.

Figure 11: Adapted by the author based on S. Dehaene et al. 1998. "Imaging Unconscious Semantic Priming." *Nature* 395: 597–600.

Figure 12: Adapted by the author from M. Pessiglione et al. 2007. "How the Brain Translates Money Into Force: A Neuroimaging Study of Subliminal Motivation." *Science* 316 (5826): 904–6. Courtesy of Mathias Pessiglione.

Figure 13: By the author.

Figure 14: By the author.

Figure 15: Adapted by the author from R. Moreno-Bote, D. C. Knill, and A. Pouget. 2011. "Bayesian Sampling in Visual Perception." *Proceedings of the National Academy of Sciences of the United States of America* 108 (30): 12491–96. Figure 1A.

Figure 16 (top): Adapted by the author from S. Dehaene et al. 2001. "Cerebral Mechanisms of Word Masking and Unconscious Repetition Priming." *Nature Neuroscience* 4 (7): 752–58. Figure 2.

Figure 16 (bottom): Adapted by the author from S. Sadaghiani et al. 2009. "Distributed and Antagonistic Contributions of Ongoing Activity Fluctuations to Auditory Stimulus Detection." *Journal of Neuroscience* 29 (42): 13410–17. Courtesy of Sepideh Sadaghiani.

Figure 17: Adapted by the author from S. van Gaal et al. 2010. "Unconscious Activation of the Prefrontal No-Go Network." *Journal of Neuroscience* 30 (11): 4143–50. Figures 3 and 4. Courtesy of Simon van Gaal.

Figure 18: Adapted by the author from C. Sergent et al. 2005. "Timing of the Brain Events Underlying Access to Consciousness During the Attentional Blink." *Nature Neuroscience* 8 (10): 1391–400.

Figure 19: Adapted by the author from A. Del Cul et al. 2007. "Brain Dynamics Underlying the Nonlinear Threshold for Access to Consciousness." *PLOS Biology* 5 (10): e260.

Figure 20: Adapted by the author from L. Fisch, E. Privman, M. Ramot, M. Harel, Y. Nir, S. Kipervasser, et al. 2009. "Neural 'Ignition': Enhanced Activation Linked to Perceptual Awareness in Human Ventral Stream Visual Cortex." *Neuron* 64: 562–74. With permission from Elsevier.

Figure 21 (top): Adapted by the author from E. Rodriguez et al. 1999. "Perception's Shadow: Long-Distance Synchronization of Human Brain Activity." *Nature* 397 (6718): 430–33. Figures 1 and 3.

Figure 21 (bottom): Adapted by the author from R. Gaillard et al. 2009. "Converging Intracranial Markers of Conscious Access." *PLOS Biology* 7 (3): e61. Figure 8.

Figure 22: Adapted by the author from R. Q. Quiroga, R. Mukamel, E. A. Isham, R. Malach, and I. Fried. 2008. "Human Single-Neuron Responses at the Threshold of Conscious Recognition." *Proceedings of the National Academy of Sciences of the United States of America* 105 (9): 3599–604. Figure 2. Copyright © 2008 National Academy of Sciences, U.S.A.

Figure 23 (right): Copyright © 2003 Neuroscience of Attention & Perception Laboratory, Princeton University.

Figure 24 (top): B. J. Baars. 1989. *A Cognitive Theory of Consciousness*. Cambridge, U.K.: Cambridge University Press. Courtesy of Bernard Baars.

Figure 24 (bottom): S. Dehaene, M. Kerszberg, and J. P. Changeux. 1998. "A Neuronal Model of a Global Workspace in Effortful Cognitive Tasks." *Proceedings of the National Academy of Sciences of the United States of America* 95 (24): 14529–34. Figure 1. Copyright © 1998 National Academy of Sciences, U.S.A.

Figure 25 (right): Courtesy of Michel Thiebaut de Schotten.

Figure 26 (bottom): G. N. Elston. 2003. "Cortex, Cognition and the Cell: New Insights into the Pyramidal Neuron and Prefrontal Function." *Cerebral Cortex* 13 (11): 1124–38. By permission of Oxford University Press.

Figure 27: Adapted by the author from S. Dehaene et al. 2005. "Ongoing Spontaneous Activity Controls Access to Consciousness: A Neuronal Model for Inattentional Blindness." *PLOS Biology* 3 (5): e141.

Figure 28: Adapted by the author from S. Dehaene et al. 2006. "Conscious, Preconscious, and Subliminal Processing: A Testable Taxonomy." *Trends in Cognitive Sciences* 10 (5): 204–11.

Figure 29: Adapted by the author from S. Laureys et al. 2004. "Brain Function in Coma, Vegetative State, and Related Disorders." *Lancet Neurology* 3 (9): 537–46.

Figure 30: Adapted by the author from M. M. Monti, A. Vanhaudenhuyse, M. R. Coleman, M. Boly, J. D. Pickard, L. Tshibanda, et al. 2010. "Willful Modulation of Brain Activity in Disorders of Consciousness." *New England Journal of Medicine* 362: 579–89. Copyright © 2010 Massachusetts Medical Society. Reprinted with permission from Massachusetts Medical Society.

Figure 31: Adapted by the author from T. A. Bekinschtein, S. Dehaene, B. Rohaut, F. Tadel, L. Cohen, and L. Naccache. 2009. "Neural Signature of the Conscious Processing of Auditory Regularities." *Proceedings of the National Academy of Sciences of the United States of America* 106 (5): 1672–77. Figures 2 and 3.

Figure 32: Courtesy of Steven Laureys.

Figure 33: Adapted by the author from J. R. King, J. D. Sitt, et al. 2013. "Long-Distance Information Sharing Indexes the State of Consciousness of Unresponsive Patients." *Current Biology* 23: 1914–19. Copyright © 2013. With permission from Elsevier.

Figure 34: Adapted by the author from G. Dehaene-Lambertz, S. Dehaene, and L. Hertz-Pannier. 2002. "Functional Neuroimaging of Speech Perception in Infants." *Science* 298 (5600): 2013–15.

Figure 35: Adapted by the author from S. Kouider et al. 2013. "A Neural Marker of Perceptual Consciousness in Infants." *Science* 340 (6130): 376–80.